Reinventing Processes for Sustainability via Process Intensification and Integration

Reinventing Processes for Sustainability via Process Intensification and Integration

Editors

Jun-Wei Lim
Worapon Kiatkittipong

MDPI

Basel • Beijing • Wuhan • Barcelona • Belgrade • Novi Sad • Cluj • Manchester

Editors

Jun-Wei Lim
Department of Fundamental and Applied Sciences
Universiti Teknologi PETRONAS
Seri Iskandar, Malaysia

Worapon Kiatkittipong
Department of Chemical Engineering
Silpakorn University
Nakhon Pathom, Thailand

Editorial Office
MDPI
St. Alban-Anlage 66
4052 Basel, Switzerland

This is a reprint of articles from the Special Issue published online in the open access journal *Processes* (ISSN 2227-9717) (available at: https://www.mdpi.com/journal/processes/special_issues/Reinventing_Processes).

For citation purposes, cite each article independently as indicated on the article page online and as indicated below:

Lastname, A.A.; Lastname, B.B. Article Title. *Journal Name* **Year**, *Volume Number*, Page Range.

ISBN 978-3-0365-9993-9 (Hbk)
ISBN 978-3-0365-9994-6 (PDF)
doi.org/10.3390/books978-3-0365-9994-6

Contents

About the Editors

Jun-Wei Lim

Dr. Jun-Wei Lim received his Ph.D. qualification in Environmental Chemistry from the Universiti Sains Malaysia in 2013. Currently, he is affiliated with Universiti Teknologi PETRONAS, serving as the Senior Lecturer and Cluster Head of the Applied Chemistry program. His major research interests are insect-based biological compounds, the bioremediation of solid wastes and wastewaters, microalgal biofuels, and green hydrogen. Accordingly, he has published more than 300 research papers, inclusive of book chapters, of late. He is a Member of The Royal Society of Chemistry (MRSC) and Associate Member of Institution of Chemical Engineers (AMIChemE) at the international level and Professional Chemist registered with Malaysian Institute of Chemistry and Professional Technologist registered with Malaysia Board of Technologists at the national level. He is also a Chartered Scientist registered with Science Council, UK, and Chartered Environmentalist registered with Society for the Environment, UK, as well as Chartered Chemist registered with The Royal Society of Chemistry, UK. He is a member of the Distinguished Adjunct Faculty at Saveetha Institute of Medical and Technical Sciences, and Visiting Professor at Prince of Songkla University.

Worapon Kiatkittipong

Dr. Worapon Kiatkittipong is an Associate Professor in the Department of Chemical Engineering, Silpakorn University, Thailand. He received a B.Eng. degree from Kasetsart University, as well as M.Eng and D.Eng. degrees from Chulalongkorn University in 2005, supported by The Royal Golden Jubilee Ph.D. Program funded by The Thailand Research Fund and supervised by Prof. Suttichai Assabumrungrat. His research focuses on the implications of biofuel and biochemical production integrated with biorefinery and the concept of process intensification. He has published more than 150 ISI/SCOPUS-indexed journals, receiving a total of 3000+ citations and an H index of 32. He has been a Guest Editor for many journals and proceedings such as *Biomass and Bioenergy*, *Energies*, and *Processes*. He previously served as the Vice Dean for Research and Academic Services, at the Faculty of Engineering and Industrial Technology, Silpakorn University (2019–2022). He is currently serving as the Associate Editor of *Science Engineering* and *Health Studies*.

processes

MDPI

Editorial

Reinventing Processes for Sustainability via Process Intensification and Integration

Worapon Kiatkittipong [1,*] and Jun Wei Lim [2,3,*]

[1] Department of Chemical Engineering, Faculty of Engineering and Industrial Technology, Silpakorn University, Nakhon Pathom 73000, Thailand
[2] HICoE-Centre for Biofuel and Biochemical Research, Institute of Self-Sustainable Building, Department of Fundamental and Applied Sciences, Universiti Teknologi PETRONAS, Seri Iskandar 32610, Perak Darul Ridzuan, Malaysia
[3] Department of Biotechnology, Saveetha School of Engineering, Saveetha Institute of Medical and Technical Sciences, Chennai 602105, India
* Correspondence: kiatkittipong_w@su.ac.th (W.K.); junwei.lim@utp.edu.my or jun_wei_85@yahoo.com (J.W.L.)

Keywords: processes; sustainability; intensification; integration; waste; environment

Citation: Kiatkittipong, W.; Lim, J.W. Reinventing Processes for Sustainability via Process Intensification and Integration. *Processes* **2024**, *12*, 63. https://doi.org/10.3390/pr12010063

Received: 19 December 2023
Accepted: 26 December 2023
Published: 27 December 2023

A waste material cannot truly be called waste when the procedures and technologies have been invented and developed to exploit and utilize it. In fact, no material should be classified as waste, especially now that various advanced approaches have been created to either recover useful products or convert waste into useful materials for a similar or other application [1]. Accordingly, lifecycle analysis has been employed in the development of various products and the fabrication of new materials in order to ensure that their impacts on the environment are sustainable in nature [2]. Furthermore, modifications and laboratory research are also being performed in order to identify the best resolutions that can offer value to humans and nature via various intensification and integration processes. This can achieve sustainability in waste production and assure that waste will always have its applications. These encompass many industrial activities, as well as sectors such as agriculture, transportation, and mining. Each of these aspects is evolving, and the integration of artificial intelligence (AI) has now begun to increase and enhance the chains' effectiveness, particularly regarding energy consumption [3]. Modelling focused on optimization and prediction has been accepted as the norm for staying abreast of the current developments in tackling the waste issues stemming from inevitable byproduct generation. A fast-moving pace within every sector can only be attained via accurate modelling and the integration of machine learning to ameliorate current processes, enabling more developing countries to shift toward development [4]. Accordingly, in this Special Issue, we intend to present essential advanced technological processes that can be applied in myriad industries in order to develop the key goal of sustainability, moving forward hand-in-hand with the Sustainable Development Goals to reach their targets by 2030 [5]. Readers can learn about the integration and intensification methods that are currently being adopted to capture the carbon stemming from waste gas generation. Novel optimization methods based on modelling, as well as numerical studies, are also reported for industries such as textiles and to forecast local disease outbreaks. Moreover, the potential use of solid organic wastes as nutrients for growing microalgae for the production of protein, carbohydrate, and lipid sources has, evidently, highlighted a major use of waste via bioconversion while avoiding landfill. Besides microalgae, bacteria are also capable of performing the composting of solid organic waste into useful green energy pellets to advocate for a waste-to-energy development plan. The potential of solar energy for process intensification has been integrated into industrial processes to partially offset energy requirements and is gaining traction regarding energy sustainability in the long term. Additionally, the importance of batteries is acknowledged, and various processes to recover lithium element from oily wastewater sources are reported. Therefore, this Special Issue, "Reinventing Processes for Sustainability via Process Intensification and Integration" (https://www.mdpi.com/journal/processes/special_issues/Reinventing_Processes, accessed on 19 December 2023), undoubtedly captures every important feature of the advanced processes in intensification and integration for permanent improvements to both human activities and the natural environment.

Conflicts of Interest: The authors declare no conflicts of interest.

List of Contributions

1. Yang, H.; Wang, Y. Research on the Path of Manufacturing Enterprises Supply Chain Integration from the Configuration Perspective. *Processes* **2021**, *9*, 1746.
2. Kingphadung, K.; Kurdkaew, P.; Siriwongwilaichat, P.; Kwonpongsagoon, S. Comparison of Performance and Economic Efficiency for Greenhouse Solar versus Hot Air Drying: A Case of Crispy Mango Production. *Processes* **2022**, *10*, 311.
3. Chia, W.; Chew, K.; Le, C.; Chee, C.; Ooi, M.; Show, P. Utilization of Aerobic Compression Composting Technology on Raw Mushroom Waste for Bioenergy Pellets Production. *Processes* **2022**, *10*, 463.
4. Tiong, Z.; Rawindran, H.; Leong, W.; Liew, C.; Wong, Y.; Kiatkittipong, W.; Abdelfattah, E.; Show, P.; Rahmah, A.; Tong, W.; et al. Impact of Various Visible Spectra on Attached Microalgal Growth on Palm Decanter Cake in Triggering Protein, Carbohydrate, and Lipid to Biodiesel Production. *Processes* **2022**, *10*, 1583.
5. Kusakabe, K.; Nagai, A.; Leong, W.; Yamasaka, K.; Nakaaki, T.; Uemura, Y.; Ikenaga, K. Dechlorination of Polyvinyl Chloride via Solvothermal Treatment with Glycerol. *Processes* **2022**, *10*, 2047.
6. Ramirez-Argaez, M.; Abreú-López, D.; Gracia-Fadrique, J.; Dutta, A. Numerical Study of Electrostatic Desalting: A Detailed Parametric Study. *Processes* **2022**, *10*, 2118.
7. Jaffar, Z.; Yunus, N.; Shaharun, M.; Allim, M.; Rahim, A. Incorporated Metal–Organic Framework Hybrid Materials for Gas Separation, Catalysis and Wastewater Treatment. *Processes* **2022**, *10*, 2368.
8. Fatima, S.; Borhan, A.; Ayoub, M.; Ghani, N. CO_2 Adsorption Performance on Surface-Functionalized Activated Carbon Impregnated with Pyrrolidinium-Based Ionic Liquid. *Processes* **2022**, *10*, 2372.
9. Othman, M.; Indawati, R.; Suleiman, A.; Qomaruddin, M.; Sokkalingam, R. Model Forecasting Development for Dengue Fever Incidence in Surabaya City Using Time Series Analysis. *Processes* **2022**, *10*, 2454.
10. Kunkhet, A.; Chudasri, D. Developing Design Approaches for Tile Pattern Designs Inspired by Traditional Textile Patterns. *Processes* **2022**, *10*, 2744.
11. Garcia, L.; Ho, Y.; Myo Thant, M.; Han, D.; Lim, J. Lithium in a Sustainable Circular Economy: A Comprehensive Review. *Processes* **2023**, *11*, 418.
12. Saenphon, T.; Phimoltares, S.; Lursinsap, C. Enhancing Mean-Variance Mapping Optimization Using Opposite Gradient Method and Interior Point Method for Real Parameter Optimization Problems. *Processes* **2023**, *11*, 465.
13. Shokrollahi, F.; Lau, K.; Partoon, B. Experimental Evaluation of Chemical Reactions Involved in Ultrasonic-Assisted Absorption of Bulk CO_2. *Processes* **2023**, *11*, 3266.

References

1. Keita, S.; Stopic, S.; Kiessling, F.; Husovic, T.V.; Kaya, E.E.; Smijanic, S.; Friedich, B. Recovery of Magnetic Particles from Wastewater Formed through the Treatment of New Polycrystalline Diamond Blanks. *Processes* **2023**, *1*, 993–1006. [CrossRef]
2. Ciacci, L.; Passarini, F. Life Cycle Assessment (LCA) of Environmental and Energy Systems. *Energies* **2020**, *13*, 5892. [CrossRef]
3. Florea, A.; Sipos, A.; Stoisor, M.C. Applying AI Tools for Modeling, Predicting and Managing the White Wine Fermentation Process. *Fermentation* **2022**, *8*, 137. [CrossRef]

4. Chandrasekar, M.; Collins, J.L.; Habibi, S.; Ong, R.G. Microfluidic Reactor Designed for Time-Lapsed Imaging Of Pretreatment And Enzymatic Hydrolysis Of Lignocellulosic Biomass. *Bioresour. Technol.* **2024**, *393*, 129989. [CrossRef] [PubMed]
5. Palka, A.; Skotnicka, M. The Health-Promoting and Sensory Properties of Tropical Fruit Sorbets with Inulin. *Molecules* **2022**, *27*, 4239. [CrossRef] [PubMed]

processes

MDPI

Review

Lithium in a Sustainable Circular Economy: A Comprehensive Review

Laura Vega Garcia [1], Yeek-Chia Ho [1,*], Maung Maung Myo Thant [2], Dong Suk Han [3] and Jun Wei Lim [4]

1 Centre for Urban Resource Sustainability, Civil and Environmental Engineering Department, Institute of Self-Sustainable Building, Universiti Teknologi PETRONAS, Seri Iskandar 32610, Perak Darul Ridzuan, Malaysia
2 Advanced Expertise Common Centre, Group Research & Technology, Project Delivery & Technology Division, Petroliam Nasional Berhad (PETRONAS), Kuala Lumpur 50088, Selangor, Malaysia
3 Center for Advanced Materials, Department of Chemical Engineering, Qatar University, Doha PO Box 2713, Qatar
4 HICoE—Centre for Biofuel and Biochemical Research, Department of Fundamental and Applied Sciences, Institute of Self-Sustainable Building, Universiti Teknologi PETRONAS, Seri Iskandar 32610, Perak Darul Ridzuan, Malaysia
* Correspondence: yeekchia.ho@utp.edu.my

Abstract: Lithium is a vital raw material used for a wide range of applications, such as the fabrication of glass, ceramics, pharmaceuticals, and batteries for electric cars. The accelerating electrification transition and the global commitment to decarbonization have caused an increasing demand for lithium. The current supply derived from brines and hard rock ores is not enough to meet the global demand unless alternate resources and efficient techniques to recover this valuable metal are implemented. In the past few decades, several approaches have been studied to extract lithium from aqueous resources. Among those studied, chemical precipitation is considered the most efficient technology for the extraction of metals from wastewater. This paper outlines the current technology, its challenges, and its environmental impacts. Moreover, it reviews alternative approaches to recover lithium via chemical precipitation, and systematically studies the effects of different operating conditions on the lithium precipitation rate. In addition, the biggest challenges of the most recent studies are discussed, along with implications for future innovation.

Keywords: lithium; lithium recovery; chemical precipitation; circular economy

Citation: Garcia, L.V.; Ho, Y.-C.; Myo Thant, M.M.; Han, D.S.; Lim, J.W. Lithium in a Sustainable Circular Economy: A Comprehensive Review. *Processes* **2023**, *11*, 418. https://doi.org/10.3390/pr11020418

Academic Editor: Mingxia Gao

Received: 15 November 2022
Revised: 15 December 2022
Accepted: 3 January 2023
Published: 30 January 2023

1. Introduction

Identified as "Energy Metal in the 21st Century", lithium is one of the main components of electric vehicles, mobile phones, and laptops. The global demand for lithium (Li) is growing exponentially, not only due to technological evolution, but also because of environmental regulations, which are committed to reducing greenhouse gas emissions caused by fossil fuel cars [1–3]. The increase in electric vehicle manufacturing has increased the global demand for Li, and it is estimated to triple by 2025 [4–6].

There is increased interest in harvesting lithium from secondary resources such as wastewater. Recapturing waste and converting it into a raw material will further diversify lithium extractive sources, increase the metal's availability, reduce material use, and make lithium production less resource-intensive. All of these factors support the circular economy, which is a popular concept being implemented around the world. A sustainable recovery process to harvest lithium from alternate resources is needed, and different approaches have been studied, including chemical precipitation, ion exchange, adsorbents, membrane-based processes, solvent extraction, and electrolysis-based systems [7–10].

Lithium extraction by adsorption and ion exchange methods has good performance and offers a simple design; however, most of the adsorbents used have poor regenerability, have low adsorption capacity, and are suitable for lithium recovery in solutions with very low concentrations of Li [11,12].

The extraction of Li using membrane technology from aqueous sources still encounters many challenges due to the poor monovalent selectivity of the conventional membranes. This method only removes divalent ions, which narrows down its application with the aim of Li concentration for recovery [13]. Noteworthy factors, such as high cost, process complexity, membrane fouling, and low permeate flux have restricted their industrial use.

On another note, solvent extraction can selectively extract lithium in an individual metal system. However, the presence of other metal ions hinders the lithium extraction efficiency [10,14]. Additionally, this method involves prolonged extraction times and requires large amounts of solvent. The use of organic solvents creates serious environmental concerns, as huge leftover volumes of solvent would be generated [11,15]

Lithium extraction by electrolysis is another technology that has been studied. Yet, it is not widely used due to a large voltage being required to increase the Li recovery rate. The cost-effectiveness of this technology for large-scale production depends on energy efficiency and reliable electricity at low cost.

Interestingly, chemical precipitation is considered the most efficient technology for the removal of dissolved metal from aqueous sources. The method is relatively simple and inexpensive to operate [12]. The overarching aim of this paper is to review the technologies studied to recover lithium via chemical precipitation, along with the novel materials used, and discuss the effects of initial pH, temperature, and reaction time on the chemical precipitation rate.

2. The Nature of Lithium

Lithium (Li) is a soft, silver–white alkali metal. It is the lightest of all metals, with an atomic weight of 6939, and the 33rd most abundant element in the Earth's crust. Li has the lowest density and electrochemical equivalent of all solids: 0.53 g/cm^3 and 0.259 g/Ah, respectively. These physical characteristics make lithium an exceptional material for batteries, as it offers high voltage, high energy density, and high heat capacity in comparison to any other metal [16–19]. Thus, Li known as an energy critical element and is being widely utilized in large-capacity rechargeable batteries, particularly Li-ion batteries (LIBs).

2.1. Sources of Lithium

Lithium can be found on Earth as a compound in igneous rocks, in brine aquifers, in lithium clay deposits, or dissolved in seawater. However, the two main economic resources of Li are brines and hard rock ores [20]. Around 40% of the global lithium production is from ores, and brines account for more than 60% of the production [18].

Worldwide ore-based lithium resources number more than 100. However, only three are commercially extracted from pegmatite ores. Pegmatite deposits contain Li minerals such as spodumene ($LiAl(Si_2O_6)$), petalite ($LiAlSi_4O_{10}$), and lepidolite ($K(Li,Al)_3(Al,Si,Rb)_4 O_{10}(F,OH)_2$), among others. Spodumene is the most important lithium mineral for the market due to its high Li content. Lepidolite is the most widespread Li mineral. Lastly, petalite, which is geographically limited, as it is mainly found in Africa, is an important source of Li [21].

Pegmatite is an igneous rock, usually of granitic composition and characterized by interlocking mineral grains [22]. Even though there are various categories for granitic pegmatites, the pegmatites of greatest economic interest are the granitic LCT pegmatites. This group take its name from their geochemical trace element composition, which consists mainly of lithium, cesium, and tantalum [23,24].

Among the Li-rich pegmatite minerals, spodumene ($LiAl(Si_2O_6)$) is currently the main source of Li from ores and the most important for the market, due to its high lithium content and widespread availability; moreover, it is considered an easily exploitable deposit. Spodumene consists of lithium aluminum silicate, and its theoretical Li content is 8.1%, as LiO_2. At present, the largest lithium ore deposit is the mine of Greenbushes, a spodumene deposit in Australia. Other spodumene deposits are in China, Russia, Canada, and Germany [25,26].

Petalite ($LiAlSi_4O_{10}$) is another commercial source of lithium. Petalite minerals contain 16.7% AL_2O_3 and 78.4% SiO_2, and their Li content can vary from 1.4 to 2.2%, or 3.0 to 4.7% as Li_2O. Petalite is considered a lithium mineral of economic importance in Zimbabwe; however, large petalite ore deposits are also found in Canada, Brazil, and Australia [18,27,28]. The literature also reports a significant amount of petalite in Namibia, China, and Russia [5,18,29].

Most of the lithium is produced from spodumene and petalite, and to a lesser extent from lepidolite. Lepidolite ($K(Li,Al)_3(Al,Si,Rb)_4O_{10}(F,OH)_2$), also known as lithium mica or zinvaldite, was one of the early ores mined for lithium, especially for salt production and glass fabrication. However, it has lost importance on the market due to its high fluorine content, which is around 9%. The theoretical Li content of lepidolite can vary from 1.39% to 3.58%, while this clue is 3.0% to 7.7% in Li_2O. Even though this mineral is not widely found in pegmatites deposits, it is exploited and commercially produced in Zimbabwe, Canada, Namibia, Brazil, Portugal, and Argentina [18,27,30,31].

Lithium can be also found in naturally occurring salt-lake brines, which are also known as continental brine or salars. Such aquifers have the vast majority of land-based lithium in the world; nearly 70% of global lithium resources are hosted in continental brines [5,23]. Lithium extraction from continental brine is considered more economical than production from hard rock ores, and is between 30% and 50% less expensive. Therefore, it is the preferred source of Li for investors [22,23,32].

Continental brines are hypersaline aquifers, with salinity values of 1.7 to 24 times that of seawater, and they are characterized by high concentrations of dissolved ions. Continental brine deposits are normally found in underground reservoirs, contained in closed basins, with the surrounding rock formations being the cause of the dissolved elements in the brine [23,26]. Only a few places in the world have the geological characteristics and arid climate that allow the economic extraction of lithium, and currently the most developed and important brines for lithium extraction are in the salars of Chile (Salar de Atacama), Argentina, and Bolivia (Salar de Uyuni), known as the lithium triangle [12,25,33].

2.1.1. Waste a New Source of Lithium

Lithium recovery from secondary resources is crucial to meet the increasing demand for the metal and ensure the sustainability of the electrification transition with electric vehicles. In terms of sustainability, recovering Li from waste such as spent Li-ion batteries and wastewater provides a potential approach to diversify Li extraction sources and make lithium production less resource intensive.

Cathode active materials contain around 5 wt.% of lithium, or 360 kt, which is equivalent to 1900 kt of Li_2CO_3 [34]. The economic worth of lithium-ion battery recycling has been estimated as USD 22,000 per ton, based on the values of lithium and cobalt [35]. Unfortunately, less than 5% of the spent LIB waste is currently recycled [34]. Hence, substantial research has been conducted on recycling Li from spent LIBs.

The recycling process of LIBs involves physical and chemical methods. The physical separation includes disassembly, crushing, screening, magnetic separation, washing, heating treatment, etc. At present, hydrometallurgical processes, which involve leaching and reduction operations, are widely used as chemical methods to recover lithium. The major industrial processes to recycle Li-ion batteries are the Toxco process, the Sony–Sumitomo process, the Umicore process, and the Recupyl process [36]. However, lithium is only recovered by the Toxco and Recupyl processes; the other processes focus primarily on the recovery of cobalt and nickel [34,37].

The Toxco process (Canada) is designed for all types of lithium-containing waste [38,39]. This process uses cryogenic conditions, where the materials are first cooled in liquid nitrogen to reduce the reactivity [22,36]. Subsequently, the batteries are shredded and submerged in an alkaline bath, where acid components are neutralized and lithium salts are dissolved. The dissolved Li salts are treated with sodium carbonate to produce lithium carbonate [40].

The Recupyl process (France) involves the mechanical separation of the batteries under an inert atmosphere to reduce the reactivity of lithium [22]. After the batteries are shredded and crushed, different fractions are obtained. The fine fraction is separated from the other and added to water. The lithium in the fine fraction dissolves in the water, and becomes rich in lithium hydroxide. The lithium is subsequently recovered by using either sodium hydroxide or phosphoric acid [36].

The literature reports additional lithium recovery processes, such as Accurec GmbH and AEA technology [22,38–40]. Accurec is a German pyrometallurgical process used to recycle batteries and recover Co–Mn alloy and lithium chloride. The method involves mechanical treatment to separate materials. After extracting electronic, plastic casing, aluminum, copper, and steel elements, the remaining electrode is agglomerated by adding a binder and is then pressed into briquettes. These pieces are then smelted in a furnace where two fractions are obtained, the Co–Mn alloy and lithium-containing slag. Lithium is extracted from the slag as either lithium chloride or lithium carbonate by acid leaching [22,38].

AEA technology is a patented method in the UK. The three main stages of the process are electrolyte extraction, electrode dissolution, and cobalt reduction. In this process, the spent LIBs are shredded in an inert and dry atmosphere. The electrolyte is extracted by an anhydrous solvent with a boiling point below the lithium salt decomposition temperature (~80 °C). Once the electrolyte is obtained, the cell pieces are immersed in solvent (N-methyl-2-pyrrolidone) at 50 °C to dissolve the binder—PVDF. The binder solution is filtered to separate other materials, such as residual copper, aluminum, steel, and plastic. The solvent is evaporated for reuse, and the residue ($LiCoO_2$ and carbon) is subjected to electrochemical reduction to obtain the lithium hydroxide solution [22,39,40].

Recent advances in the recycling of lithium-ion battery recycling are focused on hydrometallurgical processes. This method includes leaching, solvent extraction, precipitation, or electroreduction. In the leaching step, strong acids such as hydrochloric acid [41,42], sulfuric acid [41,43,44], nitric acid, and formic acid [37,45] are used, with sulfuric acid being the most common one. Hydrogen peroxide is used in the leaching process as a highly effective reducing agent, with which cobalt, nickel, and manganese are reduced to more soluble species. At the end of the recycling process, lithium is recovered and purified as a byproduct. Unfortunately, the use of inorganic acids results in the generation of a large amount of hazardous wastes with serious environmental impacts; therefore, as an alternative approach, organic acids have been explored as leaching agents in order to lessen the environmental footprint and make the recycling process more sustainable.

The use of organic acids mitigates the corrosion of equipment, minimizes the risk associated with the use of strong acids, and achieves more selective metal recovery. The most recent studies of the organic acid-assisted leaching methods included ascorbic acid [46], formic acid [47], tartaric acid [48], oxalic acid [49,50], and most recently, p-toluene sulfonic acid (PTSA), a strong organic acid [45,51]. However, the hydrometallurgical process with organic acids requires a sequence of separation and purification steps of Co and Li, making this method difficult, and resulting in product yield losses with the significant consumption of chemicals [45]. Therefore, continuous efforts are needed to find more effective organic acids, solve technical challenges, and simplify the subsequent procedures for the separation and recovery of Li.

Once the lithium is obtained in the leach liquor, the element is extracted by solvent extraction or precipitation. In first step, precipitation is used to eliminate impurities such as Cu, Al, and Fe from the leachate, followed by the solvent extraction of Co and Ni. After cobalt and nickel have been removed, lithium is precipitated by using Na_2CO_3 or CO_2 [52]. Section 3.2 contains recent studies on precipitant agents used to precipitate Li from leach liquor and wastewater from the cathode manufacturing process.

Produced water is the largest byproduct of the oil and gas sector, and thereby has significant financial and environmental implications [53,54]. Shale gas exploitation by hydraulic fracturing and horizontal drilling in unconventional reservoirs has made possible the current flourishing of gas and oil production from shale reservoirs. However, the use of these methods results in large volumes of water being produced; in fact, it is estimated that the volume will reach 499–3585 million m^3 globally by 2030 [55]. Furthermore, as wells age, there is a decrease in oil and gas production and a significant increase in the produced water. The volume of produced water could reach 98% in brownfields, leaving only 2% of oil/gas production [53]. Therefore, the growth of the market for the management and reuse of produced water is increasing.

Recent studies have revealed that produced water is a promising source of lithium, and harvesting this valuable metal from wastewater has attracted much attention from the oil and gas industry. For example, the Li concentration in Marcellus shale gas flowback and produced water is between 4 and 202 mg/L [55], Smackover brines in the U.S. contain approximately 500 mg/L of Li [7], and the Sichuan Basin (China) has an average Li concentration of 33 mg/L [55]. Oil fields such as Fox Creek and Valleyview in Canada have the potential to produce 362,000 and 385,000 metric tons of Li metal equivalents, respectively. The Smackover formation in the U.S. could potentially generate 750,000 metric tons of Li metal equivalents [7]. Therefore, produced water is a potential lithium resource. Li recovery from wastewater supports the circular economy concept by recapturing waste as a resource to manufacture a new material. Moreover, harvesting Li from produced water would not only alleviate the imbalance between supply and demand, but would also create new revenue for the oil and gas sector.

2.1.2. Global Availability

Prior to discussing the global availability of lithium, it is important to address the difference between reserves and resources and assess the available amount of lithium accordingly. Resources are geological stock that has been identified and could be mined with better technology, whereas reserves refer to the amount of lithium available for exploitation with the existing technological and socioeconomic conditions; therefore, lithium reserves are the existing sources for lithium production present in the world [30,33].

From the literature review, the consensus regarding the amount of lithium from different resources is estimated to be between 34 and 45 million tons, and the reserves available for lithium production are approximately 14 million tons. Christmann, Gloaguen [56] estimated the global lithium resources and reserves, on the basis of published data from exploration and mining companies. The compiled information contains data from 88 deposits in 22 countries, showing a total of 45.2 million tons of lithium resources and 12.22 million tons of reserves. Martin, Rentsch [26] reported above 14 million tons of available reserves and approximately 34 million tons of resources. Zhang, Hu [12] reported the resources of lithium by deposit type and its worldwide distribution. Accordingly, 26.9 million tons of lithium are found in brines, and 16.7 million tons are found in minerals, bringing the total amount of lithium resources to 43.6 million tons.

It is important to note that lithium resources have increased significantly owing to ongoing exploration, and now stand at roughly 89 million tons worldwide. New lithium resources have been identified in continental brines, geothermal brines, hectorite, oilfield brines, pegmatites, and searlesite. Moreover, recent governmental reports have revealed that over the past decade, 2021 had the largest lithium reserve volume—the total estimated reserves amounted to 22 million tons [57,58]. However, available resources need to be converted into reserves to meet the projected future consumption and ensure a reliable and diversified supply of lithium. Take into account the limited sources of Li, and we can see that it is essential to carry out Li recovery for a sustainable future. Figure 1 shows the global availability of lithium by author.

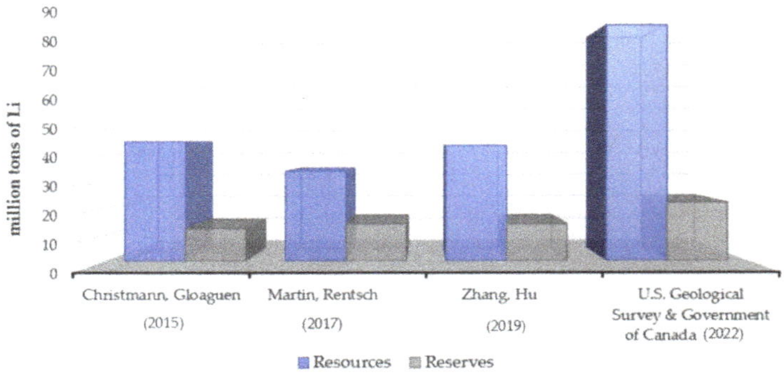

Figure 1. Lithium global availability [56–58].

2.1.3. Environmental Challenges

The processes of the extraction of lithium from brines include evaporation, crystallization, and precipitation. First, the brine is pumped out of the underground aquifer and transferred into evaporation ponds. Through solar radiation, the brine is evaporated for about one to two years [15]. During the evaporative process, some compounds such as sodium, potassium, and magnesium chlorides are removed by crystallization, at several stages of successive evaporation [5,22,59]. However, the main purpose of this stage is to increase the lithium concentration [60]. The concentrated brine is afterward brought to the carbonate precipitation phase, where residue magnesium impurities are first precipitated by calcium oxide. Subsequently, sodium carbonate is added to precipitate lithium as Li_2CO_3. At the final stage, lithium carbonate is redissolved and reprecipitated to achieve high purity in the final product [12,15,61,62].

The use of evaporation ponds in lithium processing has raised concerns about the interaction and pollution of watercourses [59]. For instance, audit reports of the Salar de Atacama and other reports of the Uyuni Salar area (Bolivia) have indicated water contamination, increased sediment loads, landscape modifications with disturbance of the native biodiversity, and changes in the ecosystem structure, which might have negative effects on the health of local people [26,59,63].

The extraction of the brine from an underground aquifer creates an imbalance in the hydrological system, as most of the water is lost during the evaporation process and only a fraction of liquid is pumped back into the aquifer. The cone of depression (which is formed during brine extraction) expands laterally and downward, causing a reduction in the pressure gradient of the aquifer. This phenomenon causes freshwater aquifers in neighboring areas to migrate towards the brine aquifer. Therefore, usable water resources are diverted away from local communities, which is a matter of concern, as the depletion of water from these resources, especially in water-scarce areas, can negatively impact the ecosystem and locals [59,60,64]. For example, The Salar del Hombre Muerto, an Argentine Lithium brine, has brought several problems to the region. Local communities have reported that Li extraction operations have contaminated the waterways used by humans, livestock, and irrigation, along with the deterioration of the soil and air quality [59].

As a further matter, fresh water is largely used for the preparation of lime solutions and the purification of lithium carbonate [61]. Approximately 5 to 50 m^3 of freshwater is used per ton of lithium carbonate [65]. Regarding the overall sustainability of the process, there is a big question about the usage of freshwater in arid and water-scarce areas.

The evaporation processes of lithium brines produce large amounts of waste. Salt-lake brines are complex systems that contain numerous ions, and nearly all of them except lithium carbonate are removed as waste, which are currently being accumulated at the verge of the Salar. As an example, the production of 20,000 tons of Li_2CO_3 per annum from a native brine, during a period of ten years, will generate a total amount of waste of 1.15 × 107 m^3, and such a volume will require a disposal site of 11.5 km^2 [61]. The increased amount of waste generation is already a major issue worldwide. Additionally, land scarcity for waste disposal is a serious and growing concern [66]. Therefore, consider the contamination of land, air, and water, along with the negative impacts on human health and the environment related to land disposal.

Currently, brine is regulated as a mineral, so there is debate over the classification of lithium brine as water or mineral. Brine is classified as a mineral mainly because it is not fit for human consumption, and it is not economically feasible to make it freshwater for agriculture. However, current studies supporting the position that brine is a type of water show that by physiochemical and thermodynamic analysis, the water molecules in brine are similar in structure to the molecules in freshwater at a pressure of 1.2 atm. The study gave a scientific perspective to persuade regulators, investors, and other stakeholders to accurately classify brine as water and adopt corresponding practices during the Li extraction process, so as to minimize the negative impacts caused by the production process resulting from a misleading categorization [64,67].

Ore deposits are another important source of lithium, but with the highest impact due to the extraction process. The common procedures used for lithium beneficiation include the concentration of the spodumene by gravity separation, magnetic separation and froth flotation, and production of Li concentrate [18,30]. Heavy media separation is a preconcentration stage to separate spodumene from gangue silicates. Spodumene is slightly heavier than gangue minerals; therefore, it sinks while the others float. Subsequently, magnetic separation is used to remove iron gangue minerals. This step can be performed before or after froth flotation. The flotation stage is the most common practice used to concentrate lithium minerals from pegmatite ores. However, the surface chemistry of the minerals, among other factors, can affect the selective recovery of Li by flotation. Thus, pre-treatment of the spodumene is necessary in the separation process, to clean or modify the surface of the mineral for its flotation to be selective [18]. In what follows, a leaching treatment using sodium hydroxide is used to enhance the flotation recovery of spodumene [68]. The lithium is finally extracted from the spodumene by either acid or alkaline digestion. The final products are lithium carbonate and lithium hydroxide. The acid digestion is performed at temperatures above 250 °C, using concentrated sulfuric acid [69], and alkaline digestion is at 1040 °C with concentrated calcium hydroxide [70]. In the case of H_2SO_4, β-spodumene reacts with the acid and lithium is extracted in the water-soluble form of lithium sulfate, at 200–300 °C [29].

Important socio-environmental impacts are caused by lithium mining operations from spodumene ores. To begin with, there are the toxic emissions caused by blasting operations, which are conducted prior to ore extraction and processing [60]. The concentration of lithium in hard rock sources requires large amounts of spodumene ore to obtain economically significant quantities, resulting in substantial volumes of waste rock [71]. Furthermore, toxic tailings are among the waste products of the extraction and processing of the ore. For example, during the processing of spodumene concentrate with sulfuric acid, large amounts of reagents are used, which lead to large waste yields [29,30].

Apart from the large amount of toxic waste generated, there is a concern about unscrupulous business practices. For instance, in 2017, a survey report showed that around 2 billion pounds of hazardous waste was thrown into watercourses by mining companies, representing a severe risk to human health and aquatic life [15]. Regarding waterways, the ore-based production of lithium involves the high consumption of water—in fact, its use of freshwater is higher than in brine-based production [25,60].

The production of lithium from spodumene ores is a significant contributor to greenhouse gas emissions, as the electricity generation required is carbon-intensive. Additionally, the energy consumption is rather high. As shown in the life cycle analysis performed by [25], one of the reasons is that the means of generating energy during the extraction process from Australian spodumene ore rely only on fossil fuels (diesel). Moreover, the subsequent processing of the spodumene concentrate into either carbonate or lithium hydroxide ($LiOH{\cdot}H_2O$) at a Chinese facility is performed using coal for heat energy.

2.2. Lithium Applications

Lithium is mainly used in the fabrication of Li-ion batteries (LIBs). These batteries are the most promising rechargeable batteries owing to their unique qualities. LIBs are not only widely used in electronics and vehicles, but also in the wheeled, naval, and aviation sectors. Lately, lithium-ion batteries have attracted attention from industry, academia, and governments, as their use in energy storage systems offers a new means of grid energy storage. Li-ion technology has the potential to eliminate the need for costly peak power plants, and at the same time incorporates the usage of renewable energy sources [72–75].

Lithium is a versatile material used for a wide range of applications. The major Li products are used in the manufacture of lubricants, ceramics, glass, synthetic plastics, rubber, pharmaceutical products, and metal alloys [19,76]. Lithium stearate is used in automotive, military, industrial, aircraft, and marine sectors as a high-temperature lubricant grease. The lithium-based grease provides high resistance to water, does not react with oxygen, and does not solidify at low temperatures.

Lithium carbonate is extensively used in the fabrication of glass and ceramics to control thermal expansion and improve thermal resistance. In the production of synthetic rubber and plastics, Li is used as a catalyst [77]. Styrene-butadiene rubber (SBR) is widely used in the production of tires. In the production of SBR, n-butyl lithium, an organic lithium compound, is used as the initiator of the polymerization reactions. n-Butyl lithium, n-butyllithium, and sec-butyl lithium are used to produce polymers with thermoplastic properties [26]. Lithium has been used in the main pharmacological treatment for bipolar disorder; it has been also used for the production of flavors and fragrances [78].

In the metallurgical industry, Li is utilized as a flux to promote the melding and soldering of metals, as it absorbs impurities and inhibits the formation of oxides. Lithium–magnesium and lithium–aluminum alloys are widely used in the manufacturing of armored vehicles and in aerospace products, as these alloys are lightweight and strong. The composites of lithium with cadmium, aluminum, copper, and manganese are used in several aircraft components. The Mg–Li alloy is the lightest, most versatile material commercially available [77]. Lithium niobate and lithium tantalate wafers are used for the production of microchips for electronics [26].

Lithium chloride is utilized in industrial dehumidifier and drying equipment. Lithium bromide is an absorbent in industrial absorption chillers. Lastly, lithium and lithium peroxide are employed to reduce carbon dioxide from the air in submarines and spacecraft [26].

Electric Vehicles Market

Global warming, or climate change, is perhaps the biggest threat to this planet. For instance, the average global temperature has increased by 0.85 °C since the pre-industrialization period [79,80]. Greenhouse gases (GHG) are the major contributing factors to climate change and air pollution, and a significant source of GHG is the transportation sector. For example, according to United States Environmental Protection Agency (EPA), in 2020 transportation accounted for the largest portion (27%) of total U.S. GHG emissions, and among cars, trucks, commercial aircraft, railroads, and other sources, light-duty vehicles emitted more greenhouse gases than any other [81].

Electric vehicles (EVs) are considered a potential solution for a sustainable and low-carbon transportation system, and an important step to reducing emissions and mitigating climate change. EVs do not produce pollutant emissions, as their internal source of energy is an electric motor powered by electric batteries. Additionally, the powertrain offers the possibility of operating with zero emissions. EVs not only offer maximum protection for the environment, but also unique features and advantages. For example, EVs are quieter and overall smoother than gas-based cars. The main reason is that the electric motor provides high torque to the wheels, resulting in smoother acceleration and deceleration, together with zero noise emissions, while operating. These vehicles have been efficiently designed to use the electric motor as a propulsion source and as a generator when decelerating or when moving by gravity. This type of "energy recovery" technology is an important feature in EVs, as it increases the overall energy efficiency of the car [82]. The electric motor of an electric car is smaller than an internal combustion engine, resulting in spacious interiors. With neither a gearbox nor a clutch, EVs make driving hassle-free.

Lithium plays an important role in the development of electric vehicles, as these are powered by Li-ion batteries. These batteries are the best available due to their superior qualities and cutting-edge technology. They have the highest energy density, volume, and mass. LIBs are the most efficient technology for energy storage, as a high density of Li ions can be stored with respect to the battery weight (energy density). By way of comparison, a Li-ion battery provides an energy density (300 to 500 Wh/kg) that is ten times greater than that of a lead-acid battery. Among other features, Li-ion batteries offer one of the highest coulombic efficiency levels (charge and discharge efficiency), a high voltage output, good high-temperature performance, light weight, easy recharging, durability, and the recycling of some of their components, which aligns with the principles of the circular economy, making this technology even more attractive [83,84].

A Li-ion battery consists of two mixed ion and electron conductors as the anode and cathode, which are separated by a separator and an ion-conducting electrolyte [85]. These batteries work on the principle of insertion reactions, where both electrons are atomic frameworks that host mobile lithium ions as charge carriers. Hence, when the battery is charging, lithium ions move from the cathode to the anode, and they move back during discharging [86]. The composition of the battery elements is an essential factor to achieve high performance in the battery, and alternative electrochemical arrangements are being developed in an effort to produce energy storage devices with a higher energy density.

The cathode is one of the critical components in LIBs and the main sources of the active lithium ions. To achieve high capacity and energy output, a significant amount of lithium is contained in this component. As a fact, LIBs in the market are referenced according to the cathode composition, and the key decisive factors in the material selection are gravimetric energy (capability of quick charging), power density, structural stability, safety, and cost [87–89]. On the other hand, the anode can enhance the inserting and escaping capacity of the lithium ions, and this electrode can be made of either carbon and lithium titanate ($Li_4Ti_5O_{12}$), Li metal, or Li (Si) alloys. The battery electrolyte is a solution that allows Li ion transfers between electrodes, and it should provide high lithium-ion transport under any operating and external conditions, such as extreme weather. The electrolyte is usually a lithium salt, such as lithium hexafluorophosphate ($LiPF_6$) or lithium perchlorate ($LiClO_4$) dissolved in a mixture of ethylene carbonate and either dimethyl carbonate, diethyl carbonate, or ethyl methyl carbonate. This organic solvent is selected for its high electrochemical stability, meaning the battery is able to operate on a higher voltage range. However, the composition could change based on the electrode material [90].

The separating membrane is a micro-porous material that allows Li-ions to pass across the pores, and is an important component for ensuring the safe operation and service life of the battery. These separators should be mechanically robust, electrochemically stable, electrolyte wettable, flexible, and offer high lithium-ion conduction. Separators should be equipped with thermal shutdown to prevent safety issues caused by the elevating temperature during operation. This feature can promptly cut off the current and prevent the electrodes from contacting [87,91,92].

2.3. Lithium Demand and Economic Perspective

According to forecasts, global lithium demand will increase significantly, and it is expected to become scarce [30,93]. Investors' reports affirm that the demand might treble by 2025 and outpace supply [94], and other sources forecast the global demand for Li will exceed 2 million metric tons by 2030 [6,95]. The accelerating electrification transition, supported by the global commitment to decarbonization, will be a strong driver of Li consumption in the next decade.

The surge in lithium demand goes alongside the exponential growth in the sale of electric vehicles. The improved range, broader model availability, and increased performance are among the factors driving sales. In addition, EVs are being promoted by policymakers through various policies and incentives, such as rebates, tax credits, discounts, zero-interest loans, and charging station deployment [26,96]. In relation to carbon neutrality targets, there is an expectation for energy storage applications to keep pace with the strong growth in renewable energy deployment around the world.

Lithium has many end use applications aside from batteries, as previously mentioned. The global consumption of the metal is mainly in the lithium battery sector: 46% of the lithium produced is used for battery production. However, glass and ceramics manufacturing consume a significant amount as well—over 27%. Lubricants and greases account for 9%, and continuous casting and polymer production consume around 5% each. Additional industrial applications, such as sanitization, organic synthesis, construction, pharmaceuticals, alloys, and alkyd resins, account for about 9%. The estimated share of the global lithium market is shown in Figure 2 [18,26].

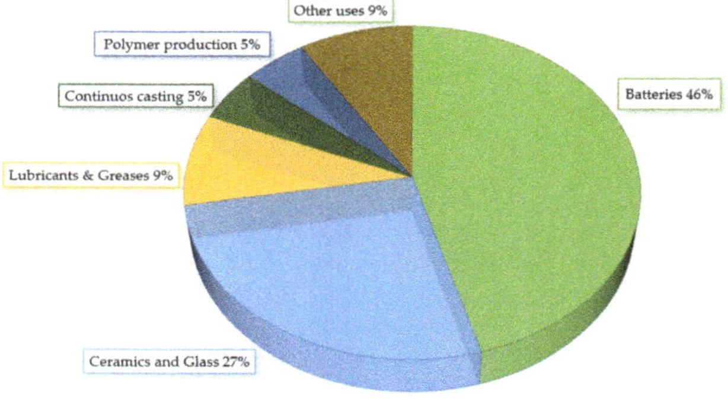

Figure 2. Global lithium end-use market shares [18,26].

The increasing demand for lithium is affecting the price of the metal. In addition, lithium deposits are concentrated in a small number of countries; therefore, there is only a limited supply of this element, a factor that also determines the price of the metal. As shown in Figure 3, in 2010, the lithium carbonate price was under 4000 USD/t, and at the end of 2015, the value rose to 6900 USD/t. This is an increase of 56% in just five years. By 2016, it amounted to 7200 USD/t [26], and in 2021 the average price was estimated at 17,000 USD/t [57]. The price of lithium in the world market will continue increasing as demand is expected to outstrip supply. Financial institutions affirm that supply could be stretched to meet the current demand; however, the response might not be enough to contain the lithium price [94].

The global demand for Li will be difficult to meet unless alternate resources and efficient techniques to recover this valuable metal are implemented. Investing in research and innovation is a critical enabler for the development of new technologies to recover lithium and secure a sustainable supply of the raw material. Funding and more engagement from electronics manufacturing companies and automakers (the biggest consumers of lithium) will increase the likelihood of novel supply coming to the market. Therefore, it is essential to look for more alternatives to cater to the demand for Li in various fields.

As noted earlier, precipitation is touted to be the most efficient, cost effective, and practical solution. It should be placed as the top consideration for Li recovery.

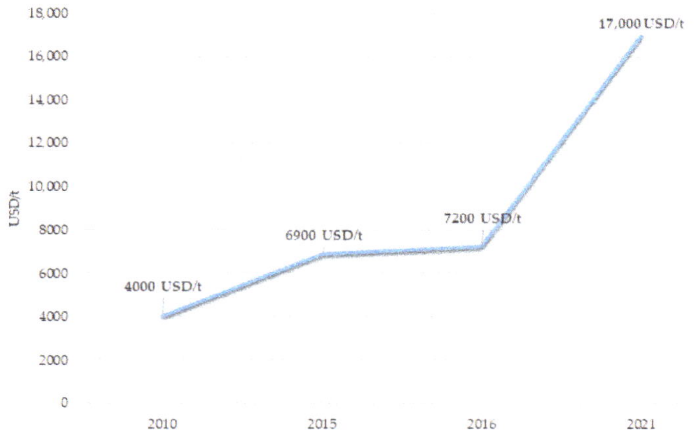

Figure 3. Lithium carbonate price (2010 to 2021) [26,57].

3. Lithium Recovery via Chemical Precipitation

Chemical precipitation is a method used in wastewater treatment to remove ionic components from aqueous wastes by adding counter-ions to reduce their solubility, thereby changing a dissolved material in water into solid particles. It is considered the most efficient technology for the removal of trace metals and rare earth elements from wastewater. The method is relatively simple and inexpensive to operate. The adjustment of pH is an essential step in the precipitation process. For instance, in some scenarios, basic conditions usually allow the dissolved metal to be converted into solid metal hydroxide, which is separated by sedimentation or filtration [97].

Lithium recovery via chemical precipitation has been the object of study in the past few decades. Different approaches have been developed, such as the carbonate precipitation method, phosphate precipitation method, and aluminate precipitation method.

3.1. Current Technology and Challenges

As previously mentioned, the traditional evaporation method requires large evaporation ponds, is time intensive, and is not suitable for all geographical locations, as it requires a dry climate and abundant sunlight. On the other hand, the efficiency of lithium recovery depends on the brine composition, as these aquifers contain high concentrations of other ions, and the co-precipitation of these impurities makes Li extraction more complicated. For example, lithium and magnesium are both alkali/alkaline earth metals and share similar ionic properties, as their ionic size is almost the same. Additionally, their radii are similar: 72 pm for Mg^{2+} and 76 pm for Li^{+}. Therefore, mining Li from brines with a high Mg/Li ratio has been a decades-long technical challenge. A high Mg/Li ratio requires large amounts of precipitant, which results in a huge amount of solid waste generation and high costs. Additionally, the method has a low recovery rate, as lithium is lost due to co-precipitation [98,99].

Carbonate precipitation is the current technology used for lithium extraction from brines at an industrial scale. However, this conventional process is ineffective for most new brine discoveries, with a low concentration of lithium. To obtain a significant level of lithium recovery, the initial Li concentration should be greater than 20 g/L [12]. In addition, the solubility of the precipitate Li_2CO_3 is still considerably high ($K_{sp} = 8.15 \times 10^{-4}$); therefore, a large amount of Na_2CO_3 is used. In addition, a high temperature (~100 °C) is required to precipitate Li as Li_2CO_3, due to its lower solubility at higher temperatures. Such processes make the mining of Li from brines challenging when it is at a low concentration [100–102].

3.2. Current Advancements

In the past few decades, the chemical precipitation of lithium has been intensively studied from an economical and scientific perspective. Different materials have been used, such as carbonate, phosphate, and aluminate, in the precipitation method.

Several studies have been conducted using sodium phosphate salts, such as Di-sodium phosphate (Na_2HPO_4), tri-sodium phosphate (Na_3PO_4), sodium pyro-phosphate ($Na_4P_2O_7$), and sodium tri-polyphosphate ($Na_5P_3O_{10}$) [103–106]. The results obtained by Alsabbagh, Aljarrah [103] show that the four types of salts studied can be used as lithium-precipitating reagents, as the lithium recovery achieved was between 23% and 27%. However, the highest percentage of lithium recovery was achieved by using tri-sodium phosphate (TSP). The research obtained a 40% recovery rate using TSP in a Dead Sea evaporated end brine with an initial lithium concentration of 40 mg/L [103]. Alsabbagh, Aljarrah [103] studied the effects of operating conditions, and the results show that the amount of precipitating reagent had a significant effect on the percentage of Li extracted. In this phase, 1 to 10 g of TSP was used, and 7 g achieved the highest recovery. The stirring speed was adjusted from 150 to 1000 rpm and the ANOVA statistical analysis showed that the stirring speed had a substantial effect. The best percentage extracted was reached at 450 rpm. The time was studied from 30 to 180 min, and the results show that the Li extraction rate did not change significantly with time, and two hours is enough for the process. Finally, the study on the influence of the temperature showed that at 40 °C, the percentage of Li extracted was slightly better than that at other temperatures (25–70 °C). Accordingly, at optimum conditions, more than 40% of the lithium was extracted from Dead Sea evaporated end brine using TSP. In this work, the pH value was neither changed nor studied.

Mulwanda, Senanayake [107] recovered lithium from an alkaline leach solution using phosphoric acid (H_3PO_4). With the purpose of increasing the efficiency of lithium extraction as Li_3PO_4, the study alternatively used H_3PO_4 instead of Na_3PO_4 to minimize the presence of sodium ions (Na^+) in the alkaline liquor. The Li-to-P molar ratio was varied from 3:1 to 3:2 to evaluate its effect on the lithium precipitation rate. The operating conditions were fixed to 90 °C, with a 2 h reaction time, initial pH of 12.5, and initial lithium concentration of 5.6 g/L. The highest precipitation efficiency (92%) was obtained at a Li:P ratio of 3:2; however, the XRD scan of the precipitate showed an additional phase in the solid, Li_2NaPO_4 (impurity). Therefore, the authors determined a molar ratio of Li:P = 3:1.6 as the most desirable, achieving around 83% recovery as Li_3PO_4.

As a new resource of lithium, Shin, Jeong [108] used the waste Li solution generated from the cathode manufacturing process. The recovery of Li as Li_3PO_4 was achieved by using phosphoric acid with the Li/PO_4 molar ratio of three. The effects of the initial Li concentration and the reaction time were studied in a synthetic solution at room temperature for three days. At various Li concentrations of 100, 250, 500, 1000, 2000, and 3000 mg/L, the results indicate that no precipitation was observed when the initial concentration was lower than 250 mg/L; once it was increased from 500 mg/L to 1000 and above, the precipitation reaction started increasing, reaching equilibrium within 12 h.

Shin, Jeong [108] concluded that the number of lithium and phosphorus ions in the solution can be a relevant factor to the precipitation rate of Li_3PO_4; therefore, at concentrations lower than 1000 mg/L, precipitation was not achieved due to a low number of Li-P nuclei. In contrast, at initial Li concentrations of 1000, 2000, and 3000 mg/L, the precipitation efficiencies achieved were 87%, 96%, and 97%, respectively, after three days of reaction. The study tested the phosphoric acid at the Li/PO_4 molar ratio of 3 in a real Li waste solution with a concentration of 2174 mg/L. The experiment was conducted at pH values of 7.0, 9.2, 11.0, and 12.4, and the precipitation efficiency achieved went from 0.2 to 81%, which indicated that the higher the pH, the higher the precipitation efficiency of Li_3PO_4. In the same study, tri-sodium phosphate was tested to evaluate the precipitation efficiency of each precipitating reagent. The experiment was conducted at a Li/PO_4 molar ratio of 3, initial Li concentration of 2000 mg/L, and initial pH of 12.9. After 24 h of reaction, the results revealed that H_3PO_4 had 92% precipitation efficiency, whereas the Na_3PO_4 recovered was 87%.

The precipitation of lithium with activated aluminum-based alloys was proposed by Li, Zhao [109]. Li was recovered as $LiCl{\cdot}Al(OH)_3{\cdot}xH_2O$. At first, Al−Ca alloy and Al−Fe alloy, each with 70% aluminum contents, were assessed to determine the lithium precipitation performance. The experiment was conducted with an initial Li concentration of 1 g/L, at 70 °C, 4:1 Al/Li ratio, and with a three-hour reaction time. The results obtained show that the precipitation rate of lithium with Al−Ca alloy was 93.6%, whereas the Al−Fe alloy achieved only 23.8%. In the subsequent experiments, the effects of the molar ratio of Al to Li, the Ca content of the Al−Ca alloy, the initial Li concentration, the reaction temperature, and the reaction time were evaluated to determine the optimum conditions.

With an initial lithium concentration of 1 g/L, a 3 h reaction time, at 70 °C, and with a 30% Ca content in the Al−Ca alloy, the results show that the molar ratio of Al/Li had a significant effect on the lithium precipitation rate. At the Al-to-Li ratio of 2:1, the recovery achieved was only 72.1%. However, the precipitation rate was improved as the molar ratio was adjusted to higher values, and at an Al/Li of 3.5:1, the process achieved 93.8% recovery. However, as the molar ratio continued to increase, an insignificant improvement in the precipitation rate was obtained. The calcium content in the Al–Ca alloy also played an important part in the precipitation process. In this phase, the Ca content was varied from 10% to 40%, and the results showed a significant increase in the Li extraction rate, from 87.1% to 94.7%, as the Ca content was adjusted from 10% to 35%. This improvement was attributed to an Al−Ca alloy with a larger surface area as the Ca content was increased.

The initial lithium concentration did not show a major effect on the precipitation rate. The recovery achieved went from 71.3% to 95.3% when the lithium-ion concentration was adjusted from 0.4 to 0.8 g/L. Still, greater values did not exhibit better recovery rates. When the temperature was evaluated, it was found that at temperatures over 70 °C, the lithium precipitation rate significantly dropped, going from 93.6% to 66% once the temperature was raised from 70 to 90 °C.

This phenomenon was attributed to thermal motion at high temperatures, where part of the $LiCl{\cdot}Al(OH)_3{\cdot}xH_2O$ decomposes and LiCl dissolves. Finally, with the previous parameters evaluated, the effect of the reaction time was studied, from 0.5 to 3 h under the conditions of a 3.5:1 Al/Li molar ratio, 35% Ca content, 8 g/L initial Li concentration, and 70 °C. The precipitation reaction was very fast, and after 1 h, 94.6% of the lithium was extracted from brine using the Al−Ca alloy.

As mentioned in the previous section, the extraction of lithium in brines with high Mg/Li mass ratio is one of the main challenges of the precipitation process. The study conducted by Liu, Zhong [99] tested aluminum-based materials in a salt-lake brine, to evaluate the influence of the Mg/Li mass ratio, among other parameters, on the Li precipitation rate. The Al-based materials showed good results, as the extraction rate achieved was 64.8% of Li and only 0.8% of Mg, in a solution containing 1 mol/L LiCl and 1 mol/L $MgCl_2$.

The following experiments were conducted using a solution containing an initial Li concentration of 1 g/L and 20 g/L of Mg. The temperature is an important factor in the precipitation process, and in this case had a significant effect when it was increased from 20 to 90 °C. The Li precipitation rate reached 65.8% at 80 °C, whereas only 18.14% was acquired at 20 °C. Interestingly, the result shows that Mg is independent of the change in temperature, as its precipitation rates at various temperatures remained below 0.16%. The reaction time was evaluated, and the results show that from 60 to 180 min, the precipitation rate improved from 34.8% to 62.8%. However, when the reaction time was above 180 min, an insubstantial change was obtained. The concentration of Mg was fixed at 20 g/L to study the effect of lithium initial concentration on the extraction of Li. As the lithium concentration was adjusted from 0.2 to 1 g/L, an evident increase in the precipitation was observed, from 30% to 64.8%. However, it dropped to 52.3% when the Li concentration was 1.5 g/L. The authors determined that as a large number of precipitates are formed at a higher Li concentration, the surface of the Al-based materials gets coated, which decreases the reaction performance. The incrementation of the initial concentration of Mg^{2+} was studied to determine its effect on the Li precipitation rate. The results reveal that an excess of Mg^{2+} in the solution had detrimental effects on the precipitation of Li. The Li precipitation rate dropped from 78.3% to 49.4% when the Mg initial concentration was increased from 15 to 40 g/L.

The production of high-purity lithium carbonate with the current technology requires re-dissolution and re-precipitation of the already precipitated Li_2CO_3. This method consumes large amounts of freshwater and chemicals. Additionally, a high Li concentration is needed to obtain a high Li recovery rate. Zhao, Zhang [101] proposed a potential approach to recover lithium from a low-concentration Li solution and obtain industrial-grade Li_2CO_3 by a one-step precipitation method, with the use of ultrasound. Initially, a comparison of ultrasound and a stirring method at different initial lithium concentrations was conducted to assess the effects of both technologies on the Li recovery rate and purity grade of Li_2CO_3. The results show that at an initial Li concentration of 10 g/L, the Li recovery rate by the ultrasound method outstripped that of the traditional technology, as more than 80% of the Li was extracted. In addition, the Li_2CO_3 produced exceeded industrial-grade purity. On the other hand, the traditional stirring method recovered over 80% of Li only when the initial Li concentration was raised to 20 g/L. Moreover, this technology did not achieve the desirable purity for industrial-grade Li_2CO_3.

Ultrasonic radiation was used to increase the efficiency of the precipitation reaction. The reason involves the formation of cavitation bubbles that provide hot spots with very high temperature and pressure gradients, and these spots can enhance mixing and particle collisions and facilitate the chemical reaction. As part of the study of the ultrasound technology, the power was varied from 90 to 300 W, at 20 KHz, to improve the recovery rate of Li. The results show an increase of more than 10% when the power was incremented from 0 until 150 W. However, at values above 150 W, minor changes were obtained.

The influences of the amount of precipitant reagent (Na_2CO_3) on the lithium precipitation rate and the purity of Li_2CO_3 were evaluated. The dosage was changed from 1 to 1.4 Na/Li (molar ratio). In this phase, the Li recovery rate increased, but the purity of Li_2CO_3 declined. The authors deduced that the increase in carbonate ions in the solution will accelerate the chemical reaction and improve the extraction of lithium. However, the presence of more ions could increase the probability of being trapped in the Li_2CO_3 complex, which increases impurities in the final product and reduces its purity. It is important to highlight that the study investigated the difference between dosing solid Na_2CO_3 versus a highly concentrated Na_2CO_3 solution in the Li recovery. The solid precipitant achieved a higher lithium recovery rate than the saturated solution. Therefore, the solid Na_2CO_3 was used throughout the whole study.

The effect of the reaction temperature was evaluated, and the results showed that the increase in the temperature accelerated the rate of the chemical reaction. When it was raised from 25 to 80 °C, the Li recovery went from 64 to 82%. Moreover, the reaction reached equilibrium in only 5 min at high temperatures. The purity of the Li_2CO_3 also improved from 96% to 99%. Such an increase was attributed to the high temperature, as Na_2CO_3 remained in solution without affecting the precipitate composition, whereas at low temperatures, the Na_2CO_3 dissolved gradually, and some of the undissolved particles were wrapped in the Li_2CO_3 precipitate.

At optimum conditions of 10 g/L of Li, solid Na_2CO_3, ultrasonic power of 150 W, and 35 min reaction time at 80 °C, the lithium recovery rate reached 82.62%, and the Li_2CO_3 was 99.01% pure. The recovery of residual lithium (2 g/L) in the filtrate was consequently extracted by the Na_3PO_4 precipitation method. Therefore, the overall recovery rate reached by the researchers was 97.4%.

3.2.1. Materials Used in the Recovery Process

In the past few decades, the chemical precipitation of lithium from aqueous sources has been intensively studied, and different materials and methods have been proposed—

in particular, carbonate precipitation, phosphate precipitation and aluminate precipitation. Table 1 shows the precipitant materials used for the extraction of lithium from various Li-containing solutions and the lithium recovery rate by precipitant at optimum operating conditions.

Table 1. Lithium recovery rate by precipitant at optimum operating conditions.

Reagents/Precipitant	Dosage	pH	T (°C)	Time (hours)	Efficiency (Li Recovery)	Reference
Sodium carbonate (Na_2CO_3)	Stoichiometric amount of solid Na_2CO_3	12	80	3	72.9%	[110]
Lime milk, NaOH, oxalic acid and carbonate	-	4.6	85	-	84%	[111]
Al−Ca alloy	Al/Li mole ratio of 3.5:1	-	70	3	94.6%	[109]
Aluminum powder and NaCl	-	6	80	3	78.3%	[99]
Tri-sodium phosphate ($Na_3PO_4{\cdot}12H_2O$)	1:1 theoretical amount	11–13	25	5	96.5%	[102]
Tri-sodium phosphate (Na_3PO_4)	PO_4^{3-}/Li^+ molar ratio of 1.3:3	-	65	2	65%	[106]
Sodium carbonate (Na_2CO_3)	Na/Li molar ratio of 1	-	80	0.6	82.62%	[101]
Sodium carbonate (Na_2CO_3)	CO_3^{-2}/Li^+ molar ratio of 1.1:2	7–8	95	-	46.5%	[112]
Al/Na_2SO_4 composite	Al/Li mole ratio of 3:1	-	70	3	89.2%	[113]
NaOH and Na_2CO_3 solution	-	-	80	1.5	85%	[100]
Aluminum chloride ($AlCl_3{\cdot}6H_2O$)	30–40 g/L	6.6–7.2	25	3	90%	[93]
Al/Na_2SO_4 composite	Al/Li mole ratio of 3:1	-	70	3	89.2%	[113]
Tri-sodium phosphate (Na_3PO_4)	Li/Na_3PO_4 mole ratio of 1	9.5	90	5	95.4%	[105]
Tri-sodium phosphate (Na_3PO_4)	5 g/L	6.3	40	0.5	40%	[103]
Facet engineered Li_3PO_4 crystal and sodium phosphate dodecahydrate ($Na_3PO_4{\cdot}12H_2O$)	40 g/L(S/L ratio) seed + 3:1 theoretical Li^+/PO_4^{3-} molar ratio of $Na_3PO_4{\cdot}12H_2O$	10–12	30	0.5	51.62%	[114]
Tri-sodium phosphate dodecahydrate ($Na_3PO_4{\cdot}12H_2O$)	Li/P mole ratio of 3:1	10–10.3	30	2	88.49%	[115]
Tri-sodium phosphate (Na_3PO_4)	1.2:1 theoretical amount	8	70	1.5	88.44%	[104]
Activated Al-Ca and Al-Fe alloys	30–40 g/L	-	70	1	94.6%	[15]
Phosphoric acid H_3PO_4	Li/P mole ratio of 3:1.6	12.5	90	2	83%	[107]
Phosphoric acid H_3PO_4	Li^+/PO_4^{3-} mole ratio of 3	12.4–13.5	25	24	81%	[108]

3.2.2. Operating Conditions and Performance

The lithium recovery rate and the operation conditions can differ considerably among precipitant materials. Additionally, the nature of the Li-bearing solution appears to be a determining factor that influences the operating conditions, as the selection of the optimum parameters for the chemical precipitation will benefit the extraction of Li rather than other ions.

Influence of pH on Lithium Precipitation

pH is a key parameter in the chemical precipitation process, and finding its optimum value is crucial to efficiently extract lithium. Li, Li [104] examined the influence of pH on the lithium precipitation rate by the phosphate precipitation method. The authors concluded that the Li extraction rate increases as the pH value reaches the basic range. When the reaction pH is above 8, the lithium extraction rate stabilizes, at nearly 80%. In acidic conditions, soluble Li compounds are formed, as the many H^+ molecules in the solution react with $PO_4{}^{3-}$ to produce $HPO_4{}^-$ and $H_2PO_4{}^-$; subsequently, Li^+ reacts with $HPO_4{}^-$ and H_2PO_4 to produce Li_2HPO_4 and LiH_2PO_4, which are highly soluble in water and therefore affect the precipitation of lithium.

High concentrations of other ions have a significant effect on the pH adjustment. To effectively precipitate Li and avoid the co-precipitation of impurities, the adjustment of pH is an essential step in the Li extraction. The recovery of Li from leach liquors could be quite challenging, as these solutions contain various dissolved species. Xiao and Zeng [102] recovered lithium from a Li-containing leach solution using Na_3PO_4. During the step of pH adjustment, the researchers noticed a substantial reduction in the concentration of ferric ions (Fe^{3+}) instead of lithium, when the pH was varied from 0 to 5.5. The rationale behind the change involves the formation of $FePO_4$, as phosphate preferentially precipitates ferric ions rather than Li ions at pH values from 0 to 5.5. From this study, the authors concluded that the extraction of lithium from a solution containing Li^+ and Fe^{3+} using phosphate materials should be kept in the pH range of 5.5 to 14.0. The optimum pH window to recover lithium as Li_3PO_4 was on the scale of 11.0–13.0. At pH values higher than 13.5, the study found that Li was being precipitated as lithium hydroxide (LiOH) rather than lithium phosphate. This finding further validated the optimum pH window of 11.0 to 13.0. Lastly, the verification experiment achieved a 96.5% lithium recovery rate at a pH of 11.

Shin, Jeong [108] recovered lithium from a lithium waste solution using phosphoric acid. The effect of the pH on the Li precipitation rate was evaluated using pH levels of 7.0, 9.2, and 12.4. The results show a significant improvement in the lithium precipitation rate as the pH became more basic. The precipitation efficiency went from 0.2 to 81% as the pH of the solution increased from 6.4 to 12.4.

Increasing the pH might not be a critical step when using aluminum-based materials. Liu, Zhong's [99] study is a case in point. They achieved a 78.3% lithium recovery rate by adding aluminum-based materials, and the pH value (5.5) of the brine remained unchanged throughout the precipitation process. The authors highlighted that a change in pH in the brine could lead to negative effects within the salt-lake ecosystem.

Influence of Temperature on Lithium Precipitation

According to the collision theory of reactivity, chemical reactions occur when reactant particles "effectively collide". Moreover, an effective molecular collision requires a minimum amount of kinetic energy in the molecule. A high temperature increases the average kinetic energy of the reactant molecules and causes molecules to move faster, increasing the rate of intermolecular collisions. These collisions promote more molecules to interact in the reaction and increase the reaction rate. Thus, in theory, the precipitation rate of lithium should increase with the increase in the temperature.

For example, Shin, Joo [105] used Na_3PO_4 to study the effect of the temperature on the Li precipitation rate, at 30, 60, and 90 °C. The results showed that at 90 °C, the precipitation rate obtained was over 90% in only 5 min of reaction time, regardless of the initial Li concentration (1.7, 4.0 and, 7 g/L). At 30 °C, the lithium extracted was only 70% over the same period of time, and high recovery was only achieved at the highest Li concentration.

The maximum precipitation rate of Li, Li [104] was reached at 70 °C, using Na_3PO_4 as the precipitant, in a Li solution containing 262.43 mg/L of lithium. When the temperature was increased from 50 to 70 °C, the extraction increased from about 77% to 86.57%. The authors observed that the higher the temperature, the better the Na_3PO_4 precipitation of lithium. However, once the temperature was raised from 70 to 90 °C, the extraction rate plateaued.

Li, Zhao [109] extracted lithium as $LiCl{\cdot}Al(OH)_3{\cdot}xH_2O$ from a salt-lake brine using aluminum-based alloys. In this study, the increase in the temperature (50 to 90 °C) did not have a positive effect on the lithium recovery rate. First, there was no significant difference in the Li extraction efficiency when the temperature was raised from 50 to 70 °C. As the temperature was increased from 70 to 90 °C, a significant decline in the Li precipitation rate was observed, dropping from 93.6 to 66%. The authors stated that at more than 70 °C, $LiCl{\cdot}Al(OH)_3{\cdot}xH_2O$ disintegrates and LiCl dissolves in water, decreasing the extraction rate of Li.

About 90% of Li was extracted from Dead Sea brine as lithium aluminate, using aluminum chloride ($AlCl_3{\cdot}6H_2O$). The highest yields were achieved at room temperature, and as the temperature was increased, the lithium recovery declined [93,116]. Theoretically, a high temperature increases the rate of a reaction; however, temperature does not exhibit the same behavior in all precipitation methods. High lithium recovery can be achieved at room temperature and prolonged reaction times when the initial Li concentration is high.

The solubility product of the precipitate plays an important part in the selection of the reaction temperature. For example, the solubility product of Li_2CO_3 is relatively high, 8.15×10^{-4} (pKsp = 2.2), in comparison with Li_3PO_4, 2.37×10^{-11} (pKsp = 10.63) [104]. Therefore, the carbonate precipitation method is usually conducted in the range of 80 to 100 °C, as the solubility of Li_2CO_3 decreases as the temperature rises [110,112].

Influence of Reaction Time on Lithium Precipitation

The reaction time may perhaps be linked to the initial Li concentration of the solution and the reaction temperature: high Li concentrations and high temperatures shorten the time required to achieve efficient lithium extraction. Shin, Joo [105] studied the precipitation efficiency of lithium over time at different temperatures and initial lithium concentrations. In only five minutes, more than 90% of the lithium was extracted at 90 °C, despite the initial Li concentration. At 30 °C, only 70% was extracted in the first five minutes; however, the precipitation efficiency of lithium increased gradually, given more time, and at higher Li concentrations. The authors indicated that to obtain the critical size in the crystallization of Li into Li_3PO_4, a longer period of time is needed. Therefore, the extraction rate of lithium as Li_3PO_4 is slow at low temperatures, and the higher the temperature, the faster the rate.

Xiao and Zeng [102] used Na_3PO_4 to extract lithium from a synthetic LiCl solution; the precipitation of Li reached 96.5% after 5 h of reaction time at 25 °C. Shin, Jeong [108] used phosphoric acid to recover Li from a LiOH solution. The experiment achieved around 92% precipitation efficiency after 24 h of reaction time at room temperature, with an initial Li concentration of 2000 mg/L. Evidently, a low reaction temperature requires a longer period of time to extract lithium as Li_3PO_4, and a high lithium concentration is key to obtaining a significant extraction rate.

3.3. Challenges and Outlook

Although much progress has been made in lithium recovery via chemical precipitation, there are still some challenges to overcome. Many studies have been conducted to separate lithium from magnesium in salt-lake brines; however, a high Mg/Li ratio requires large amounts of precipitant, which results in huge amount of solid waste generation and high costs [98]. The study of the aluminate precipitation method showed high lithium recovery. However, increasing the magnesium concentration to a Mg/Li mass ratio of 20:1 was detrimental to the precipitation process, and the lithium precipitation efficiency decreased [113].

To improve the precipitation separation of lithium from magnesium, the effects of various parameters on the precipitation process must be entirely and systematically studied.

The efficiency of lithium recovery depends on the Li-bearing composition. High concentrations of dissolved ions result in the co-precipitation of these impurities, making the Li extraction process more complicated. To efficiently harvest lithium from waste solutions, the characteristics of impurity removal and the interactions between the present ions must be understood. Once the removal of impurities is achieved, Li can be recovered in a one-step precipitation process with a high recovery rate and high purity of the final Li product [117].

The phosphate precipitation method has shown promising results in recovering lithium from aqueous sources. However, the direct recovery of lithium from solutions with low concentrations of lithium faces serious challenges, such as low induction period and low efficiency. Additionally, the formation of a Li_3PO_4 precipitate requires large supersaturation and a high nucleation energy barrier [114]. To achieve a high recovery rate, the reaction temperature should be raised to 70 °C or higher, and the initial lithium concentration should be higher than 2 g/L [105,114,118]. Thus, a high recovery rate is still conditional on a high initial lithium concentration.

Even though the proposed chemical precipitation technologies are technically feasible, the majority of the reported materials are still at the bench scale. To prove the concept and evaluate the feasibility of the proposed methods, pilot-scale tests would be an important step.

Lanthanum compounds have received lots of attention in various applications because of their flexibility and multifunctionality. There is indeed significant evidence of La^{3+} ions being extensively used to remove toxic pollutants from wastewater [119–121]. Lanthanum reagents have shown several advantages over traditional precipitation methods, such as high performance with and without pH adjustment. Although it is called a rare earth element (RE), La is relatively abundant, which means it offers a potentially cost-effective option. RE technology for wastewater treatment does not have negative effects on the ecosystem. Moreover, the La precipitation method has favorable characteristics for closed-loop technology, as it can form insoluble complexes with carbonate (CO_3^{2-}), hydroxide (OH^-), and fluoride (F^-) [122], favoring its recovery for reuse. In summary, the RE technology exhibits promising features for precipitating Li from wastewater with closed-loop technology, which aligns with the circular economy principles.

4. Conclusions

Current extraction methods are water- and energy-intensive. The conventional process for extracting lithium from brines requires a dry climate, abundant sunlight, and large evaporation ponds that are harmful to the environment. The inherent slowness of the process, and the demands for both land and freshwater made by the method, raise questions regarding the sustainability of the process. Lithium-brine developers need new technology to quickly and efficiently achieve the intended production levels.

The accelerating electrification transition, supported by the global commitment to decarbonization, is a strong driver of Li consumption. In addition, there is an expectation for energy storage applications to keep pace with the strong growth in renewable energy deployment around the world. The growing trend is projected to continue, to the point that Li availability may become scarce.

The increasing demand for lithium is affecting the price of the metal. In fact, the price of lithium on the world market has significantly increased since 2010, and this is not expected to be temporary.

The global demand for Lithium will be difficult to meet unless alternate resources and efficient techniques to recover this valuable metal are implemented.

Chemical precipitation is a mature technology, and is considered the most efficient method for the removal of trace metals and rare earth elements from wastewater. In the past few decades, the chemical precipitation of lithium has been intensively studied, and it has shown great potential due to the high lithium recovery rates (>98%) and high purity of the final product (more than 99%). At the optimum operating conditions, lithium can be precipitated in quite a short time, and the method can selectively precipitate lithium and avoid co-precipitation of impurities. Different methods and materials used to recover lithium from aqueous solutions were critically reviewed above.

Mining Li from brines with high Mg/Li ratios is still challenging, as these elements share similar ionic properties. Aluminate precipitation can effectively precipitate lithium in brines of this type; however, the method involves high alkali consumption and a high cost of production.

Further research should be conducted on harvesting lithium from new brines with lower grades of Li (<100 mg/L). The vast majority of studies have focused on the precipitation of lithium from solutions with above 1 g/L of Li.

Investing in research and innovation is a critical enabler for the development of new technologies to recover lithium and secure a sustainable supply of the raw material. Funding and more engagement from the industrial sector will increase the likelihood of novel supplies coming into market.

In summary, the development of a sustainable process to harvest lithium from Li waste solutions, such as produced water, may result in a new revenue stream for companies. At the same time, it may contribute to the circular economy by converting hazardous waste into a raw material. Moreover, by recovering lithium from wastewater, Li production can be bolstered to meet the growing demands and power the energy economy.

Author Contributions: Conceptualization, L.V.G. and M.M.M.T.; methodology, L.V.G. and Y.-C.H.; validation, Y.-C.H. and J.W.L.; formal analysis, L.V.G.; investigation, L.V.G.; resources, Y.-C.H., M.M.M.T. and D.S.H.; writing—original draft preparation, L.V.G.; writing—review and editing, L.V.G.; supervision, Y.-C.H., M.M.M.T. and D.S.H.; project administration, Y.-C.H.; funding acquisition, Y.-C.H., M.M.M.T. and D.S.H. All authors have read and agreed to the published version of the manuscript.

Funding: This research was funded by PETRONAS, grant number 015LC0-340.

Institutional Review Board Statement: Not applicable.

Informed Consent Statement: Not applicable.

Data Availability Statement: Not applicable.

Acknowledgments: The authors would like to express our deepest gratitude to Norhayama Bt Ramli for technical assistance and Universiti Teknologi PETRONAS for providing laboratory facilities. The authors would also like to thank the reviewers for all their comments, along with Juan Diego Merchan for his support and encouragement throughout the study.

Conflicts of Interest: The authors declare no conflict of interest.

References

1. Qiao, D.; Wang, G.; Gao, T.; Wen, B.; Dai, T. Potential impact of the end-of-life batteries recycling of electric vehicles on lithium demand in China: 2010–2050. *Sci. Total Environ.* **2021**, *764*, 142835. [CrossRef] [PubMed]
2. Pavón, S.; Kahl, M.; Hippmann, S.; Bertau, M. Lithium recovery from production waste by thermal pre-treatment. *Sustain. Chem. Pharm.* **2022**, *28*, 100725. [CrossRef]
3. Shankar Naik, S.; Lee, S.J.; Yu, Y.; Al-Mohaimeed, A.M.; Theerthagiri, J.; Choi, M.Y. Novel approach for the synthesis and recovery of lithium carbonate using a pulsed laser irradiation technique. *Mater. Lett.* **2022**, *308*, 131218. [CrossRef]
4. Siekierka, A.; Bryjak, M. Selective sorbents for recovery of lithium ions by hybrid capacitive deionization. *Desalination* **2021**, *520*, 115324. [CrossRef]
5. Alessia, A.; Alessandro, B.; Maria, V.-G.; Carlos, V.-A.; Francesca, B. Challenges for sustainable lithium supply: A critical review. *J. Clean. Prod.* **2021**, *300*, 126954. [CrossRef]
6. Zhou, G.; Chen, L.; Li, X.; Luo, G.; Yu, Z.; Yin, J.; Fan, L.; Chao, Y.; Jiang, L.; Zhu, W. Construction of truncated-octahedral $LiMn_2O_4$ for battery-like electrochemical lithium recovery from brine. *Green Energy Environ.* 2022, *in press*. [CrossRef]
7. Kumar, A.; Fukuda, H.; Hatton, T.A.; Lienhard, J.H. Lithium Recovery from Oil and Gas Produced Water: A Need for a Growing Energy Industry. *ACS Energy Lett.* **2019**, *4*, 1471–1474. [CrossRef]

8. Lee, J.; Chung, E. Lithium recovery by solvent extraction from simulated shale gas produced water—Impact of organic compounds. *Appl. Geochem.* **2020**, *116*, 104571. [CrossRef]
9. Zante, G.; Trébouet, D.; Boltoeva, M. Solvent extraction of lithium from simulated shale gas produced water with a bifunctional ionic liquid. *Appl. Geochem.* **2020**, *123*, 104783. [CrossRef]
10. Seip, A.; Safari, S.; Pickup, D.M.; Chadwick, A.V.; Ramos, S.; Velasco, C.A.; Cerrato, J.M.; Alessi, D.S. Lithium recovery from hydraulic fracturing flowback and produced water using a selective ion exchange sorbent. *Chem. Eng. J.* **2021**, *426*, 130713. [CrossRef]
11. Bazrgar Bajestani, M.; Moheb, A.; Dinari, M. Preparation of lithium ion-selective cation exchange membrane for lithium recovery from sodium contaminated lithium bromide solution by electrodialysis process. *Desalination* **2020**, *486*, 114476. [CrossRef]
12. Zhang, Y.; Hu, Y.; Wang, L.; Sun, W. Systematic review of lithium extraction from salt-lake brines via precipitation approaches. *Miner. Eng.* **2019**, *139*, 105868. [CrossRef]
13. Razmjou, A.; Eshaghi, G.; Orooji, Y.; Hosseini, E.; Korayem, A.H.; Mohagheghian, F.; Boroumand, Y.; Noorbakhsh, A.; Asadnia, M.; Chen, V. Lithium ion-selective membrane with 2D subnanometer channels. *Water Res.* **2019**, *159*, 313–323. [CrossRef]
14. Steven Kurniawan, Y.; Rao Sathuluri, R.; Ohto, K.; Iwasaki, W.; Kawakita, H.; Morisada, S.; Miyazaki, M. A rapid and efficient lithium-ion recovery from seawater with tripropyl-monoacetic acid calix[4]arene derivative employing droplet-based microreactor system. *Sep. Purif. Technol.* **2019**, *211*, 925–934. [CrossRef]
15. Khalil, A.; Mohammed, S.; Hashaikeh, R.; Hilal, N. Lithium recovery from brine: Recent developments and challenges. *Desalination* **2022**, *528*, 115611. [CrossRef]
16. Nishio, K. Primary Batteries—Nonaqueous Systems | Lithium–Manganese Dioxide. In *Encyclopedia of Electrochemical Power Sources*; Garche, J., Ed.; Elsevier: Amsterdam, The Netherlands, 2009; pp. 83–92.
17. Choubey, P.K.; Kim, M.-S.; Srivastava, R.R.; Lee, J.-C.; Lee, J.-Y. Advance review on the exploitation of the prominent energy-storage element: Lithium. Part I: From mineral and brine resources. *Miner. Eng.* **2016**, *89*, 119–137. [CrossRef]
18. Tadesse, B.; Makuei, F.; Albijanic, B.; Dyer, L. The beneficiation of lithium minerals from hard rock ores: A review. *Miner. Eng.* **2019**, *131*, 170–184. [CrossRef]
19. Wang, J.; Yue, X.; Wang, P.; Yu, T.; Du, X.; Hao, X.; Abudula, A.; Guan, G. Electrochemical technologies for lithium recovery from liquid resources: A review. *Renew. Sustain. Energy Rev.* **2022**, *154*, 111813. [CrossRef]
20. Sterba, J.; Krzemień, A.; Riesgo Fernández, P.; Escanciano García-Miranda, C.; Fidalgo Valverde, G. Lithium mining: Accelerating the transition to sustainable energy. *Resour. Policy* **2019**, *62*, 416–426. [CrossRef]
21. Oliazadeh, M.; Aghamirian, M.; Ali, S.; Legault, E.; Gibson, C. Flowsheet Development for Benefication of Lithium Minerals from Hard Rock Deposits. In *Extraction*; Springer International Publishing: Cham, Switzerland, 2018.
22. Meshram, P.; Pandey, B.; Mankhand, T. Extraction of lithium from primary and secondary sources by pre-treatment, leaching and separation: A comprehensive review. *Hydrometallurgy* **2014**, *150*, 192–208. [CrossRef]
23. Bowell, R.J.; Lagos, L.; de los Hoyos, C.R.; Declercq, J. Classification and Characteristics of Natural Lithium Resources. *Elements* **2020**, *16*, 259–264. [CrossRef]
24. Grew, E.S. The Minerals of Lithium. *Elements* **2020**, *16*, 235–240. [CrossRef]
25. Kelly, J.C.; Wang, M.; Dai, Q.; Winjobi, O. Energy, greenhouse gas, water life cycle analysis of lithium carbonate and lithium hydroxide monohydrate from brine and ore resources and their use in lithium ion battery cathodes and lithium ion batteries. *Resour. Conserv. Recycl.* **2021**, *174*, 105762. [CrossRef]
26. Martin, G.; Rentsch, L.; Höck, M.; Bertau, M. Lithium market research—Global supply, future demand and price development. *Energy Storage Mater.* **2017**, *6*, 171–179. [CrossRef]
27. Bulatovic, S.M. Beneficiation of Lithium Ores. In *Handbook of Flotation Reagents: Chemistry, Theory and Practice*; Bulatovic, S.M., Ed.; Elsevier: Amsterdam, The Netherlands, 2015; Chapter 28; pp. 41–56.
28. Sitando, O.; Crouse, P.L. Processing of a Zimbabwean petalite to obtain lithium carbonate. *Int. J. Miner. Process.* **2012**, *102–103*, 45–50. [CrossRef]
29. Yelatontsev, D.; Mukhachev, A. Processing of lithium ores: Industrial technologies and case studies—A review. *Hydrometallurgy* **2021**, *201*, 105578. [CrossRef]
30. Vikström, H.; Davidsson, S.; Höök, M. Lithium availability and future production outlooks. *Appl. Energy* **2013**, *110*, 252–266. [CrossRef]
31. Grosjean, C.; Miranda, P.H.; Perrin, M.; Poggi, P. Assessment of world lithium resources and consequences of their geographic distribution on the expected development of the electric vehicle industry. *Renew. Sustain. Energy Rev.* **2012**, *16*, 1735–1744. [CrossRef]
32. Pistilli, M. Investing News Network. 2022. Available online: https://investingnews.com/how-to-invest-in-lithium/\T1 \textquotedblright (accessed on 26 October 2022).
33. Cabello, J. Lithium brine production, reserves, resources and exploration in Chile: An updated review. *Ore Geol. Rev.* **2021**, *128*, 103883. [CrossRef]
34. Kumar, R.; Liu, C.; Ha, G.-S.; Park, Y.-K.; Ali Khan, M.; Jang, M.; Kim, S.-H.; Amin, M.A.; Gacem, A.; Jeon, B.-H. Downstream recovery of Li and value-added metals (Ni, Co, and Mn) from leach liquor of spent lithium-ion batteries using a membrane-integrated hybrid system. *Chem. Eng. J.* **2022**, *447*, 137507. [CrossRef]

35. Zhou, L.F.; Yang, D.; Du, T.; Gong, H.; Luo, W.B. The Current Process for the Recycling of Spent Lithium Ion Batteries. *Front. Chem.* **2020**, *8*, 578044. [CrossRef]
36. Sonoc, A.; Jeswiet, J.; Soo, V. Opportunities to Improve Recycling of Automotive Lithium Ion Batteries. *Procedia CIRP* **2015**, *29*, 752–757. [CrossRef]
37. Rouquette, L.M.J.; Lemaître, T.; Vieceli, N.; Petranikova, M. Intensification of lithium carbonation in the thermal treatment of spent EV Li-ion batteries via waste utilization and selective recovery by water leaching. *Resour. Conserv. Recycl. Adv.* **2022**, *17*, 200125. [CrossRef]
38. Espinosa, D.C.R.; Bernardes, A.; Tenório, J. An overview on the current processes for the recycling of batteries. *J. Power Sources* **2004**, *135*, 311–319. [CrossRef]
39. Lain, M.J. Recycling of lithium ion cells and batteries. *J. Power Sources* **2001**, *97–98*, 736–738. [CrossRef]
40. Al-Thyabat, S.; Nakamura, T.; Shibata, E.; Iizuka, A. Adaptation of minerals processing operations for lithium-ion (LiBs) and nickel metal hydride (NiMH) batteries recycling: Critical review. *Miner. Eng.* **2013**, *45*, 4–17. [CrossRef]
41. Joulié, M.; Laucournet, R.; Billy, E. Hydrometallurgical process for the recovery of high value metals from spent lithium nickel cobalt aluminum oxide based lithium-ion batteries. *J. Power Sources* **2014**, *247*, 551–555. [CrossRef]
42. Guzolu, J.S.; Gharabaghi, M.; Mobin, M.; Alilo, H. Extraction of Li and Co from Li-ion Batteries by Chemical Methods. *J. Inst. Eng. India Ser. D* **2017**, *98*, 43–48. [CrossRef]
43. Chen, Y.; Chang, D.; Liu, N.; Hu, F.; Peng, C.; Zhou, X.; He, J.; Jie, Y.; Wang, H.; Wilson, B.P.; et al. Biomass-Assisted Reductive Leaching in H_2SO_4 Medium for the Recovery of Valuable Metals from Spent Mixed-Type Lithium-Ion Batteries. *JOM* **2019**, *71*, 4465–4472. [CrossRef]
44. Yang, Y.; Xu, S.; He, Y. Lithium recycling and cathode material regeneration from acid leach liquor of spent lithium-ion battery via facile co-extraction and co-precipitation processes. *Waste Manag.* **2017**, *64*, 219–227. [CrossRef]
45. Liu, J.; Mak, T.Y.; Meng, Z.; Wang, X.; Cao, Y.; Lu, Z.; Suen, D.W.-S.; Lu, X.-Y.; Tang, Y. Efficient recovery of lithium as Li_2CO_3 and cobalt as Co_3O_4 from spent lithium-ion batteries after leaching with p-toluene sulfonic acid. *Hydrometallurgy* **2022**, *216*, 106012. [CrossRef]
46. Nayaka, G.P.; Pai, K.V.; Manjanna, J.; Keny, S.J. Use of mild organic acid reagents to recover the Co and Li from spent Li-ion batteries. *Waste Manag.* **2016**, *51*, 234–238. [CrossRef] [PubMed]
47. Gao, W.; Zhang, X.; Zheng, X.; Lin, X.; Cao, H.; Zhang, Y.; Sun, Z. Lithium Carbonate Recovery from Cathode Scrap of Spent Lithium-Ion Battery: A Closed-Loop Process. *Environ. Sci. Technol.* **2017**, *51*, 1662–1669. [CrossRef]
48. He, L.-P.; Sun, S.-Y.; Mu, Y.-Y.; Song, X.-F.; Yu, J.-G. Recovery of Lithium, Nickel, Cobalt, Manganese from Spent Lithium-Ion Batteries Using l-Tartaric Acid as a Leachant. *ACS Sustain. Chem. Eng.* **2017**, *5*, 714–721. [CrossRef]
49. Verma, A.; Kore, R.; Corbin, D.R.; Shiflett, M.B. Metal Recovery Using Oxalate Chemistry: A Technical Review. *Ind. Eng. Chem. Res.* **2019**, *58*, 15381–15393. [CrossRef]
50. Zeng, X.; Li, J.; Shen, B. Novel approach to recover cobalt and lithium from spent lithium-ion battery using oxalic acid. *J. Hazard. Mater.* **2015**, *295*, 112–118. [CrossRef]
51. Yadav, P.; Jie, C.J.; Tan, S.; Srinivasan, M. Recycling of cathode from spent lithium iron phosphate batteries. *J. Hazard. Mater.* **2020**, *399*, 123068. [CrossRef]
52. Ekberg, C.; Petranikova, M. Lithium Batteries Recycling. In *Lithium Process Chemistry*; Chagnes, A., Światowska, J., Eds.; Elsevier: Amsterdam, The Netherlands, 2015; Chapter 7; pp. 233–267.
53. Al-Ghouti, M.A.; Al-Kaabi, M.A.; Ashfaq, M.Y.; Da'na, D.A. Produced water characteristics, treatment and reuse: A review. *J. Water Process. Eng.* **2019**, *28*, 222–239. [CrossRef]
54. He, Y.; Flynn, S.L.; Folkerts, E.J.; Zhang, Y.; Ruan, D.; Alessi, D.S.; Martin, J.W.; Goss, G.G. Chemical and toxicological characterizations of hydraulic fracturing flowback and produced water. *Water Res.* **2017**, *114*, 78–87. [CrossRef]
55. Tian, L.; Liu, Y.; Tang, P.; Yang, Y.; Wang, X.; Chen, T.; Bai, Y.; Tiraferri, A.; Liu, B. Lithium extraction from shale gas flowback and produced water using H1.33Mn1.67O4 adsorbent. *Resour. Conserv. Recycl.* **2022**, *185*, 106476. [CrossRef]
56. Christmann, P.; Gloaguen, E.; Labbé, J.-F.; Melleton, J.; Piantone, P. Global Lithium Resources and Sustainability Issues. In *Lithium Process Chemistry*; Chagnes, A., Światowska, J., Eds.; Elsevier: Amsterdam, The Netherlands, 2015; Chapter 1; pp. 1–40.
57. U.S. Geological Survey. Mineral commodity summaries 2022. In *Mineral Commodity Summaries*; U.S. Geological Survey: Reston, VA, USA, 2022; pp. 100–101.
58. Canada, G.O. Lithium Facts. Available online: https://www.nrcan.gc.ca/our-natural-resources/minerals-mining/minerals-metals-facts/lithium-facts/24009 (accessed on 10 March 2022).
59. Garcés, I.; Álvarez, G. Water mining and extractivism of the Salar de Atacama, Chile. *WIT Trans. Ecol. Environ.* **2020**, *245*, 189–199.
60. Chordia, M.; Wickerts, S.; Nordelöf, A.; Arvidsson, R. Life cycle environmental impacts of current and future battery-grade lithium supply from brine and spodumene. *Resour. Conserv. Recycl.* **2022**, *187*, 106634. [CrossRef]
61. Flexer, V.; Baspineiro, C.; Galli, C. Lithium recovery from brines: A vital raw material for green energies with a potential environmental impact in its mining and processing. *Sci. Total Environ.* **2018**, *639*, 1188–1204. [CrossRef]
62. Stamp, A.; Lang, D.; Wäger, P. Environmental impacts of a transition toward e-mobility: The present and future role of lithium carbonate production. *J. Clean. Prod.* **2012**, *23*, 104–112. [CrossRef]
63. Wanger, T.C. The Lithium future—Resources, recycling, and the environment. *Conserv. Lett.* **2011**, *4*, 202–206. [CrossRef]
64. Ejeian, M.; Grant, A.; Shon, H.K.; Razmjou, A. Is lithium brine water? *Desalination* **2021**, *518*, 115169. [CrossRef]

65. Baspineiro, C.F.; Franco, J.; Flexer, V. Performance of a double-slope solar still for the concentration of lithium rich brines with concomitant fresh water recovery. *Sci. Total Environ.* **2021**, *791*, 148192. [CrossRef]
66. Feo, G.D.; Gisi, S.D. Using MCDA and GIS for hazardous waste landfill siting considering land scarcity for waste disposal. *Waste Manag.* **2014**, *34*, 2225–2238. [CrossRef]
67. Morse, I. Water or Mineral? In Chile, a Debate Over Lithium Brine. Available online: https://undark.org/2020/12/21/chile-debate-over-lithium-brine/ (accessed on 21 December 2020).
68. Yu, F.; Wang, Y.; Zhang, L. Effect of spodumene leaching with sodium hydroxide on its flotation. *Physicochem. Probl. Miner. Process.* **2015**, *51*, 745–754.
69. Sverdrup, H.U. Modelling global extraction, supply, price and depletion of the extractable geological resources with the LITHIUM model. *Resour. Conserv. Recycl.* **2016**, *114*, 112–129. [CrossRef]
70. Rosales, G.D.; Ruiz, M.d.C.; Rodriguez, M.H. Novel process for the extraction of lithium from β-spodumene by leaching with HF. *Hydrometallurgy* **2014**, *147–148*, 1–6. [CrossRef]
71. Prior, T.; Wäger, P.A.; Stamp, A.; Widmer, R.; Giurco, D. Sustainable governance of scarce metals: The case of lithium. *Sci. Total Environ.* **2013**, *461–462*, 785–791. [CrossRef] [PubMed]
72. Breeze, P. Power System Energy Storage Technologies. In *Power Generation Technologies*, 3rd ed.; Breeze, P., Ed.; Newnes: Oxford, UK, 2019; Chapter 10; pp. 219–249.
73. Shen, K.; Xu, X.; Tang, Y. Recent progress of magnetic field application in lithium-based batteries. *Nano Energy* **2022**, *92*, 106703. [CrossRef]
74. Bandini, G.; Caposciutti, G.; Marracci, M.; Buffi, A.; Tellini, B. Characterization of lithium-batteries for high power applications. *J. Energy Storage* **2022**, *50*, 104607. [CrossRef]
75. Gutsch, M.; Leker, J. Global warming potential of lithium-ion battery energy storage systems: A review. *J. Energy Storage* **2022**, *52*, 105030. [CrossRef]
76. Das, D.; Abarajitha, R.; Kay, P.; Ramamurthy, V.; Goycoolea, F.M.; Das, N. Selective recovery of lithium from spent coin cell cathode leachates using ion imprinted blended chitosan microfibers: Pilot scale studies provide insights on scalability. *J. Hazard. Mater.* **2022**, *431*, 128535. [CrossRef]
77. Aral, H.; Vecchio-Sadus, A. Lithium: Environmental Pollution and Health Effects. In *Encyclopedia of Environmental Health*, 2nd ed.; Nriagu, J., Ed.; Elsevier: Oxford, UK, 2011; pp. 116–125.
78. Gao, K.; Calabrese, J.R. The mechanisms of action of lithium in bipolar disorder. In *Neurobiology of Bipolar Disorder*; Quevedo, J., Carvalho, A.F., Vieta, E., Eds.; Academic Press: Cambridge, MA, USA, 2021; Chapter 31; pp. 357–364.
79. Callery, S. *NASA's Scientific Visualization Studio*; NASA: Washington, DC, USA, 2022.
80. Lau, L.C.; Lee, K.; Mohamed, A. Global warming mitigation and renewable energy policy development from the Kyoto Protocol to the Copenhagen Accord—A comment. *Renew. Sustain. Energy Rev.* **2012**, *16*, 5280–5284. [CrossRef]
81. E.P.A. U.S. Fast Facts on Transportation Greenhouse Gas Emissions. Available online: https://www.epa.gov/greenvehicles/fast-facts-transportation-greenhouse-gas-emissions (accessed on 20 May 2022).
82. Manzetti, S.; Mariasiu, F. Electric vehicle battery technologies: From present state to future systems. *Renew. Sustain. Energy Rev.* **2015**, *51*, 1004–1012. [CrossRef]
83. Benveniste, G.; Sánchez, A.; Rallo, H.; Corchero, C.; Amante, B. Comparative life cycle assessment of Li-Sulphur and Li-ion batteries for electric vehicles. *Resour. Conserv. Recycl. Adv.* **2022**, *15*, 200086. [CrossRef]
84. Renault. The Electric Car: How Does Its Lithium-Ion Battery Work? Available online: https://www.renaultgroup.com/en/news-on-air/news/the-electric-car-how-does-its-lithium-ion-battery-work/ (accessed on 18 October 2019).
85. Office of Energy Efficiency & Renewable Energy. How Does a Lithium-Ion Battery Work? Available online: https://www.energy.gov/eere/articles/how-does-lithium-ion-battery-work (accessed on 14 September 2017).
86. Li, Y. Electrochemical Ion Insertion: Mechanisms and Applications in Energy Storage and Computing. In Proceedings of the 2018 AIChE Annual Meeting, Pittsburgh, PA, USA, 28 October–2 November 2018.
87. Hannan, M.A.; Hoque, M.M.; Hussain, A.; Yusof, Y.; Ker, P.J. State-of-the-Art and Energy Management System of Lithium-Ion Batteries in Electric Vehicle Applications: Issues and Recommendations. *IEEE Access* **2018**, *6*, 19362–19378. [CrossRef]
88. Iskandar Radzi, Z.; Helmy Arifin, K.; Zieauddin Kufian, M.; Balakrishnan, V.; Rohani Sheikh Raihan, S.; Abd Rahim, N.; Subramaniam, R. Review of spinel $LiMn_2O_4$ cathode materials under high cut-off voltage in lithium-ion batteries: Challenges and strategies. *J. Electroanal. Chem.* **2022**, *920*, 116623. [CrossRef]
89. Jeevanantham, B.; Shobana, M.K. Enhanced cathode materials for advanced lithium-ion batteries using nickel-rich and lithium/manganese-rich LiNixMnyCozO2. *J. Energy Storage* **2022**, *54*, 105353.
90. Or, T.; Gourley, S.W.D.; Kaliyappan, K.; Yu, A.; Chen, Z. Recycling of mixed cathode lithium-ion batteries for electric vehicles: Current status and future outlook. *Carbon Energy* **2020**, *2*, 6–43. [CrossRef]
91. Li, J.; Zhang, Y.; Shang, R.; Cheng, C.; Cheng, Y.; Xing, J.; Wei, Z.; Zhao, Y. Recent advances in lithium-ion battery separators with reversible/irreversible thermal shutdown capability. *Energy Storage Mater.* **2021**, *43*, 143–157. [CrossRef]
92. Feng, X.; Ouyang, M.; Liu, X.; Lu, L.; Xia, Y.; He, X. Thermal runaway mechanism of lithium ion battery for electric vehicles: A review. *Energy Storage Mater.* **2018**, *10*, 246–267. [CrossRef]
93. Samadiy Murodjon, X.Y.; Li, M.; Duo, J.; Deng, T. Lithium Recovery from Brines Including Seawater, Salt Lake Brine, Underground Water and Geothermal Water. In *Thermodynamics and Energy Engineering*; InTech Open: London, UK, 2020; pp. 1–39. [CrossRef]

94. Treadgold, T. Lithium Price Tipped To Rise After Warning of 'Perpetual Deficit'. Available online: https://www.forbes.com/sites/timtreadgold/2021/07/02/lithium-price-tipped-to-rise-after-warning-of-perpetual-deficit/?sh=178e8dcc4ab7 (accessed on 2 July 2021).
95. Garside, M. Statista. 2022. Available online: https://www.statista.com/ (accessed on 28 October 2022).
96. Wu, Y.-C.; Kontou, E. Designing electric vehicle incentives to meet emission reduction targets. *Transp. Res. Part D Transp. Environ.* **2022**, *107*, 103320. [CrossRef]
97. Galhardi, J.A.; Luko-Sulato, K.; Yabuki, L.N.M.; Santos, L.M.; da Silva, Y.J.A.B.; da Silva, Y.J.A.B. Rare earth elements and radionuclides. In *Emerging Freshwater Pollutants*; Dalu, T., Tavengwa, N.T., Eds.; Elsevier: Amsterdam, The Netherlands, 2022; Chapter 17; pp. 309–329.
98. Sun, Y.; Wang, Q.; Wang, Y.; Yun, R.; Xiang, X. Recent advances in magnesium/lithium separation and lithium extraction technologies from salt lake brine. *Sep. Purif. Technol.* **2021**, *256*, 117807. [CrossRef]
99. Liu, X.; Zhong, M.; Chen, X.; Zhao, Z. Separating lithium and magnesium in brine by aluminum-based materials. *Hydrometallurgy* **2018**, *176*, 73–77. [CrossRef]
100. Celso Quintero, J.M.D.; Fierro, F.; Thennis, T.; Zhang, Y.; Videla, Á.; Rojas, R. Development of a co-precipitation process for the preparation of magnesium hydroxide containing lithium carbonate from Li-enriched brines. *Hydrometallurgy* **2020**, *198*, 105515. [CrossRef]
101. Zhao, C.; Zhang, Y.; Cao, H.; Zheng, X.; Van Gerven, T.; Hu, Y.; Sun, Z. Lithium carbonate recovery from lithium-containing solution by ultrasound assisted precipitation. *Ultrason. Sonochem.* **2019**, *52*, 484–492. [CrossRef]
102. Xiao, C.; Zeng, L. Thermodynamic study on recovery of lithium using phosphate precipitation method. *Hydrometallurgy* **2018**, *178*, 283–286. [CrossRef]
103. Alsabbagh, A.; Aljarrah, S.; Almahasneh, M. Lithium enrichment optimization from Dead Sea end brine by chemical precipitation technique. *Miner. Eng.* **2021**, *170*, 107038. [CrossRef]
104. Li, H.; Li, H.; Liang, J.; Yan, H.; Cai, Z. Study on the Synergistic Extraction of Lithium from Spent Lithium Cobalt Oxide Batteries by Molten Salt Electrolysis and Two-Step Precipitation Method. *Crystals* **2021**, *11*, 1163. [CrossRef]
105. Shin, D.J.; Joo, S.-H.; Lee, D.; Shin, S.M. Precipitation of lithium phosphate from lithium solution by using sodium phosphate. *Can. J. Chem. Eng.* **2021**, *100*, 3760–3767. [CrossRef]
106. Song, Y.; Zhao, T.; He, L.; Zhao, Z.; Liu, X. A promising approach for directly extracting lithium from α-spodumene by alkaline digestion and precipitation as phosphate. *Hydrometallurgy* **2019**, *189*, 105141. [CrossRef]
107. Mulwanda, J.; Senanayake, G.; Oskierski, H.; Altarawneh, M.; Dlugogorski, B.Z. Leaching of lepidolite and recovery of lithium hydroxide from purified alkaline pressure leach liquor by phosphate precipitation and lime addition. *Hydrometallurgy* **2021**, *201*, 105538. [CrossRef]
108. Shin, J.; Jeong, J.-M.; Lee, J.B.; Cho, H.-J.; Kim, Y.H.; Ryu, T. Preparation of lithium carbonate from waste lithium solution through precipitation and wet conversion methods. *Hydrometallurgy* **2022**, *210*, 105863. [CrossRef]
109. Li, Y.-H.; Zhao, Z.-W.; Liu, X.-H.; Chen, X.-Y.; Zhong, M.-L. Extraction of lithium from salt lake brine by aluminum-based alloys. *Trans. Nonferrous Met. Soc. China* **2015**, *25*, 3484–3489. [CrossRef]
110. An, J.W.; Kang, D.J.; Tran, K.T.; Kim, M.J.; Lim, T.; Tran, T. Recovery of lithium from Uyuni salar brine. *Hydrometallurgy* **2012**, *117*, 64–70. [CrossRef]
111. Xu, Z.; Zhang, H.; Wang, R.; Gui, W.; Liu, G.; Yang, Y. Systemic and Direct Production of Battery-Grade Lithium Carbonate from a Saline Lake. *Ind. Eng. Chem. Res.* **2014**, *53*, 16502–16507. [CrossRef]
112. Wu, S.; Tao, W.; Zheng, Y.; Yang, Y.; Yu, J.; Cui, J.; Lu, Y.; Shi, Z.; Wang, Z. Novel process for the extraction of lithium carbonate from spent lithium-containing aluminum electrolytes by leaching with aluminum nitrate and nitric acid. *Hydrometallurgy* **2020**, *198*, 105505. [CrossRef]
113. Liu, X.; Zhong, M.; Chen, X.; Li, J.; He, L.; Zhao, Z. Enriching lithium and separating lithium to magnesium from sulfate type salt lake brine. *Hydrometallurgy* **2020**, *192*, 105247. [CrossRef]
114. Liu, D.; Li, Z.; He, L.; Zhao, Z. Facet engineered Li3PO4 for lithium recovery from brines. *Desalination* **2021**, *514*. 115186. [CrossRef]
115. Liu, D.; Zhao, Z.; Xu, W.; Xiong, J.; He, L. A closed-loop process for selective lithium recovery from brines via electrochemical and precipitation. *Desalination* **2021**, *519*, 115302. [CrossRef]
116. Swain, B. Recovery and recycling of lithium: A review. *Sep. Purif. Technol.* **2017**, *172*, 388–403. [CrossRef]
117. Zhao, C.; He, M.; Cao, H.; Zheng, X.; Gao, W.; Sun, Y.; Zhao, H.; Liu, D.; Zhang, Y.; Sun, Z. Investigation of solution chemistry to enable efficient lithium recovery from low-concentration lithium-containing wastewater. *Front. Chem. Sci. Eng.* **2020**, *14*, 639–650. [CrossRef]
118. Song, Y.-J. Recovery of Lithium as Li3PO4 from Waste Water in a LIB Recycling Process. *Korean J. Met. Mater.* **2018**, *56*, 755–762. [CrossRef]
119. Kajjumba, G.W.; Marti, E.J. A review of the application of cerium and lanthanum in phosphorus removal during wastewater treatment: Characteristics, mechanism, and recovery. *Chemosphere* **2022**, *309*, 136462. [CrossRef]
120. Kajjumba, G.W.; Fischer, D.; Risso, L.; Koury, D.; Marti, E.J. Application of cerium and lanthanum coagulants in wastewater treatment—A comparative assessment to magnesium, aluminum, and iron coagulants. *Chem. Eng. J.* **2021**, *426*, 131268. [CrossRef]

121. Karthikeyan, P.; Banu, H.; Meenakshi, S. Removal of phosphate and nitrate ions from aqueous solution using La^{3+} incorporated chitosan biopolymeric matrix membrane. *Int. J. Biol. Macromol.* **2019**, *124*, 492–504. [CrossRef]
122. Han, K.N. Characteristics of Precipitation of Rare Earth Elements with Various Precipitants. *Minerals* **2020**, *10*, 178. [CrossRef]

processes MDPI

Review

Incorporated Metal–Organic Framework Hybrid Materials for Gas Separation, Catalysis and Wastewater Treatment

Zahirah Jaffar [1], Normawati M. Yunus [1,*], Maizatul Shima Shaharun [2], Muhammad Faizadmesa Allim [3] and Asyraf Hanim Ab Rahim [1]

1 Institute of Contaminant Management, Centre for Research in Ionic Liquid (CORIL), Department of Fundamental and Applied Sciences, Universiti Teknologi PETRONAS, Seri Iskandar 32610, Malaysia
2 Institute of Contaminant Management, Centre for Contaminant Control & Utilization (CenCoU)), Department of Fundamental and Applied Sciences, Universiti Teknologi PETRONAS, Seri Iskandar 32610, Malaysia
3 Department of Fundamental and Applied Sciences, Universiti Teknologi PETRONAS, Seri Iskandar 32610, Malaysia
* Correspondence: normaw@utp.edu.my; Tel.: +605-368-7689

Abstract: The special features of metal–organic frameworks (MOFs), namely, tunable porosity, exceptional structure, high surface area and high adsorption capability enable them to be widely studied in many applications including carbon capture and storage (CCS), biomedical engineering, catalysis and pollutant treatment. Despite these remarkable properties, MOFs are known to be moisture-sensitive, hardly recyclable and expensive in fabrication cost which limits their breakthrough performance in more efficient uses. Recently, extensive studies have been devoted to counter those shortcomings by embedding MOFs with support materials using various series of synthetic designs to yield incorporated MOF hybrid materials to counter their limitations. In view of this interest, this review summarizes the latest developments of incorporated MOFs with various materials, namely, ionic liquids (ILs), membranes and metal species. Pre-synthetic and post-synthetic synthesis methods are also discussed. This review also aims to highlight the factors associated with incorporated MOF performance such as materials selection and mass ratio which could have favorable effects in gas separation, catalysis and wastewater treatment applications. The data indicate that incorporated MOF hybrid materials exhibit exceptional properties including excellent robustness and stability. Correspondingly, in comparison to pristine MOFs, incorporated MOF hybrid materials significantly improve, among others, the gas selectivity, catalyst activity and dye removal efficiency in gas separation, catalysis and wastewater treatment, respectively. In addition, the challenge related to the utilization of this newly incorporated material is mentioned.

Keywords: MOF; hybrid; incorporated; gas separation; catalysis; wastewater

Citation: Jaffar, Z.; Yunus, N.M.; Shaharun, M.S.; Allim, M.F.; Rahim, A.H.A. Incorporated Metal–Organic Framework Hybrid Materials for Gas Separation, Catalysis and Wastewater Treatment. *Processes* **2022**, *10*, 2368. https://doi.org/10.3390/pr10112368

Academic Editor: Farooq Sher

Received: 11 September 2022
Accepted: 9 November 2022
Published: 11 November 2022

1. Introduction

Metal–organic frameworks (MOFs) are non-porous materials made by combining organic linkers and metal inorganic units via strong bonds [1–3]. These materials were initially discovered in the 1990s with 1,3,5-benzenetricarboxylate (BTC) MOFs that consisted of three equal spaces of carboxylate groups (Figure 1) and were amongst the first MOFs to emerge in the research [4]. Over the years, MOFs have captured researchers' attention and due to their facile tunability, numerous new MOFs structures have been identified and synthesized [5]. To date, almost 100,000 of MOFs have been synthesized and the number is expected to keep on increasing with time [6]. Examples of common MOFs are Cu-BTC, HKUST-1, MIL-101, UiO-66, MOF-5, MOF-74, CPO-27 and MIL-53 [2,7,8]. All MOFs own different crystal topologies and coordinating sites, depending on their organic linker. In 2018, Pu et al. synthesized Mg-MOF-74 and compared its performance with commercial zeolite in ethylene/ethane separation [9]. The work reported on better adsorptive ability of

Mg-MOF-74 in comparison to the zeolite used. The breakthrough experiments suggested the molded Mg-MOF-74 was robust and stable for industrial application. Other than that, Yao et al. synthesized MIL-101(Fe)-C_4H_4 to encapsulate doxorubicin hydrochloride (DOX), a drug used as anthracycline anticancer [10]. The encapsulation is required as DOX cannot be directly introduced as it can be toxic to normal cells and cause irreversible cardiotoxicity. The study showed that, due to the modification of C_4H_4 at the MOF structure, the MIL-101(Fe)-C_4H_4 exhibited high stability in aqueous solution. The MOF was also capable of high-loading DOX and thus provides a promising result for multifunctional nanosystems in cancer therapy.

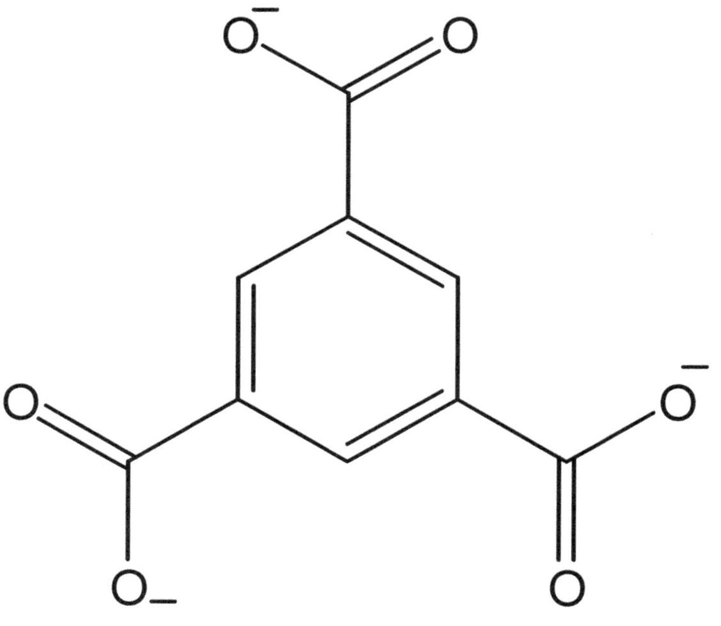

Figure 1. Structure of 1,3,5-benzenetricarboxylate.

MOFs possess some very important properties such as high surface area, tunable pore size and structure, high CO_2 affinity and high thermal and chemical stabilities [1,11–15]. However, MOFs suffer some limitations such as instability in water and acidic conditions; thus many attempts have been made to overcome and increase MOF performance [16,17]. For example, the fabrication of MOFs by introducing various materials such as ionic liquids (ILs) [18,19], graphene [20], biomolecules [21] and polyoxometalates (POM) [22] had produced newly adsorbents known as incorporated MOF hybrid materials. Generally, an incorporated MOF can be defined as a material consisting of an MOF as the main compound and other substance as shown in Figure 2. An incorporated MOF has different physical and chemical properties than its pristine MOF. Alternatively, incorporated MOFs could also be known as hybrid MOFs. The incorporation of other materials with MOFs has been attempted to encounter MOF limitations while enhancing MOF efficiency and capability [23].

At present, MOFs and incorporated MOF hybrid materials have been used in various applications including drug delivery, gas separation, catalysis and luminescent sensors [25,26]. In addition, incorporated MOFs hybrid materials have also been widely used in numerous applications such as gas separation [18,27], wastewater treatment [28,29], catalytic processes [22]. This review specifically discusses several synthesis methods of incorporated MOF hybrid materials, and the application of incorporated MOF hybrid materials in gas separation, catalysis and wastewater treatment.

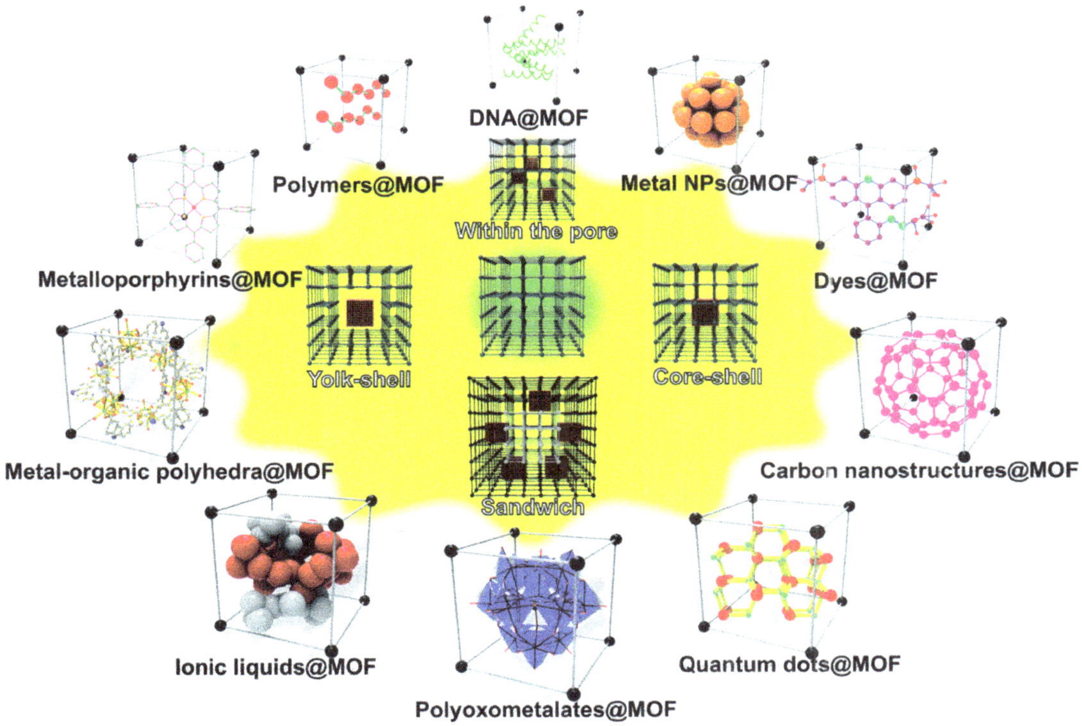

Figure 2. The schematic of incorporated MOF with various materials (reproduced with permission from ref [24]).

2. Incorporated MOFs Synthesis Method

In recent years, researchers have improvised the function of MOFs by incorporation with other materials to increase their efficiency in certain applications while overcoming the problem faced by MOFs. The incorporated MOFs are generally functionalized via two methods: (1) pre-synthetic and (2) post-synthetic [30], where the former's building units are modified before the synthesis of a porous framework and the latter is carried out after the synthesis of targeted parent MOFs. Thus, this section will discuss the details of both syntheses of incorporated MOF techniques.

In the pre-synthetic technique, the earliest strategy in an attempt to prepare the MOF hybrid was through the ionothermal synthesis method. Ionothermal is a solvothermal synthesis technique for MOF hybrid synthesis in which ILs act as solvents [9]. Xu et al. reported the ionothermal synthesis of five MOF hybrids consisting of $NI(OAc)_2$-H_3BTC in various IL solvents with different alkyl chains (n-ethyl, n-propyl and n-butyl) combined with halide anions (I^-, Br^- and Cl^-) [31]. Analysis reveals the changes in structural relative thermodynamic stability were affected by IL cation due to size matching existing between guest and host cavity. Meanwhile, halide anions had significantly affected the structural formation of MOF hybrids because of their differences in nucleophilicity and basicity. In 2019, Lu and co-workers reported the synthesis of polyoxometalate-MOFs (PMOFs) to be used as photocatalysts [32]. The work reported a successful synthesis of five PMOFs with active, high proton conductivity and stable photocatalytic properties for H_2 evolution and rhodamine-B (RhB) degradation under visible-light exposure. Though ionothermally synthesized MOF hybrids produce promising outcomes, due to the complex matrix of ligand functionalization, the synthesis of incorporated MOFs utilizing a pre-synthetic route becomes more challenging. Due to the strong binding of IL cations onto the MOF structure, the ionothermally MOF hybrid did not possess the same properties owned by bulk ILs. Furthermore, pre-synthetic functionalization of the MOFs also deals with issues such as the limited functional group scope for application [33]. This approach may result in incompatible physicochemical properties of MOFs products due to mismatched ligands in combination with synthetic conditions in the synthesis process [30]. Several organic ligands with functional groups are thermally unstable, thus leading to decomposition during the pre-synthesis process. Moreover, several parameters inclusive of weight percentage of the loaded MOFs and the targeted average particle size must be critically considered during the MOF synthesis to achieve the application end-uses of the incorporated MOFs [34]. This is principally true because the modification is targeted to be carried out before the MOF synthesis [30,34,35].

Generally, post-synthetic modification (PSM) is opted for by many researchers due to the new final functionalized MOF product being yielded, without transfiguring the parent structural and topological build-up [33]. Otal et al. mentioned in their report that MOFs' parent properties such as crystallinity, structure and porosity are those unique traits that can be preserved in the event where the PSM synthesis route is utilized [36]. It is undeniably a facile yet exhibiting series of complex approaches including covalent modifications, non-covalent interactions and MOFs used as a precursor and sacrificial template, as reported by Sosa et al. [37]. In 1999, Kiang et al. compared 20 crystals of porous phenylacetylene silver salts with stable hexagonal channel crystal structures of MCM-41. The study reported on having coordinatively replaced bond crystals from the initial covalent bonds, producing a porous solid with chemical stability in a well-crystallized form [38]. This is one of the earliest PSM utilizations related to MOF study. While there are many doctrines of PSM being designed, researched, explored and reported, there are two main approaches that are widely being used: (1) covalent PSM and (2) coordinative PSM [35,36]. Kalaj et al. reported that covalent PSM is carried out by functionalizing organic linkers of the solid framework with a reagent to produce an incorporated MOF product possessing new functionality, while coordinative PSM is conducted by the introduction of new organic molecules containing metal ligating groups onto the inorganic building units of the MOF after the solid, porous framework synthesis [35].

3. Application of Incorporated MOFs

MOFs have been recognized in various applications such as in the areas of separation, catalysis and wastewater treatment. However, incorporated MOFs are still new or unrecognized to many fields of application. Thus, the aim of this paper is to extensively review the use of hybrids or incorporated MOFs in several applications, namely, gas separation, catalysis and wastewater treatment.

3.1. Gas Separation Applications

Natural gas (NG) is widely used as an energy source other than crude oil and coal and it is formed deep under the earth's surface [39]. NG usually consists of alkanes ranging from methane (CH_4) to pentane (C_5H_{12}), carbon dioxide (CO_2), hydrogen sulfide (H_2S), nitrogen(N_2) and helium (He) [40]. NG is commonly used as a fuel and a starting material to produce other valuable products. In addition, NG is always seen as an alternative green approach to reduce the greenhouse effect. Thus, the demand for NG will continue to increase until other alternatives of energy sources are found. However, the other components in NG besides CH_4 may also be classified as impurities that need to be removed via purification processes as these impurities have detrimental effects on the quality of NG. Amongst all the impurities, CO_2 is the most crucial to be eliminated during gas processing besides hydrocarbons and water because of the kinetic diameter size of CO_2 is almost similar with water and methane [41]. Several gas processing technologies used by NG industries to remove CO_2 are absorption, distillation, membrane separation and hydrates [42]. The adsorption process is another potential technique to capture CO_2 because of its low energy requirement and operational expenses [43]. There are many available adsorbents in industry to capture CO_2 including MOFs. One study has stated that MOFs were considered as adsorbents in CO_2 capture due to their stability performance, bigger pore volume and high surface area which are characteristics of excellent adsorbents. This leads to the increasing use of MOFs in the current technology of CO_2 adsorption. Another study has shown that some MOFs such as MOF-199, MOF-2, MOF-3 and MOF-74 were successfully used as adsorbents in CO_2 removal [44]. All these MOFs have various kinds of metals and ligands in their structures. Even though they are produced from various kind of metals and ligands, their synthesis process is considered as the simplest synthesis route when compared to other porous materials such as zeolites. However, the crucial limitation of MOFs, which is similar to other adsorbents, is their stability towards water. In view of this drawback, an alternative by incorporation of other materials in MOFs was suggested to overcome the limitations of MOFs. Therefore, this section will be focusing on the synthesis and utilization of incorporated MOF with ionic liquids (ILs) or membranes for CO_2 removal.

Ionic liquids (ILs) are defined as molten salts that have melting points below 100 °C. ILs are seen as an immense potential absorbent in CO_2 separation which are low vapour pressure, excellent thermal stability and high CO_2 solubility. Besides, the cation and anion of ILs can be modified along with their functionalities to suit the needs. Because of this uniqueness, ILs can be combined with MOFs to complement each other to improve the performance of MOFs. Zeeshan et al. synthesized new MOF/IL composite which was a combination of 1-(2-hydroxyethyl)-3-methylimidazolium dicyanamide, [HEMIM] [DCA] and zeolitic imidazolate framework, ZIF-8 [44]. The MOF/IL was synthesized by using the wet impregnation method with 40% weight loading of IL in MOFs. ZIF-8 was activated at 105 °C in the vacuum oven overnight to remove moisture and impurities. Next, the [HEMIM] [DCA] was dissolved in acetone and stirred at room temperature for 1 h. Then, ZIF-8 was added to the solution. The solution was stirred for 6 h and then dried overnight in a vacuum oven. The resulting MOF/IL which was in powder form was characterized by using X-ray diffraction (XRD), scanning electron microscopy (SEM), Fourier transform infrared spectroscopy (FTIR), transmission electron microscopy (TEM), X-ray photoelectron spectroscopy (XPS), Brunauer−Emmett−Teller (BET) and thermogravimetric analysis (TGA). Characterization results of the ZIF-8/[HEMIM] [DCA] showed that IL was deposited on the external surface of MOF. At 1 mbar and 25 °C, the ZIF-8/[HEMIM] [DCA] exhibited increments of 5.7- and 45-fold of CO_2 uptake and CO_2/CH_4 selectivity, respectively, compared to the single MOF. These remarkable results indicate that, by simply performing a straightforward post-synthesis alteration, a high level of improvement in CO_2 capture and CO_2/CH_4 selectivity compared to the parent MOF could be achieved. The core-shell type of ZIF-8/[HEMIM] [DCA] composites described here operates as a new platform for high-performance gas separation applications due to the existence of a wide number of suitable MOF/IL pairs for the application.

Ferreira and his teams studied the IL-impregnated MOF composite materials by using a combination of ZIF-8 with numerous ILs for CO_2/CH_4 separation [18]. They used the wet impregnation method to synthesize the MOF/IL. In the general procedure, ILs were dissolved in acetone for 15 min at room temperature and the resulting mixture was added into a vial containing ZIF-8. The vial was capped, and the mixture was stirred overnight at room temperature. The mixture was further stirred without cap for 4–5 h at 65 °C. Then, the sample was dried at 100 °C for 3–4 h. The resulting MOF/IL was characterized by using TGA, N_2 adsorption−desorption equilibrium at 77 K, powder XRD(PXRD), FTIR and STEM. A total of 10 imidazolium-based ILs were used in their study, namely, 1-ethyl-3-methylimidazolium bis(trifluoromethyl) sulfonylimide [C_2MIM] [NTf_2], 1-(2-hydroxyethyl)-3-methylimidazolium bis(trifluoromethyl)sulfonylimide [C_2OHMIM] [NTf_2], 1-hexyl-3-methylimidazolium bis(trifluoromethyl)sulfonylimide [C_6MIM] [NTf_2],

1-benzyl-3-methylimidazolium bis(trifluoromethyl)sulfonylimide [BzMIM] [NTf2], 1-decyl-3-methylimidazolium bis(trifluoromethyl)sulfonylimide [C_{10}MIM] [NTf_2], trihexyl (tetradecyl)phosphonium bis(trifluoromethyl)sulfonylimide [$P_{6\,6\,6\,14}$] [NTf_2], 1-hexyl-3- methylimidazolium dicyanamide[C_6MIM] [$N(CN)_2$], 1-hexyl-3-methylimidazolium tricyanomethanide [C_6MIM] [$C(CN)_3$], 1-hexyl-3-methylimidazolium chloride [C_6MIM] [Cl] and 1-ethyl-3-methylimidazolium acetate [C_2MIM] [Ac]. The results of the characterization confirmed that the impregnation of IL in the MOF was successful with evident of interactions between the IL and the MOF. However, for both CO_2 and CH_4, all ZIF-8/ILs composites exhibit lower high-pressure adsorption capabilities than that of pure ZIF-8. This is because of the reduction in the overall pore volume losses due to the blocked MOF pores by IL. In contrast, at 0.5 bar, ZIF-8 /[C_2MIM] [Ac] exhibited higher adsorption capability than pure ZIF-18 and this could be attributed to strong attractions between the anion of the IL and the CO_2 gas.

Sezginel and colleagues used 1-butyl-3-methylimidazolium tetrafluoroborate [BMIM] [BF_4] IL and a commercially available MOF, Cu-BTC, for the study of gas separation using MOF impregnated with IL [45]. In this wet impregnation, IL was dissolved in acetone prior to the addition of Cu-BTC and the mixture was stirred at 30 °C for 6 h before drying in a vacuum oven at 105 °C overnight. The Cu-BTC/[BMIM] [BF_4] underwent characterizations, namely, by using SEM, XRD, TGA, FTIR, BET and elemental analysis before the high-pressure CO_2 analysis. On a separate note, prior to the MOF-hybrid preparation, two solvents that are capable of dissolving [BMIM] [BF_4] were selected, which are ethanol and acetone. The selectivity of the solvent for preparation of MOF hybrid is crucial in order to ensure an undisturbed crystal structure of MOF after the IL impregnation process. Different Cu-BTC samples were then soaked on ethanol and acetone before both solvents were removed by evaporation. Analysis on SEM and XRD revealed ethanol had caused a deformation of crystal structure. In contrary, SEM and XRD analysis showed that the crystal structure of Cu-BTC soaked in acetone is almost undisturbed, thus making it the most suitable solvent for the preparation of MOF hybrid. The cause of solvent effect remains unclear in the production of MOF hybrid; however, it could be speculated that some solvents may interrupt the crystal lattice structure of MOF by acting as guest molecules, thus causing structural disruption as seen in this study. In the meantime, BET results indicated that the ILs loading increases and the surface area of the composite materials decreases due to the penetration of bulky [BMIM] [BF_4] in pores Cu-BTC. Similarly, the thermal stability of the Cu-BTC/[BMIM] [BF_4] by TGA showed the reduction in the decomposition temperature as the IL loading increases as compared to pure IL and CuBTC. Meanwhile, the incorporation of IL in Cu-BTC increases the CH_4 selectivity over CO_2, H_2, and N_2 gases by 1.5-fold in contrast to pristine Cu-BTC.

Almost four decades later, membrane technology has been applied to separate and capture gas such as other conventional separation methods which are absorption, adsorption and distillation [46,47]. Membranes have been used in gas separation due to low energy consumption, low environmental impact, high efficiency and simple operation. Castarlenas et al. studied gas separation by using mixed matrix membrane (MMM) and UiO-66_GO hybrid materials [48]. UiO-66_GO hybrids were obtained by hydrothermal synthesis of MOF UiO-66 on graphite oxide (GO). In this study, the UiO-66-GO was used as a filler in preparing the MMM with two different polymers, namely, polysulfone and polyimide. The filler was dispersed in chloroform after the addition of the polymer. The mixture was stirred overnight and treated in ultrasonic bath for about 15 min. The membrane was set on a petri dish to allow the chloroform to naturally evaporate at room temperature overnight. After drying, the film was placed in a vacuum oven to remove remaining solvent at a certain temperature based on the polymer used. XRD, nitrogen adsorption–desorption, TGA, DSC and SEM were used to characterize the membrane. The barrier effect in the MMMs created with the hybrids was compensated by the presence of UiO-66. The H_2/CH_4 mixture, whose separation is mostly based on variations in kinetic diameters, was nevertheless significantly affected by this effect, and the best H_2/CH_4 separation results in terms of both permeability and selectivity were achieved with a polyimide-based MMM containing solely UiO-66 as filler. Results indicate that the UiO-66_GO hybrids had produced the best separation results in the case of the CO_2/CH_4 separation, where both CO_2 diffusion and adsorption are encouraged by the effective action of the MOF which is well disseminated on the GO sheets.

Chen and his teams analyzed gas separation using hybrid membranes fabricated by using PI-TB (Polyimide-Tröger's Base) microporous polymer matrix and UiO-66-NH_2 MOF as the filler [47]. PI-TB and UiO-66-NH_2 nanoparticles were vacuum-dried overnight at 120 °C for membrane fabrication. The pure gas permeabilities and ideal selectivity for CO_2/CH_4, CO_2/N_2, and O_2/N_2 systems were investigated at a temperature of 35 °C and 1 bar pressure. The experimental results revealed that hybrid membrane containing 30 wt% UiO-66-NH_2 displayed the best gas separation performance an outstanding permeabilities improvement of 166% for CO_2 and O_2. Meanwhile, comparable gas pair selectivities were observed for CO_2/CH_4, CO_2/N_2 and O_2/N_2 with pure PI-TB membrane. It should be noted that its CO_2 separation performance exhibits a tendency of approaching the 2008 Robeson upper bound, indicating a potential application prospect for CO_2 removal from natural gas. Additionally, the hybrid membranes developed in this work performed better at gas separation at a low temperature.

Nabais et al. studied the performances of mixed matrix membranes (MMMs) prepared from the combination of poly (ionic liquid)s (PILs), an IL comprising the same anion as of the PIL and an MOF. Three different MOFs which act as fillers in the MMM system have been used, namely, Al-MIL-53, $Cu_3(BTC)_2$ and ZIF-8 [49]. PIL/IL was dissolved in DMF and stirred for 8 h. Meanwhile, different weight loading of MOFs in DMF solution were also prepared in separate vials before the solutions were sonicated followed by constant stirring for 8 h. Then, the solutions were mixed and stirred and left overnight. Lastly, the resulting membrane solution was casted on plates and heated at 343 K for 7–8 h. The method that was used to synthesize MOFs and membrane was the solvent evaporation method. This membrane underwent several characterizations including FTIR, SEM-EDS, TGA, mechanical properties and pure gas permeation. For TGA results, except for MMMs containing ZIF-8, the prepared MMMs generally exhibit a weight loss stage when the temperature was increased to 473 K, presumably as a result of the evaporation of some remaining solvent. At around 654 K, the PIL/IL membrane's thermal decomposition (T_d) begun. It has been noted that the T_d drops as the MOF loading increases, for MMMs with 30 wt% loadings of Al-MIL-53, $Cu_3(BTC)_2$, and ZIF-8, respectively, the T_d drops to 639 K, 598 K and 615 K. The acquired TGA data demonstrated that the suggested MMMs may be appropriate for use in CO_2/H_2 gas separation given that syngas streams have a temperature range of 313 to 523 K; also, no significant weight losses were seen in this temperature range. As a result, the pure gas permeation measurements showed that both CO_2 permeability and CO_2/H_2 ideal selectivity were improved. The high ionic content of the PIL/IL composite matrix and the high CO_2 adsorption capabilities of the MOFs were primarily responsible for these outcomes. Additionally, the intrinsic properties of the included MOFs, specifically their porosity volume, cavity shape, and BET surface area, had a significant impact on the membranes' permeability. Membranes based on ZIF-8 achieved the highest results for CO_2 permeability, up to 97.2 barrier with 30 wt% loading, whereas membranes based on Al-MIL-53 demonstrated the highest improvement in ideal selectivity (up to 13.3). The summary for all hybrid materials for gas separation in this review is tabulated in Table 1.

Table 1. Summary of hybrid materials for gas separation.

Hybrid Materials	Synthesis Method	Separation	Efficiency	Reference
ZIF-8/[HEMIM] [DCA]	Wet impregnation	CO_2, CH_4	Increase in CO_2 uptake and decrease in CH_4 uptake.	[44]
ZIF-8/[C_2MIM] [NTf_2] ZIF-8/[C_2OHMIM] [NTf_2] ZIF-8/[C_6MIM] [NTf_2] ZIF-8/[BzMIM] [NTf_2] ZIF-8/[C_{10}MIM] [NTf_2] ZIF-8/[$P_{6\,6\,6\,14}$] [NTf_2] ZIF-8/[C_6MIM] [$N(CN)_2$] ZIF-8/[C_6MIM] [$C(CN)_3$] ZIF-8/[C_6MIM] [Cl] ZIF-8/[C_2MIM] [Ac]	Wet impregnation	CO_2/CH_4	Better ideal selectivity between CO_2/CH_4.	[18]
CuBTC/[BMIM] [BF_4]	Wet impregnation	CO_2/CH_4, CO_2/N_2, CO_2/H_2, CH_4/N_2, CH_4/H_2, N_2/H_2	CH_4 selectivity over CO_2, H, and N_2 gases increase by 1.5-fold.	[45]
UiO-66_GO	Hydrothermal	H_2/CH_4, CO_2/CH_4	UiO-GO hybrids produced the best separation results in the case of the CO_2/CH_4 separation.	[48]
UiO-66-NH_2/Tröger's Base Polyimide Hybrid Membranes	Solution casting method	CO_2/CH_4, CO_2/N_2, O_2/N_2	Exhibits ~166% increases in CO_2 and O_2 permeabilities. CO_2 separation performance shows a trend in approaching the 2008 Robeson upper bound, showing a potential application prospect for CO_2 removal from NG.	[47]
MIL-53(Al)/poly[Pyr11] [Tf_2N] + [C_4mpyr] [Tf_2N] (MMM) $Cu_3(BTC)_2$/MMM ZIF-8/MMM	Solvent evaporation	CO_2/H_2	Improvement in both CO_2 permeability and CO_2/H_2 ideal selectivity.	[49]

3.2. Catalysis Applications

Catalysis is important to speed up the reaction rate and save energy utilities because catalysts have active sites that can make the reaction faster and lower the activation energy. There are a few types of catalysts that have been used in many fields including homogeneous, heterogeneous and biological catalysts. Each of these catalysts have their own advantages and drawbacks. MOFs fall under the category of heterogeneous catalysts which will be the discussion in this sub section. MOFs have been exposed as potential catalysts due to their properties of having high densities in catalytic sites in addition to their ability to withstand high temperature conditions. In addition to the tunability of MOF catalytic sites, these materials have high catalytic activity [50–52], better recyclability [53] and high selectivity features [54]. Because of these remarkable properties possessed by MOFs, the efficacy of MOFs in catalysis can be improved by combining them with other species or materials as discussed in this sub section.

Sun and colleagues tested on an MOF-incorporated IL as a heterogeneous catalyst in the cycloaddition of CO_2 with epichlorohydrin (ECH) [55]. An acid–base neutralization route synthesis of MOF/IL was conducted by reacting the amino-functionalized IL, 1-methyl-3-(2-amino-ethyl)imidazolium bromide, [2-aemim] [Br], with MIL-101-SO_3H. Both precursor synthesis steps were detailed in their report. Using water as a solvent, atmospheric pressure environment, and room temperature conditions, the MOF-incorporated IL system was fabricated. A high crystallinity with initial MOF parent imitation indicated that introduction of [2-aemim] [Br] did not affect the original structure of MIL-101-SO_3H framework, as resulted from XRD diffractograms. FTIR and X-ray photoelectron spectroscopy (XPS) analyses supported the presence of an amino group at the MIL-101-SO_3H framework where the former result showed an imidazole ring stretching vibration as 1575 cm^{-1} correlating with lean [2-aemim] [Br] spectrum and the latter revealed the presence of an N element in the MIL-101-SO_3H/IL sample that was absent in the original MOF sample, supporting the claim of an acid–base reaction between the N and S components of IL and MOF, respectively.

In order to evaluate the catalytic performance of synthesized MIL-101-SO_3H/IL, Sun and co-workers conducted the cyclization of CO_2 with epichlorohydrin to compare the conversion ratio of epoxide to carbonate via ^1H NMR analysis from an integration ratio of peak. The high conversion yield of epichlorohydrin to chloropropene carbonate where the proposed mechanism can be referred from [55] was obtained with the use of MIL-101-SO_3H/IL (N, 3.5 wt%) catalyst (88% yield) compared to a control (0% yield) and non-IL-appended MIL-101-SO_3H (58% yield) under reaction conditions: epichlorohydrin (10.8 mmol), catalyst (80 mg), CO_2 pressure (1 atm), reaction temperature (90 °C) and reaction time (36 h) in a 25 mL Schlenk tube with a loaded stir bar. The MOF/IL catalyst also demonstrated good recyclability after a cycloaddition reaction being conducted three times with the same reaction conditions. It was also noted by Sun and teammates where the powder XRD comparative analysis of fresh MOF-incorporated IL catalysts and recovered IL@MIL-101-SO_3H showed a well-preserved framework integrity after reusing three times, indicating good reusability and stability performance. This self-assembly, facile fabrication of MOF-incorporated IL as heterogeneous catalyst in CO_2 chemical fixation for the epichlorohydrin catalytic conversion supported the utilization of incorporated MOF in carbon management and catalysis application.

In addition, MOF can also be incorporated with metal species to be derived as a catalyst. Veisi and associates published their study on the Suzuki–Miyaura coupling reaction using UiO-66-biguanidine/Pd nanocomposites as the catalyst [56]. It is noted in their report, where there is no research reported so far on this convention, that related to Suzuki–Miyaura coupling reactions which utilize Pd functionalized UiO-66 modified MOF; hence, this intrigued Veisi and co-workers to fill in the research gap. The incorporated MOF was synthesized by metal species loading via the addition of sodium tetrachloropalladate (Na_2PdCl_4) aqueous solution in a uniformly dispersed solution of UiO-66-biguanidine (water as solvent) for 12 h. The final solid products were centrifuged, washed with deionized water followed by ethanol by two thresholds, and lastly vacuum dried at 60 °C for 24 h, yielding UiO-66-biguanidine/Pd. Detailed stepwise synthesis procedures of UiO-66-biguanidine from precursors UiO-66-NH_2 and dicyandiamide can be referred to in their previous report [56]. Elemental analysis of UiO-66-biguanidine/Pd revealed the presence of Pd and C species as an addition to the precursor elemental profile of UiO-66-NH_2 that displayed signals of Zr, N and O only. The additional element signals are correlated with the organic linker attachment (biguanidine) and Pd that are immobilized all over the framework. Upholding the post-synthetic modification synthesis method, the conservation of framework integrity for both UiO-66-biguanidine/Pd and UiO-66-NH_2 was proven due to having a roughly similar XRD pattern despite having poor crystallinity. Moreover, Langmuir surface area of final metal-incorporated MOF (629 m^2/g) in this study is lower than the original parent MOF (831 m^2/g), which further confirms the immobilization of Pd nanoparticles, resulting partial pore blocking on the MOF surface as investigated from data of nitrogen adsorption–desorption characterization. In addition, the structural nature is concluded to be a microporous framework due to type I isotherm being observed.

In the cross-coupling reaction of phenyl boronic acid and 4-boromotoluene adopted in this study by Veisi and colleagues, optimized conditions with high yield for the conversion were found from the matrix screening of solvent, base, catalyst load and reaction temperature—3 mL solvent [ethanol/water (1:1)], 2 mmol of K_2CO_3 as base, 0.1 mol% Pd loading, and 50 °C reaction temperature in an equimolar reaction of aforementioned reactants. To consider the aspect of green catalytic for this Suzuki–Miyaura coupling reaction, the good recyclability of UiO-66-biguanidine/Pd as the catalyst was recorded with seven successive cycles of high yield performance, whereas reaching the eighth and ninth cycle, the yield reduced to 90% and 86%, respectively. Veisi and co-workers explained the phenomenon might be due to the oxidation of active species from ambient air, agglomeration of metal loadings or deactivation from organic species deposits [56]. This new MOF-incorporated metal catalyzed reaction, specifically the Pd species, can be a breakthrough for the Suzuki–Miyaura coupling reaction for carbon-carbon bond exploration, with a greener reaction condition design.

The difficulty in recycling 2-methylimidazole (2MeIm) motivated Wu and co-workers to explore a new hybrid material to be used as heterogenous catalyst for CO_2 conversion into cyclic carbonate [57]. As 2MEIm can easily be coordinated with Co ions to form ZIF-67, Co-BTC MOF was selected for the purpose of producing a new catalyst. In their work, 2MEIm was grafted onto the surface of Co-BTC through the solvothermal method. Initially, the pre-prepared Co-BTC was dispersed in the mixture containing methanol (MeOH) and various mass ratios of 2MeIm to produce Co-BTC/2MEIm-0.2, Co-BTC/2MEIm-0.5, Co-BTC/2MEIm-1.0, Co-BTC/2MEIm-1.5 and Co-BTC/2MEIm-2.0. The mixture was stirred for 30 min and transferred to Teflon-lined stainless-steel autoclave where the reaction was conducted at 120°C for 24 h. Collected precipitate was centrifuged and dried at 60°C. XRD analysis showed the presence of peaks representing Co-BTC and ZIF-67. While the intensity of Co-BTC peak decreased, the peak intensity of ZIF-67 increased progressively with the mass ratio of 2MEIm. Meanwhile, the presence of a new crystal phase at 12.22° on modified material could be related to the coordination existing between 2MEIm and Co ions. A new formation of hexagonal prism crystal shape was also observed on FESEM images of modified Co-BTC. Moreover, BET analysis revealed the presence of micro- and mesopores on Co-BTC/2MEIm. The BET surface area showed an increasing trend as the 2MEIm mass ratio increased due to the presence of mesoporous in the modified sample. The same trend was also observed in total pore volume. Co-BTC/2MEIm was then used for the catalyst in CO_2 conversion and the result was evaluated based on ECH conversion, CPC selectivity and CPC yield. The presence of 2MEIm ligand on Co-BTC had improved the ECH conversion, CPC selectivity and conversion up to 98.25%, 98.72% and 95.21%, respectively. According to authors, the presence of 2MEIm provided more Lewis acid site exposure to the reactants, thus improved the catalytic activity of Co-BTC/2MEIm compared to pristine Co-BTC. As Co-BTC/2MEIm-1.0 produced the highest yield of CPC, this heterogenous catalyst was then selected for optimization study on CO_2 conversion in which almost 100% ECH conversion and CPC selectivity was able to be achieved at optimal conditions. The recyclability on Co-BTC/2MEIm-1.0 was also conducted to evaluate the catalytic performance. In general, the ECH conversion was slightly decreased after three times of usage. In the meantime, the CPC selectivity showed an increasing trend after the fourth cycle recovery of the catalyst. This could be due to the slow release of water molecules during the recovery reaction, thus slightly impacting ECH conversion and CPC selectivity. However, there were no significant changes occurring between fresh and recycled catalyst which indicates the stability of Co-BTC/2MEIm-1.0 during the reaction. Additional study on CO_2 conversion coupling with epoxides was also conducted. The result revealed the newly modified catalyst was able to assist in the transformation of epoxides into their corresponding carbonates.

Ten and co-workers [58] demonstrated the combination of UiO-66 with Ag-Pd nanoparticles (NPs) to be applied as the catalyst for the conversion of propylene glycol (PG) to lactic acid (LA). In their study, the UiO-66/AgPd was synthesized through different methods which are incipient wetness impregnation (IWI) and double solvent impregnation (DS). In the IWI method, two types of solvent were used which were water (IWI–water) and acetonitrile (IWI–acetonitrile). In the IWI–water procedure, the solution containing Ag, $PdNO_3$ and water was added dropwise into UiO-66. The reduction of Ag and Pd was conducted under an argon environment at a temperature of 200 °C for 2 h. Then, the resulting impregnated material (UiO-66/AgPd-H_2O) was dried overnight at room temperature. According to the authors, the impregnation protocol with AgPd NPs in IWI-acetonitrile was conducted using a similar procedure as in IWI–water procedure but utilizing acetonitrile as the impregnation solution. In this method, silver nitrate and palladium acetylacetonate were dissolved in acetonitrile and slowly added via dropwise procedure into Uio-66 before drying to produce UiO-66/AgPd (ACN). In the meantime, the DS method was carried out by suspending UiO-66 in n-heptane. The mixture was sonicated for 30 min. After that, hydrophilic aqueous solution of Ag and $PdNO_3$ was added dropwise while being vigorously stirred thus resulting in a suspension. Then, 62wt% of aqueous solution of hydrazine was slowly added and stirred for 1 h. The solid, UiO-66/AgPd (DS), was decanted from n-heptane and dried at 150 °C. Analysis on textural properties of all incorporated materials revealed that UiO-66/AgPd (H_2O) and UiO-66/AgPd (ACN) maintained the BET surface area and micropore volume as pristine MOF. Meanwhile, a reduction in both parameters were observed in UiO-66/AgPd (DS) which was due to pore filling by Ag clusters. The study on catalytic activity of all materials showed that the UiO-66/AgPd (DS) recorded the highest PG conversion with a value of 7%. This could be due to a large formation of bimetallic NPs on UiO-66/AgPd (DS). Moreover, the study further explored the utilization of different ratios of Ag for sample preparation through the DS method. It was identified that the selectivity towards LA increased as the concentration of Ag increased. The presence of active and accessible active sites of Ag/Pd NPs on MOF, particularly on incorporated materials prepared through the DS method, improved PG conversion. The discussion on the catalysis activity of incorporated MOFs is summarized in Table 2.

Table 2. Summary of catalysis applications in hybrid materials.

Hybrid Materials	Synthesis Method	Application	Efficiency	Reference
MIL-101−SO_3H/IL	Acid−base attraction	CO_2 Chemical Fixation	High catalytic activity for cyclizing CO_2 with epichlorohydrin under atmospheric pressure	[55]
UiO-66-biguanidine/Pd	Hydrothermal	Suzuki–Miyaura coupling	Better recyclability	[56]
Co-BTC/2MEIm	Solvothermal	CO_2 Chemical Fixation	Higher selectivity and higher conversion	[57]
UiO-66/Bimetallic AgPd	Incipient Wetness Impregnation Double solvent impregnation	PG into LA	Higher PG conversion	[58]

3.3. *Wastewater Treatment*

Almost 80% of our Earth's surface is covered with water. In addition, up to 60% of the human body consists of water. However, due to human activities such as plantation, farming and industrial development, our water sources have been polluted. Industrial activity such as quarrying, glove production, pulp and poultry processing industries [59] have produced large amounts of wastewater, thus making water treatment one of the main environmental issues worldwide apart from global warming, deforestation, air pollution and the depletion of natural resources. The presence of water pollutant threatens the aquatic ecosystem, thus disturbing the food chain. An example of water pollutants are insecticides, pesticides, heavy metals, volatile organic compounds and chemical waste [60]. According to Shojaei and co-worker, the wastewater treatment can be conducted through three methods which are physical, chemical and biological [61]. However, the physical adsorption process by using various adsorbents such as zeolite, chitosan and carbons has been reportedy much more efficient due to their affordable cost, simple and fast procedure [62]. In the meantime, the usage of pristine and hybrid MOF in wastewater treatment has attracted researchers' attention due to interesting properties such as various functionalities and porosities that can be tuned according to the application of interest. For example, the presence of functionality in MOF structure such as NH_2, -OH, -SO_3OH and -COOH from ILs is one of important strategies to improve adsorptive application [63]. Moroever, a wider availability of MOF and functional groups could produce an MOF hybrid with porosity that suited targeted compounds. Therefore, this subsection will be focusing on wastewater treatment by using MOF and hybrid MOF material.

Bisphenol A (BPA) is one of endocrine-disrupting compounds (EDCs), which commonly presents in the aquatic environment due to its extensive usage, mainly in the food packaging industry. While BPA is acutely toxic to aquatic organisms, it is also potentially damaging to the human reproductive system and thyroid glands and has been linked to cancer due to its ability to mimic the estrogen hormone. The leaching of BPA could occur due to incomplete removal during the wastewater treatment process which may interfere with our water source. Thus, Badra and co-workers [62] synthesized the Bio-MOF-1 derived carbon (BDMC) which consists of Zn and adenine interconnected with biphenyldicarboxylate linkers. In their study, the Bio-MOF-1 was prepared by adding 1:2:3 mol ratio of adenine: bisphenyl-4,4′-dicarboxylic acid: zinc acetate dihydrate into Teflon containing DMF. The mixture was mixed for 10 min and placed in an autoclave container following by heating at 130 °C for 24 h. The Bio-MOF-1, which appeared as colorless crystal, was filtered and washed with DMF before being dried at 100 °C overnight. Meanwhile, the prepared Bio-MOF-1 was carbonized at 1000 °C for 6, 12 and 24 h producing BDMC-6, BDMC-12 and BDMC-24, respectively. The finishing products were collected and washed with 2.0 M hydrochloric acid (HCl) and dried at 100 °C under vacuum overnight. Analysis on the XRD pattern of all samples revealed the presence of new broad peaks at 20–25° and 40–45° at carbonized Bio-MOF-1 that are associated with the carbonization process. Further analysis on carbonized Bio-MOF-1 revealed the BMDC-12 has the highest surface area and mesopore volume. The adsorption of BPA by using BDMCs and activated carbon (AC) was conducted at 25 °C in which the result was analyzed by using UV–Visible spectrophotometer. The BPA adsorbed amount by all BDMC samples was observed to be higher than AC. The study on BPA adsorption kinetics and isotherms showed the order of efficiency is AC < BDMC-6 < BDMC-24 < BDMC-12 indicates the remarkable performance of BDMC-12 in BPA removal. The pH study showed negligible effect on BPA removal. In the meantime, the removal of BPA by using BDMC-12 is efficient up to four cycles with no reduction in its porosity thus confirming the robustness of this porous material.

Furthermore, heavy metals such as arsenic, mercury, selenium and lead [64] could exert severe effects towards environment and human health. According to Rehman et al., the consumption of water containing heavy metal will lead to many health issues such as cardiovascular disorders, neuronal damage and renal injuries [65]. Many methods have been developed to treat heavy metal contamination in which nanofiltration (NF) has shown promising results due to its low cost with high effectiveness. Gnanasekaran and co-workers synthesized Zn-MOF-5 incorporated with three polymeric membranes which are cellulose acetate (CA), polyethersulfone (PES) and polyvinylidene fluoride (PVDF) for heavy metal removal as it has been reported that the embedding of MOFs on polyimide support showed good performance in terms of solvent permeability [37]. In their work, Zn-MOF-5 was prepared by loading a specific mass of terephthalic acid and zinc nitrate hexahydrate into containers containing DMF. The mixture was stirred and heated at 150 °C for 4 h. The white precipitate was then collected through filtration and washed with acetone before being dried at 60 °C for 3 h. Meanwhile, the fabrication of Zn-MOF-5 with polymeric membranes was conducted by using immersion precipitation method. Approximately, 0.5% of Zn-MOF-5 was loaded into DMF and sonicated to maximize the homogeneity. The polymers were dissolved in the mixture of Zn-MOF-5 and stirred for 3 h. The solution was cast to form a film which then was immersed in distilled water for 24 h. The functional group analysis of Zn-MOF-5 incorporated polymeric membranes (Zn-MOF-5/PM) revealed the presence of peaks that confirmed the presence of MOF. In XRD patterns, the peak occurrence at 9.8° of all Zn-MOF-5/PM materials were slightly shifted, probably due to the interaction between MOF and polymeric membranes. The morphology analysis showed an improvement in Zn-MOF-5/PES, Zn-MOF-5/CA and Zn-MOF-5/PVDF in term of pore radius, pore size and porosity, respectively. The introduction of Zn-MOF-5 had simultaneously increased the membrane's hydrophilicity and rejection capability. Analysis on heavy metal removal application showed the promising performance of modified membranes as all materials recorded more than 50% rejection efficiency for Cu (II). For Co (II), the Zn-MOF/CA recorded the highest rejection efficiency as 77.0%.

Textile industry is the main contribution in generating effluent containing hazardous dyes [66]. Synthetic dyes contained large amount of azo group which are amine and benzidine that are carcinogenic. Furthermore, some water-soluble dyes are non-biodegradable, thus could threaten the marine ecosystem by impairing the photosynthesis process and reducing the dissolved oxygen level which will disrupt our food source [67]. Kavak and co-workers [68] demonstrated the usage of Al-MIL-53/[BMIM] [PF_6] for dye removal in the water purification process. In this work, 70 wt% of dehydrated Al-MIL-53 was loaded into a container comprising of dissolved [BMIM] [PF_6]. The mixture was stirred at 35 °C until acetone was completely evaporated. The newly hybrid porous material was collected and dried overnight at 105 °C. Incorporating [BMIM] [PF_6] into the structure of Al-MIL-53 had reduced the BET surface area and calculated pore volume as compared to pristine Al-MIL-53. Meanwhile, SEM analysis indicates the crystal structure of Al-MIL-53 remained intact after [BMIM] [PF_6] incorporation. This result was also supported by XRD analysis. Moreover, the change in XRD intensity spectrum was also observed corresponding to molecular level interaction of IL with MOF. The study Al-MIL-53/[BMIM] [PF_6] as adsorbent for dye removal was performed in 10 ppm cationic (methylene blue) and anionic dye (methyl orange) solution. A comparative study on the performance of pristine Al-MIL-53 and Al-MIL-53/[BMIM] [PF_6] revealed 23.3% and 82.3% of methylene blue removal, respectively, within 1 min of reaction time. In 3 h of reaction time, the removal efficiency of methylene blue using Al-MIL-53/[BMIM] [PF_6] increased to 99.3%. The same trend was observed for methyl orange in which Al-MIL-53/[BMIM] [PF_6] recorded 99.2% removal efficiency compared to Al-MIL-53 (53.69%). In the binary mixture of methylene blue and methyl orange, the performance of Al-MIL-53/[BMIM] [PF_6] did not show significant difference compared to single dye mixture. This indicates that there is no competition existed between dyes with different charge. On the other hand, the study on the kinetic model showed the removal of methylene blue and methyl orange by Al-MIL-53/[BMIM] [PF_6] follows the pseudo-second order. Further study on recyclability Al-MIL-53/[BMIM] [PF_6] demonstrated the successful regeneration of this hybrid porous material without significant loss of its original mass.

In the meantime, Younis and co-workers chose to study the performance of heterojunction photocatalysts prepared via in situ encapsulation of 5% zinc doped titanium oxynitride ($Zn_{0.05}$TiOxNy) into MOF-5 solid support for dye removal. The material was prepared by using the microwave technique. The research on the usage of a photocatalyst, mainly titanium oxide (TiO_2), had been extensively carried out due to its easy preparation, recyclability and robustness towards alkali/acid environments [69]. However, due to its poorly visible light absorption and nano particle size [70], which may be challenging during the recovering process, the combination of various material is required. In this work, the usage of MOF-5 with ZnO4 cluster was applied due to its UV-driven property. The support material, $Zn_{0.05}$TiOxNy @MOF-5 was prepared in a microwave in which the crystal growth of MOF-5 was conducted in the presence of 33 wt% of $Zn_{0.05}$TiOxNy nanoparticles. Analysis on XRD patterns of $Zn_{0.05}$TiOxNy, MOF-5 and $Zn_{0.05}$TiOxNy@MOF-5 showed the presence of peaks corresponding to each component, thus indicating a successful synthesis process. Although the surface area and pore volume of MOF-5 hybridized with $Zn_{0.05}$TiOxNy decreased, the encapsulation of $Zn_{0.05}$TiOxNy with MOF-5 resulted in an increasing value for both parameters. The study on optical property of MOF-5/$Zn_{0.05}$TiOxNy showed an improvement in visible light absorption compared to pristine material. The utilization of MOF-5/$Zn_{0.05}$TiOxNy for dye removal was conducted in methylene blue solution. It was determined that, under dark conditions, MOF-5 and MOF-5/$Zn_{0.05}$TiOxNy own the fastest methylene blue adsorption rate under 60 min reaction conditions. Additional study on antimicrobial activity towards *E. coli*, *S. aureus* and *C. albicans* indicates MOF-5/$Zn_{0.05}$TiOxNy was among the photocatalyst that had the highest ability in antimicrobial inhibition growth under visible light. The performance of MOF-5/$Zn_{0.05}$TiOxNy was also tested by using real industrial wastewater that contaminated with methylene blue, bacteria/yeast pathogens under dark/light conditions It has been determined that MOF-5/$Zn_{0.05}$TiOxNy showed good performance in terms of chemical oxygen demand (COD) adsorption removal. According to the authors, the hybridization of $Zn_{0.05}$TiOxNy with MOF-5 improved the recyclability of adsorbent up to five cycles. To conclude, all four adsorbates as discussed above are summarized in Table 3.

Table 3. Summary of wastewater treatment applications used hybrid materials as adsorbate.

Hybrid Materials	Synthesis Method	Pollutant Removal	Efficiency	Reference
Bio-MOF-1-derived carbons (BMDC)	Carbonization	Bisphenol A(BPA)	BMDC-12h adsorbent showed an efficiency ~5 times that of a commercial activated carbon in BPA capture.	[62]
Embedded polymeric membrane • MOF-5/PES • MOF-5/CA • MOF-5/PVDF	Immersion precipitation	Metal ions (copper Cu (II) and cobalt Co (II))	More than 50% rejection efficiency for Cu(II). CA/MOF-5 showed 77% rejection efficiency on Co(II).	[71]
Al-MIL-53/[BMIM] [PF_6]	Wet impregnation	Dye removal	increased dye removal efficiency and the maximum dye adsorption capacity for both methylene blue (MB) and methyl orange (MO).	[68]
MOF-5/$Zn_{0.05}$TiOxNy	Microwave	Organic dye and microbial contaminants	Shows highest ability in inhibit microbial growth. Can be recyled up to five times.	[72]

4. Conclusions

In this review, we have discussed in detail the development of incorporated MOF hybrid materials and the latest update of their application in gas separation, catalysis and wastewater treatment. Each of the work was discussed separately and summarized in order to provide a better understanding, mainly in terms of MOF hybrid performance on the mentioned applications. There are numerous techniques opted for to modify MOFs and post-synthetic modification (PSM) is the most popular choice due to the wider possibility of MOF and other material combinations available. Incorporation of other materials in MOFs to form MOF hybrid materials has shown the increase in the materials' efficiency and performance in gas separation, catalysis and wastewater treatment applications mainly in term of recyclability. It is inevitable in the future that more incorporated MOF hybrid materials with better features, high stability and facile recyclability will be developed and studied for various industrial applications due to the limitless availability of possible combinations. Nevertheless, it is important to take into consideration the synthesis protocol such as solvent suitability to ensure minimal damage on MOF structures. Furthermore, it can be concluded that the mass ratio of other materials to be combined with MOF is a key factor for incorporated MOF performance. In the meantime, although incorporated MOFs exhibit a spectacular result in the mentioned applications, a few challenges have been identified. For example, direct introduction of MOF into a polymeric membrane could possibly cause adhesion on the membrane surface, thus allowing gas molecules to escape. Furthermore, the selection of materials to be coupled with MOF should be highly considered and must be well suited with applications. For instance, several ILs are hydrophilic in nature; thus, they are not suitable to be used in applications such as wastewater treatment as they can easily leach into the environment. The capability of this hydrophilic ILs to easily absorb water molecules could also jeopardize the cage structure of MOF. Therefore, further studies are still required to improve the capability of this hybrid material. As for recommendations, a study related to toxicity of this hybrid material should be conducted as its application may involve exposure in our environment. In addition, the combination of MOF with biodegradable material could also be considered.

Author Contributions: Conceptualization, N.M.Y.; resources, Z.J. and N.M.Y.; writing—original draft preparation, Z.J., A.H.A.R. and M.F.A.; writing—review and editing, M.S.S. and N.M.Y.; supervision, N.M.Y. and M.S.S.; project administration, N.M.Y.; funding acquisition, N.M.Y. All authors have read and agreed to the published version of the manuscript.

Funding: This research was funded by the Ministry of Higher Education (MOHE), Malaysia, under the Fundamental Research Grant Scheme (FRGS/1/2020/STG04/UTP/02/4) entitled: *The effects of ionic liquid structures on the stability and CO_2 adsorption of hydrophobic ionic liquid-metal organic framework (IL-MOF) composite materials* and the APC was also funded by the FRGS/1/2020/STG04/UTP/02/4 (cost centre 015MA0-128).

Institutional Review Board Statement: Not applicable.

Informed Consent Statement: Not applicable.

Data Availability Statement: Not applicable.

Acknowledgments: The authors would like to acknowledge the Ministry of Higher Education (MOHE), Malaysia for providing financial assistance under Fundamental Research Grant Scheme (FRGS/1/2020/STG04/UTP/02/4), Universiti Teknologi PETRONAS and Centre for Research in Ionic Liquid (CORIL) for providing the required facilities to conduct this research work.

Conflicts of Interest: The authors declare no conflict of interest.

References

1. Zhang, X.; Zhang, S.; Tang, Y.; Huang, X.; Pang, H. Recent advances and challenges of metal–organic framework/graphene-based composites. *Compos. Part B Eng.* **2021**, *230*, 109532. [CrossRef]
2. Seth, S.; Matzger, A.J. Metal–organic frameworks: Examples, counterexamples, and an actionable definition. *Cryst. Growth Des.* **2017**, *17*, 4043–4048. [CrossRef]
3. Ding, M.; Cai, X.; Jiang, H.-L. Improving MOF stability: Approaches and applications. *Chem. Sci.* **2019**, *10*, 10209–10230. [CrossRef] [PubMed]
4. Zhang, X.; Chen, Z.; Liu, X.; Hanna, S.L.; Wang, X.; Taheri-Ledari, R.; Maleki, A.; Li, P.; Farha, O.K. A historical overview of the activation and porosity of metal–organic frameworks. *Chem. Soc. Rev.* **2020**, *49*, 7406–7427. [CrossRef] [PubMed]
5. Anumah, A.; Louis, H.; Zafar, S.-U.; Hamzat, A.T.; Amusan, O.O.; Pigweh, A.I.; Akakuru, O.U.; Adeleye, A.T.; Magu, T.O. Metal-Organic Frameworks (MOFs): Recent Advances in Synthetic Methodologies and Some Applications. *Chem. Methodol.* **2018**, *3*, 283–305. [CrossRef]
6. Moosavi, S.M.; Nandy, A.; Jablonka, K.M.; Ongari, D.; Janet, J.P.; Boyd, P.G.; Lee, Y.; Smit, B.; Kulik, H.J. Understanding the diversity of the metal-organic framework ecosystem. *Nat. Commun.* **2020**, *11*, 4068. [CrossRef] [PubMed]
7. Kaur, R.; Kaur, A.; Umar, A.; Anderson, W.A.; Kansal, S.K. Metal organic framework (MOF) porous octahedral nanocrystals of Cu-BTC: Synthesis, properties and enhanced adsorption properties. *Mater. Res. Bull.* **2018**, *109*, 124–133. [CrossRef]
8. Cota, I.; Martinez, F.F. Recent advances in the synthesis and applications of metal organic frameworks doped with ionic liquids for CO_2 adsorption. *Coord. Chem. Rev.* **2017**, *351*, 189–204. [CrossRef]
9. Pu, S.; Wang, J.; Li, L.; Zhang, Z.; Bao, Z.; Yang, Q.; Yang, Y.; Xing, H.; Ren, Q. Performance Comparison of Metal–Organic Framework Extrudates and Commercial Zeolite for Ethylene/Ethane Separation. *Ind. Eng. Chem. Res.* **2018**, *57*, 1645–1654. [CrossRef]
10. Yao, L.; Tang, Y.; Cao, W.; Cui, Y.; Qian, G. Highly Efficient Encapsulation of Doxorubicin Hydrochloride in Metal–Organic Frameworks for Synergistic Chemotherapy and Chemodynamic Therapy. *ACS Biomater. Sci. Eng.* **2021**, *7*, 4999–5006. [CrossRef]
11. Furukawa, H.; Cordova, K.E.; O'Keeffe, M.; Yaghi, O.M. The Chemistry and Applications of Metal-Organic Frameworks. *Science* **2013**, *341*, 1230444. [CrossRef] [PubMed]
12. Briggs, L.; Newby, R.; Han, X.; Morris, C.G.; Savage, M.; Krap, C.P.; Easun, T.L.; Frogley, M.D.; Cinque, G.; Murray, C.A. Binding and separation of CO_2, SO_2 and C_2H_2 in homo-and hetero-metallic metal–organic framework materials. *J. Mater. Chem. A* **2021**, *9*, 7190–7197. [CrossRef]
13. Ghanbari, T.; Abnisa, F.; Daud, W.M.A.W. A review on production of metal organic frameworks (MOF) for CO_2 adsorption. *Sci. Total Environ.* **2020**, *707*, 135090. [CrossRef] [PubMed]
14. Vikrant, K.; Kumar, V.; Kim, K.-H.; Kukkar, D. Metal–organic frameworks (MOFs): Potential and challenges for capture and abatement of ammonia. *J. Mater. Chem. A* **2017**, *5*, 22877–22896. [CrossRef]
15. Salehi, S.; Anbia, M.J. High CO_2 adsorption capacity and CO_2/CH_4 selectivity by nanocomposites of MOF-199. *Energy Fuels* **2017**, *31*, 5376–5384. [CrossRef]
16. Yuan, S.; Feng, L.; Wang, K.; Pang, J.; Bosch, M.; Lollar, C.; Sun, Y.; Qin, J.; Yang, X.; Zhang, P.J. Stable metal–organic frameworks: Design, synthesis, and applications. *Adv. Mater.* **2018**, *30*, 1704303. [CrossRef]
17. Ding, M.; Jiang, H.-L. Improving Water Stability of Metal–Organic Frameworks by a General Surface Hydrophobic Polymerization. *CCS Chem.* **2021**, *3*, 2740–2748. [CrossRef]

18. Ferreira, T.J.; Ribeiro, R.P.; Mota, J.P.; Rebelo, L.P.; Esperança, J.M.; Esteves, I.A. Ionic Liquid-Impregnated Metal–Organic Frameworks for CO_2/CH_4 Separation. *ACS Appl. Nano Mater.* **2019**, *2*, 7933–7950. [CrossRef]
19. Zhang, Y.-Y.; Xu, W.; Cao, J.-F.; Shu, Y.; Wang, J.-H. Ionic liquid modification of metal-organic framework endows high selectivity for phosphoproteins adsorption. *Anal. Chim. Acta* **2021**, *1147*, 144–154. [CrossRef]
20. Dao, X.; Nie, M.; Sun, H.; Dong, W.; Xue, Z.; Li, Q.; Liao, J.; Wang, X.; Zhao, X.; Yang, D.; et al. Electrochemical performance of metal-organic framework MOF(Ni) doped graphene. *Int. J. Hydrogen Energy* **2022**, *47*, 16741–16749. [CrossRef]
21. An, H.; Li, M.; Gao, J.; Zhang, Z.; Ma, S.; Chen, Y. Incorporation of biomolecules in Metal-Organic Frameworks for advanced applications. *Coord. Chem. Rev.* **2019**, *384*, 90–106. [CrossRef]
22. Sun, J.; Abednatanzi, S.; Van Der Voort, P.; Liu, Y.-Y.; Leus, K. POM@MOF Hybrids: Synthesis and Applications. *Catalysts* **2020**, *10*, 578. [CrossRef]
23. Xiang, W.; Zhang, Y.; Lin, H.; Liu, C.-J. Nanoparticle/Metal–Organic Framework Composites for Catalytic Applications: Current Status and Perspective. *Molecules* **2017**, *22*, 2103. [CrossRef] [PubMed]
24. Chen, L.; Luque, R.; Li, Y. Controllable design of tunable nanostructures inside metal–organic frameworks. *Chem. Soc. Rev.* **2017**, *46*, 4614–4630. [CrossRef] [PubMed]
25. Kuppler, R.J.; Timmons, D.J.; Fang, Q.-R.; Li, J.-R.; Makal, T.A.; Young, M.D.; Yuan, D.; Zhao, D.; Zhuang, W.; Zhou, H.-C. Potential applications of metal-organic frameworks. *Co-ord. Chem. Rev.* **2009**, *253*, 3042–3066. [CrossRef]
26. Pettinari, C.; Marchetti, F.; Mosca, N.; Tosi, G.; Drozdov, A. Application of metal−organic frameworks. *Polym. Int.* **2017**, *66*, 731–744. [CrossRef]
27. Ferreira, T.J.; Vera, A.T.; De Moura, B.A.; Esteves, L.M.; Tariq, M.; Esperança, J.M.; Esteves, I.A. Paramagnetic ionic liquid/metal organic framework composites for CO_2/CH_4 and CO_2/N_2 separations. *Front. Chem.* **2020**, *8*, 590191. [CrossRef]
28. Ming, M.; Yin, S.; Shi, J. Poly(ionic liquids)-Impregnated UiO-66 composites for efficient sequestration of dichromate. *J. Solid State Chem.* **2022**, *310*, 123091. [CrossRef]
29. Abbasi, Z.; Cseri, L.; Zhang, X.; Ladewig, B.P.; Wang, H. Metal–Organic Frameworks (MOFs) and MOF-Derived Porous Carbon Materials for Sustainable Adsorptive Wastewater Treatment. In *Sustainable Nanoscale Engineering*; Elsevier: Amsterdam, The Netherlands, 2019; pp. 163–194. [CrossRef]
30. Dutta, A.; Pan, Y.; Liu, J.-Q.; Kumar, A. Multicomponent isoreticular metal-organic frameworks: Principles, current status and challenges. *Co-ord. Chem. Rev.* **2021**, *445*, 214074. [CrossRef]
31. Xu, L.; Yan, S.; Choi, E.-Y.; Lee, J.Y.; Kwon, Y.-U. Product control by halide ions of ionic liquids in the ionothermal syntheses of Ni–(H)BTC metal–organic frameworks. *Chem. Commun.* **2009**, *23*, 3431–3433. [CrossRef]
32. Lu, K.; Peláez, A.L.; Wu, L.-C.; Cao, Y.; Zhu, C.-H.; Fu, H. Ionothermal Synthesis of Five Keggin-Type Polyoxometalate-Based Metal–Organic Frameworks. *Inorg. Chem.* **2019**, *58*, 1794–1805. [CrossRef]
33. Mandal, S.; Natarajan, S.; Mani, P.; Pankajakshan, A. Post-Synthetic Modification of Metal–Organic Frameworks Toward Applications. *Adv. Funct. Mater.* **2021**, *31*, 2006291. [CrossRef]
34. Yoo, D.K.; Woo, H.C.; Jhung, S.H. Removal of particulate matter with metal–organic framework-incorporated materials. *Co-ord. Chem. Rev.* **2020**, *422*, 213477. [CrossRef]
35. Kalaj, M.; Cohen, S.M. Postsynthetic modification: An enabling technology for the advancement of metal–organic frameworks. *ACS Cent. Sci.* **2020**, *6*, 1046–1057. [CrossRef] [PubMed]
36. Otal, E.H.; Kim, M.L.; Hattori, Y.; Kitazawa, Y.; Hinestroza, J.P.; Kimura, M. Versatile Covalent Postsynthetic Modification of Metal Organic Frameworks via Thermal Condensation for Fluoride Sensing in Waters. *Bioengineering* **2021**, *8*, 196. [CrossRef]
37. Sosa, J.D.; Bennett, T.F.; Nelms, K.J.; Liu, B.M.; Tovar, R.C.; Liu, Y. Metal–Organic Framework Hybrid Materials and Their Applications. *Crystals* **2018**, *8*, 325. [CrossRef]
38. Kiang, Y.-H.; Gardner, G.B.; Lee, S.; Xu, A.Z.; Lobkovsky, E.B. Variable Pore Size, Variable Chemical Functionality, and an Example of Reactivity within Porous Phenylacetylene Silver Salts. *J. Am. Chem. Soc.* **1999**, *121*, 8204–8215. [CrossRef]
39. Li, H.; Li, L.; Lin, R.-B.; Zhou, W.; Zhang, Z.; Xiang, S.; Chen, B. Porous metal-organic frameworks for gas storage and separation: Status and challenges. *Energy Chem.* **2019**, *1*, 100006. [CrossRef]
40. Faramawy, S.; Zaki, T.; Sakr, A.-E. Natural gas origin, composition, and processing: A review. *J. Nat. Gas Sci. Eng.* **2016**, *34*, 34–54. [CrossRef]
41. Elhenawy, S.E.M.; Khraisheh, M.; AlMomani, F.; Walker, G. Metal-organic frameworks as a platform for CO_2 capture and chemical processes: Adsorption, membrane separation, catalytic-conversion, and electrochemical reduction of CO_2. *Catalysts* **2020**, *10*, 1293. [CrossRef]
42. Rufford, T.; Smart, S.; Watson, G.; Graham, B.; Boxall, J.; da Costa, J.D.; May, E. The removal of CO_2 and N_2 from natural gas: A review of conventional and emerging process technologies. *J. Pet. Sci. Eng.* **2012**, *94*, 123–154. [CrossRef]
43. Grande, C.A.; Roussanaly, S.; Anantharaman, R.; Lindqvist, K.; Singh, P.; Kemper, J. CO_2 Capture in Natural Gas Production by Adsorption Processes. *Energy Procedia* **2017**, *114*, 2259–2264. [CrossRef]
44. Zeeshan, M.; Nozari, V.; Yagci, M.B.; Isık, T.; Unal, U.; Ortalan, V.; Keskin, S.; Uzun, A. Core–shell type ionic liquid/metal organic framework composite: An exceptionally high CO_2/CH_4 selectivity. *J. Am. Chem. Soc.* **2018**, *140*, 10113–10116. [CrossRef] [PubMed]
45. Sezginel, K.B.; Keskin, S.; Uzun, A. Tuning the Gas Separation Performance of CuBTC by Ionic Liquid Incorporation. *Langmuir* **2016**, *32*, 1139–1147. [CrossRef] [PubMed]

46. Bernardo, P.; Clarizia, G. 30 years of membrane technology for gas separation. *Chem. Eng. Trans.* **2013**, *32*, 1999–2004.
47. Chen, Z.; Hong, Z.; Wu, H.; Li, C.; Jiang, Z. Tröger's Base Polyimide Hybrid Membranes by Incorporating UiO-66-NH_2 Nanoparticles for Gas Separation. *Ind. Eng. Chem. Res.* **2022**, *61*, 3418–3427. [CrossRef]
48. Castarlenas, S.; Téllez, C.; Coronas, J. Gas separation with mixed matrix membranes obtained from MOF UiO-66-graphite oxide hybrids. *J. Membr. Sci.* **2017**, *526*, 205–211. [CrossRef]
49. Nabais, A.R.; Martins, A.P.; Alves, V.D.; Crespo, J.G.; Marrucho, I.M.; Tomé, L.C.; Neves, L.A. Poly (ionic liquid)-based engineered mixed matrix membranes for CO_2/H_2 separation. *Sep. Purif. Technol.* **2019**, *222*, 168–176. [CrossRef]
50. Pascanu, V.; Miera, G.G.; Inge, A.K.; Martín-Matute, B. Metal–Organic Frameworks as Catalysts for Organic Synthesis: A Critical Perspective. *J. Am. Chem. Soc.* **2019**, *141*, 7223–7234. [CrossRef]
51. Liu, Y.; Chen, G.; Chen, J.; Niu, H. Excellent Catalytic Performance of Ce–MOF with Abundant Oxygen Vacancies Supported Noble Metal Pt in the Oxidation of Toluene. *Catalysts* **2022**, *12*, 775. [CrossRef]
52. Chughtai, A.H.; Ahmad, N.; Younus, H.A.; Laypkov, A.; Verpoort, F. Metal–organic frameworks: Versatile heterogeneous catalysts for efficient catalytic organic transformations. *Chem. Soc. Rev.* **2015**, *44*, 6804–6849. [CrossRef] [PubMed]
53. Chatterjee, A.; Hu, X.; Lam, F.L.-Y. Towards a recyclable MOF catalyst for efficient production of furfural. *Catal. Today* **2018**, *314*, 129–136. [CrossRef]
54. Gangu, K.K.; Jonnalagadda, S.B. A review on metal-organic frameworks as congenial heterogeneous catalysts for potential organic transformations. *Front. Chem.* **2021**, *9*, 747615. [CrossRef] [PubMed]
55. Sun, Y.; Huang, H.; Vardhan, H.; Aguila, B.; Zhong, C.; Perman, J.A.; Al-Enizi, A.M.; Nafady, A. Ma, S. Facile Approach to Graft Ionic Liquid into MOF for Improving the Efficiency of CO_2 Chemical Fixation. *ACS Appl. Mater. Interfaces* **2018**, *10*, 27124–27130. [CrossRef]
56. Veisi, H.; Abrifam, M.; Kamangar, S.A.; Pirhayati, M.; Saremi, S.G.; Noroozi, M.; Tamoradi, T.; Karmakar, B. Pd immobilization biguanidine modified Zr-UiO-66 MOF as a reusable heterogeneous catalyst in Suzuki–Miyaura coupling. *Sci. Rep.* **2021**, *11*, 1–14. [CrossRef]
57. Wu, Y.; Song, X.; Xu, S.; Zhang, J.; Zhu, Y.; Gao, L.; Xiao, G. 2-Methylimidazole Modified Co-BTC MOF as an Efficient Catalyst for Chemical Fixation of Carbon Dioxide. *Catal. Lett.* **2019**, *149*, 2575–2585. [CrossRef]
58. Ten, S.; Torbina, V.V.; Zaikovskii, V.I.; Kulinich, S.A.; Vodyankina, O.V. Bimetallic AgPd/UiO-66 Hybrid Catalysts for Propylene Glycol Oxidation into Lactic Acid. *Materials* **2020**, *13*, 5471. [CrossRef]
59. Muralikrishna, I.V.; Manickam, V. Industrial wastewater treatment technologies, recycling, and reuse. *Environ. Manag.* **2017**, 295–336.
60. Muralikrishna, I.V.; Manickam, V.J.E.M. Chapter One—Introduction. In *Environmental Management*; Elsevier: Amsterdam, The Netherlands, 2017; pp. 1–4, ISBN 978-0-12-811989-1.
61. Shojaei, S.; Shojaei, S. Optimization of process. In *Soft Computing Techniques in Solid Waste and Wastewater Management*; Elsevier: Amsterdam, The Netherlands, 2021; pp. 381–392.
62. Bhadra, B.N.; Lee, J.K.; Cho, C.-W.; Jhung, S.H. Remarkably efficient adsorbent for the removal of bisphenol A from water: Bio-MOF-1-derived porous carbon. *Chem. Eng. J.* **2018**, *343*, 225–234. [CrossRef]
63. Russo, V.; Hmoudah, M.; Broccoli, F.; Iesce, M.R.; Jung, O.-S.; Di Serio, M. Applications of Metal Organic Frameworks in Wastewater Treatment: A Review on Adsorption and Photodegradation. *Front. Chem. Eng.* **2020**, *2*, 581487. [CrossRef]
64. Masindi, V.; Muedi, K.L. Environmental contamination by heavy metals. *Heavy Met.* **2018**, *10*, 115–132.
65. Rehman, K.; Fatima, F.; Waheed, I.; Akash, M.S.H. Prevalence of exposure of heavy metals and their impact on health consequences. *J. Cell. Biochem.* **2017**, *119*, 157–184. [CrossRef]
66. Maheshwari, K.; Agrawal, M.; Gupta, A. Dye Pollution in Water and Wastewater. In *Novel Materials for Dye-Containing Wastewater Treatment*; Springer: Berlin/Heidelberg, Germany, 2021; pp. 1–25.
67. Lellis, B.; Fávaro-Polonio, C.Z.; Pamphile, J.A.; Polonio, J.C. Effects of textile dyes on health and the environment and bioremediation potential of living organisms. *Biotechnol. Res. Innov.* **2019**, *3*, 275–290. [CrossRef]
68. Kavak, S.; Durak, Ö.; Kulak, H.; Polat, H.M.; Keskin, S.; Uzun, A. Enhanced Water Purification Performance of Ionic Liquid Impregnated Metal–Organic Framework: Dye Removal by [BMIM][PF6]/MIL-53 (Al) Composite. *Front. Chem.* **2021**, *8*, 622567. [CrossRef] [PubMed]
69. Kumar, S.; Ahlawat, W.; Bhanjana, G.; Heydarifard, S.; Nazhad, M.; Dilbaghi, N. Nanotechnology-Based Water Treatment Strategies. *J. Nanosci. Nanotechnol.* **2014**, *14*, 1838–1858. [CrossRef]
70. Iervolino, G.; Zammit, I.; Vaiano, V.; Rizzo, L. Limitations and Prospects for Wastewater Treatment by UV and Visible-Light-Active Heterogeneous Photocatalysis: A Critical Review. In *Heterogeneous Photocatalysis*; Springer: Berlin/Heidelberg, Germany, 2019; pp. 225–264. [CrossRef]
71. Gnanasekaran, G.; Balaguru, S.; Arthanareeswaran, G.; Das, D.B. Removal of hazardous material from wastewater by using metal organic framework (MOF) embedded polymeric membranes. *Sep. Sci. Technol.* **2018**, *54*, 434–446. [CrossRef]
72. Younis, S.A.; Serp, P.; Nassar, H.N. Photocatalytic and biocidal activities of ZnTiO2 oxynitride heterojunction with MOF-5 and g-C3N4: A case study for textile wastewater treatment under direct sunlight. *J. Hazard. Mater.* **2020**, *410*, 124562. [CrossRef] [PubMed]

processes

MDPI

Article

Research on the Path of Manufacturing Enterprises Supply Chain Integration from the Configuration Perspective

Hongxiong Yang and Yunpeng Wang *

School of Management, Tianjin University of Technology, Tianjin 300384, China; yanghongxiong@email.tjut.edu.cn
* Correspondence: Wangyunpeng0926@163.com; Tel.: +86-152-3050-1075

Abstract: In digital transformation and development, supply chain integration has become a key strategy to improve supply chain synergy efficiency and enhance enterprise competitiveness. Based on the survey data of 185 manufacturing enterprises in Tianjin, the fuzzy set qualitative comparative analysis (fs QCA) method is used to explore the synergistic mechanism of government policy, supply chain partnership, information sharing, risk avoidance, and intelligence degree on supply chain integration and the interaction among them. The results show that: (1) a single factor does not constitute a necessary condition for promoting supply chain integration, but the formation and development of supply chain partnership plays a universal role in promoting supply chain integration; (2) the "multiple concurrent" of five factors constitute the diversified configuration of a driving supply chain integration path, that is, the driving supply chain integration path has the characteristic of "all roads lead to the same destination"; (3) there is a "sharing" type, a "cooperative" type, and a "cooperative-sharing" type three equivalent path, whereby the formation of supply chain partnership can enhance the trust between manufacturing enterprises and suppliers and customers, increase the transaction frequency of upstream and downstream enterprises, and improve the cooperation efficiency. The utility model can effectively reduce the "long whip" problem caused by the information asymmetry, and improve the operational efficiency and stability of the whole supply chain. The purpose of this study is to inspire manufacturing enterprises in the context of digital supply chain integration to improve the collaborative efficiency of the supply chain.

Keywords: manufacturing enterprise; supply chain integration; integration path; collaborative efficiency; qualitative comparative analysis of fuzzy sets

Citation: Yang, H.; Wang, Y. Research on the Path of Manufacturing Enterprises Supply Chain Integration from the Configuration Perspective. *Processes* **2021**, *9*, 1746. https://doi.org/10.3390/pr9101746

Academic Editors: Jun-Wei Lim and Worapon Kiatkittipong

Received: 2 September 2021
Accepted: 26 September 2021
Published: 29 September 2021

1. Introduction

In today's digital trend of information development, enterprises must keep up with the degree of unceasing enhancement, and improvement of the supply chain in the development of intelligent collaborative efficiency has become a hot spot in the industry. For enterprises to keep up with current developments, it is necessary to explore more agile and efficient supply chain integration pathways to promote the intelligent coordination of the supply chain efficiency. Based on this, more and more enterprises are paying attention to supply chain integration. Supply chain integration can make the enterprise more collaborative via cooperation between the customers and suppliers, and to some extent, can realize the resource sharing. To enhance the competitiveness of the supply chain and promote the improvement of enterprise operating performance [1], this is the key to gain a competitive edge in the complex market environment. However, in a dynamic and complex environment, the primary concerns of enterprises are which supply chain integration strategy to adopt and how to integrate it [2], as the study of manufacturing in the supply chain integration path is of extremely practical significance.

In recent years, many types of studies on supply chain integration have been carried out in academia and have achieved fruitful results. To sum up, supply chain integration is a management strategy of manufacturing enterprises that enables enterprises to carry out higher-level cooperation through cooperation and coordination to improve the competitiveness and operating benefits of enterprises in the supply chain [2–5]. There are many factors affecting supply chain integration, including government support, supply chain partnership, information sharing, risk avoidance, etc. Different scholars have also carried out relevant studies on how these factors affect supply chain integration. Zeng Mingang et al. [6] investigated 180 manufacturing enterprises in the Pearl River Delta region and found that government policies can indirectly promote supply chain integration by influencing environmental uncertainty. Pan Ancheng et al. [7] studied a project team based on the cultural background of reason and found that a close partnership is an effective means to improve the integration of the supply chain. From the perspective of reducing the bullwhip effect, Zhang Wei et al. [8] pointed out that in the supply chain, information transmission and sharing can help improve the performance of the whole supply chain. Wang Li [9] believes that the optimization and integration of the manufacturing supply chain can provide effective help for enterprises to avoid risks, improve their profitability and speed up capital turnover. However, at present, Chinese enterprises are gradually developing towards the direction of digitalization, and the degree of intelligence of enterprises is deepening, so the degree of intelligence of enterprises in the supply chain has become a new factor that may affect the integration of supply chain. Huo Yanfang et al. [10] pointed out that if manufacturing enterprises want to gain advantages in market competition, the most important thing is to grasp the current intelligent manufacturing environment and design a convenient and effective supply chain. Throughout the past, the research of supply chain integration found that most still focus on the integration of various factors of a single probe. The relative lack of systematic analysis of the supply chain integration factors and the mechanism of interaction between various factors and the path of research make it hard to get an effectively suitable supply chain enterprise integration path to attract manufacturing enterprises in the supply chain integration. In particular, few scholars introduce how to integrate the supply chain of manufacturing enterprises under the background of digitalization so that it can better adapt to the requirements of intelligent manufacturing on the supply chain function. Therefore, this article introduces integrity research; the integration of the existing factors includes intelligent degree study, the integrated use of literary analysis, a questionnaire survey method, and a fuzzy qualitative comparative analysis method to build the research model of manufacturing enterprise supply chain integration, providing a combination of quantitative and qualitative analysis to explore the integration of the supply chain paths. To improve the competitiveness of the supply chain of manufacturing enterprises and the overall collaborative efficiency under the background of digitization, the optimal combination of factors conducive to supply chain integration is sought.

2. Literature Review

2.1. The Nature of Supply Chain Integration

Supply chain integration is the focus of supply chain management. It improves the operational performance of the supply chain by optimizing the behavior process, organization, business process, and management links related to the supply chain by core enterprises [11]. Scholars at home and abroad have conducted a lot of studies on supply chain integration. Malonii and Benton [3] defined supply chain integration as a higher level of cooperative management between supply chain partners to provide customers with higher value and improve competitive advantages, and divided supply chain integration into internal integration and external integration for the first time. Huo et al. [4] pointed out in their studies that supply chain integration reflects enterprises' ability to formulate strategic alliances, restructuring resources, build seamless processes, and exchange and share information. Xu Ke [5] believes that supply chain integration is strategic cooperation of information integration, resource coordination, inter-organization interconnection, and process management. Generally, a supply chain is composed of suppliers, manufacturers and customers. Therefore, Flynn et al. [2] divides supply chain integration into three dimensions: customer integration, supplier integration, and internal integration. Among them, the integration of customers and suppliers is called external integration, which refers to the synchronization of strategic cooperation and process cooperation among enterprises by establishing partnerships with core enterprises in the supply chain. Internal integration refers to the synchronization of strategy, operation, and process through internal collaboration to meet customer needs.

The comprehensive research of scholars found that although different scholars put different emphasis on the nature of supply chain integration, they all pointed to its role in enterprise strategic cooperation and process management. Few scholars have systematically explored and studied the complex process of synergistic action among factors promoting supply chain integration. Therefore, based on previous studies on influencing factors of supply chain integration and the current status of digital transformation of Chinese manufacturing enterprises, this paper comprehensively studies the influence mechanism of government policy, supply chain partnership, information sharing, risk avoidance and intelligence degree on the supply chain integration path of manufacturing enterprises. In order to enhance the competitive advantage of manufacturing enterprises in the process of digital development and improve the efficiency of cooperation between manufacturing enterprises and other enterprises in the supply chain, this paper studies the supply chain integration path from the two aspects of internal integration and external integration.

2.2. Internal Factors of Supply Chain Integration

By analyzing many studies, we found that integration of internal factors of supply chain manufacturing facilitates better cooperation with external suppliers to meet customer demand and between different departments using the integration methods. This allows faster interaction of internal information and the optimizing of enterprise coordination to actively avoid risks, improve their stability and maintain competitive advantages. First of all, from the perspective of information integration, Xu et al. [12] explained the importance of information exchange and sharing between partners by comparing whether information sharing has fluctuations in orders and changes in safe stock between manufacturers and sellers. Zhou Sihua and Wan Guohua [13] found that the improvement of information capability in supply chain enterprises can significantly positively affect supply chain integration. Phillip and Wendell [14] argue that information communication is the key to maintaining supply chain partnerships, and the existence of such relationships can reduce supply chain integration costs and improve supply chain integration efficiency. Moreover, information exchange between enterprises can have a positive impact on organizational capability and supply chain operational performance through supply chain integration [15]. Secondly, from the perspective of risk avoidance, risk avoidance refers to the decision-making behavior that enterprises choose to avoid risks because they are afraid of risks. Pure risks and environmental fluctuations will reduce the overall benefits of the supply chain [16], so it is necessary to study the impact of risk aversion on supply chain integration. Based on the study of supply chain risk management on the performance of commercial circulation enterprises, Wu Qiang et al. [17] found that risk avoidance, risk sharing, and risk transfer in supply chain risk management have a significant positive impact on enterprise performance. Weingarten et al. [18] discussed the role of risk and risk management practices in promoting the success of supply chain integration from a relational perspective and made it clear that the adoption of risk avoidance behavior can promote the success of supply chain integration. The goal of risk avoidance is to reduce the complexity and uncertainty of the supply chain system through supply chain integration, to significantly improve the effectiveness of the supply chain system [19]. Finally, from the perspective of intelligent degree, current scholars do not study the relationship between the degree of intelligent manufacturing and the supply chain integration path, but with the advent of the era of the digital economy, the Internet and the rapid development of artificial intelligence, as the intelligent mode of production is more popular with consumers, intelligent manufacturing transformation has become the trend of the times. Zhou Jiajun et al. [20] observed from the perspective of technological foundations that intelligent manufacturing technology is a manufacturing behavior that realizes product design, manufacturing, management, and maintenance through humanoid behavior operations—such as perception and human–computer interaction based on information technology and artificial intelligence, to realize the integration of the two. The development of the intelligent manufacturing industry also requires the corresponding improvement of intelligent supply chain management systems to a certain extent. Barton et al. [21] pointed out that the key to creating a good supply chain is the improvement of reaction ability and adaptation to high-level intelligent system integration and management ability. It can be seen that the introduction of artificial intelligence technology into supply chain management, the creation of intelligent supply chain management systems, and the use of artificial intelligence and big data analysis technology to explore supply chain management data can improve management efficiency and reduce management costs. Therefore, this paper takes information sharing, risk avoidance, and intelligence degree as internal factors driving supply chain integration.

2.3. External Factors of Supply Chain Integration

The research on external influencing factors of supply chain integration can help enterprises get more help in the new development pattern of double circulation and improve the cooperation quality of supply chain partner enterprises. Reviewing previous studies, it can be seen that both government support and supply chain partners have a certain impact on supply chain integration [22,23]. Governments can often influence the industrial structure by allocating resources, and biased government policies will have a lasting impact on the business environment of enterprises [24]. Zeng Mingang, Wu Shaobo et al. [23] selected 251 manufacturing supply chains in Guangdong province to study their integration relationship and found that government policies can play a role in supply chain integration by affecting environmental uncertainty, and it has a significant positive impact on supply chain integration. Zeng Mingang [6] used empirical research to show that customer trust, supplier trust, customer integration, and supplier integration are all significantly affected by government support. Dou Chao et al. [25] pointed out that big customers with government backgrounds can be recognized by the capital market, which can significantly improve the valuation level of enterprises and show their competitive advantages and value. Liu Minghua et al. [26] studied the data of 218 upstream and downstream enterprises of the supply chain and found that partnership has a significant positive impact on supply chain integration. Ye Fei and Xu Xuejun [27] studied the degree of trust, relationship commitment, information sharing, and operational performance among enterprises in the supply chain, and found that trust can positively affect the relationship between relationship commitment, information sharing, and operational performance. Feng Hua et al. [28] believe that sharing information, complementary resources, and creating a reliable transaction environment among supply chain partners are conducive to the integration of nodal enterprises' business activities, to reduce opportunistic behaviors and transaction costs. To sum up the above research findings: the formation and development of any enterprise cannot be separated from the support and assistance of the government. The formation of a good mutual assistance relationship between enterprises and the government can make it easier for enterprises to attract partners, enhance their competitive advantages and promote the integration of the supply chain. In addition, by allying with other enterprises in the supply chain, enterprises can interact and coordinate their resources and information based on mutual trust, which is conducive to the cooperation among enterprises in the supply chain. Therefore, this paper takes government policy and supply chain partnership as external factors driving supply chain integration.

2.4. Marginal Contribution of Supply Chain Integration

Based on the analysis of the above factors affecting supply chain integration, this paper holds that the influence of government policies, supply chain partnerships, information sharing, risk avoidance, and intelligence degree on supply chain integration is not independent but complex. The synergistic effect among all factors makes the traditional regression analysis or structural equation method unable to solve the diversified practical problems of the supply chain integration path. The combination of different integration factors will promote supply chain integration from different aspects. However, the combination of different integration paths will improve the synergy efficiency of the supply chain from different aspects. In this paper, through a questionnaire to collect data and information sharing, we focus on risk aversion and intellectualized degree as internal driving factors of supply chain integration, and government policy and the supply chain partnership as the external driving factors of supply chain integration, constructing the theoretical research framework shown in Figure 1, using fuzzy qualitative research methods of comparative analysis (fs QCA). It reveals how government policy, supply chain partnership, information sharing, risk avoidance, and intelligence drive manufacturing enterprises to conduct supply chain integration, thus providing a theoretical reference for manufacturing enterprises to conduct supply chain integration practice.

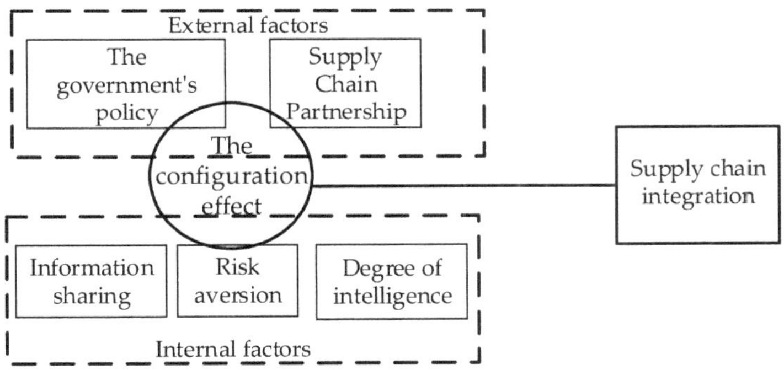

Figure 1. The research framework of supply chain integration theory.

This paper has three marginal contributions to the study of supply chain integration path: First, in terms of integration factors, the paper studies the influence of intelligence degree on supply chain integration under the background of digital development, links this factor with supply chain integration, and discusses the influence of the former on the latter and its mechanism. Second, in terms of research methods, traditional statistical analysis methods such as linear regression and structural equation can only deal with completely corresponding correlation, while the fs QCA method can deal with all kinds of complex asymmetric relations, ensuring the effectiveness of research results. Thirdly, based on the perspective of configuration, this paper adopts the holistic and systematic analysis to empirically test how the configuration of five factors—namely government policy, supply chain partnership, information sharing, risk avoidance and intelligence degree—can better promote the integration of supply chain, and makes a reasonable explanation for the corresponding configuration.

3. Research Process

3.1. Data Collection

This paper takes the supply chain of manufacturing enterprises in Tianjin as an example, based on the following two points: (1) Tianjin is one of the largest industrial cities in China, and its industrial structure has been constantly upgrading, which is an important carrier of coordinated industrial development in The Beijing–Tianjin–Hebei region;

(2) In the context of the new normal of economy, the manufacturing industry in Tianjin is developing towards information and intellectualization, with diversified industrial structures and rapid development of emerging industries. The study on the supply chain integration path of manufacturing enterprises in Tianjin will provide a reference for the supply chain integration of manufacturing enterprises in the whole country. In addition, suppliers and distributors are closely linked in the manufacturing industry, which is the industry with the most obvious characteristics of the supply chain and is more typical in the integration of the supply chain. Therefore, this paper takes the supply chain of manufacturing in Tianjin as an example, adopts a questionnaire survey to collect relevant data, and takes 216 manufacturing enterprises in Tianjin as samples. The questionnaire survey started from Mary 2021 to May 2021. A total of 216 questionnaires were sent out, of which 205 were recovered with a recovery rate of 94.91%. After removing invalid questionnaires, 185 valid questionnaires were collected with effective recovery of 90.24%. The main characteristics of the samples are shown in Table 1:

Table 1. Distribution of basic features of samples (N = 185).

Sample Characteristics	The Basic Information	Sample Size	Specific Gravity (%)
The enterprise-scale	A large enterprise	43	23.24%
	Medium-sized enterprise	87	47.03%
	Small and micro enterprises	55	29.73%
The enterprise type	Agriculture and sideline, food, tobacco, and alcohol	15	8.11%
	Textiles and Apparel	13	7.03%
	Mineral, metal processing, and products	21	11.35%
	The pharmaceutical industry	17	9.19%
	Machinery, transportation equipment	32	17.29%
	Communications, electronics, electrical appliances	39	21.08%
	Other manufacturing	48	25.95%
Sources of funding	Domestic enterprises	163	88.11%
	The foreign capital enterprise	7	3.78%
	The joint venture	15	8.11%
Enterprise establishment year	5 years or less	48	25.95%
	6 to10 years	59	31.89%
	11 to 15 years	45	24.32%
	16 years or more	33	17.84%

3.2. *Qualitative Comparative Analysis Method Based on Fuzzy Sets*

The selection of the fs QCA method is mainly based on two considerations: (1) The manufacturing enterprise supply chain integration shared with multiple factors, such as enterprise risk response, the government's policy support, and the information sharing degree of several different fronts due to the different combinations of conditions that form different paths of manufacturing enterprise supply chain integration; each path [29] has different characteristics, corresponding to different types of manufacturing enterprises. Compared with traditional quantitative analysis methods, the fs QCA analysis method can deal with the influence of more antecedent variables on the result variables, so it can better analyze the influence path of the five antecedent variables on the integration results. (2) In manufacturing enterprise supply chain integration, the antecedent and consequent variables are not necessarily sufficient and necessary relationships, such as information sharing in the supplier data sharing, although they help an enterprise to master the supplier's stock information, price, quality, and so on, which is advantageous for the enterprise to make comparisons of different suppliers, choose the best suppliers, and promote enterprises of supply chain integration. However, it is also possible that some suppliers' data information may be faked, making enterprises make wrong choices, resulting in corresponding economic losses, affecting the normal operation of the supply chain, and thus reducing the overall collaborative efficiency of the supply chain. Compared with the traditional statistical analysis method which can only deal with the complete corresponding correlation, the fs QCA method can deal with all kinds of complex asymmetric relations. The analysis software used below is fs QCA3.0.

3.3. Measurement of Variables

The Likert five-point scale was selected to measure all variables. To ensure the reliability and validity of the scale, the measurement of variables was based on the existing mature scale, and a few items in the questionnaire were revised and adjusted according to the needs of the study in the process of the questionnaire predictive test. Among them, the supply chain integration scale of outcome variable comes from Flynn et al. [2], which measures supplier integration, internal integration, and customer integration from three dimensions, including 10 items such as "we track major customers to get feedback". Among the conditional variables, government policies are mainly referred to and adapted from the research of Cai et al. [30] and Chen et al. [31], and measured from three dimensions of policy support, policy adjustment and political relations, including four questions such as "the government helps our enterprises to obtain resources (capital, labor, key raw materials, etc.)". Partnership mainly adopts the views of Morgan and Hunt [32], Kumar and Dissel [33], and is measured from two dimensions of organizational trust and relationship commitment, including nine questions such as "major suppliers share their available inventory level information with us". Risk avoidance mainly adopts the research theories of Li Mingqiang [34] and Yan Cunyan [35], which are measured from three dimensions of risk system, management attitude and coping methods, including three questions such as "the intensity of risk aversion of enterprises". Information sharing comes from Flynn et al. [2], which measures demand, inventory and, supplier data from three dimensions, including five questions such as "We share demand prediction information of related products with major supply chain partners". The degree of intelligence mainly adopts the research of Meng Fansheng and Zhao Gang [36], and is measured from three dimensions of a product, production mode, and intelligent management, including four items such as "intelligent level of enterprise operation and management". Descriptive statistics of outcome variables and conditional variables are shown in Table 2.

Table 2. Descriptive statistical analysis results.

Condition Variables	The Mean	The Standard Deviation	GP	SCP	IS	RA	DOI	SCI
Government policy	2.7297	0.8652	1					
Supply chain partnership	3.0775	1.0858	−0.444 **	1				
Information sharing	2.9319	0.9789	0.308 **	−0.576 **	1			
Risk aversion	2.8378	0.9787	−0.376 **	0.284 **	0.218 **	1		
Degree of intelligence	3.5608	0.9430	−0.437 **	0.156 *	−0.147 *	0.244 **	1	
Supply chain integration	3.5130	0.7953	−0.055	0.390 **	0.133	0.529 **	0.218 **	1

Notes: **. At the level 0.01 (double-tailed), the correlation is significant *. At the level of 0.05 (double-tailed), the correlation was significant. "GP" means "government policy", "SCP" means "supply chain partnership", "IS" means "information sharing", "RA" means "risk aversion", "DOI" means "degree of intelligence", "SCI" means "supply chain integration".

3.4. Reliability and Validity Test

SPSS25.0 was used to test the reliability and validity of the scale [37], and the results are shown in Table 3. Cronbach's α and composite reliability (CR) ranged from 0.897 to 0.938 and 0.763 to 0.973, respectively. The reliability indexes were all above 0.7, which met the internal consistency of the scale, indicating the high reliability of the questionnaire [38]. Exploratory factor analysis (EFA) was conducted on the questionnaire, the statistical test (Kaiser Meyer Olkin, KMO) ranged from 0.749 to 0.942 [39], which passed the significance level test of 0.001. At the same time, six factors with characteristic roots greater than 1 were analyzed, explaining 78.56% (higher than 50%) variance, indicating that the study had good structural validity. Confirmatory factor analysis (CFA) of AMOS23.0 scale showed that the model fit degree was high ($\chi^2/DF = 2.134$, RMSEA = 0.0785, IFI = 0.912, TLI = 0.900,

CFI = 0.911) [40,41], and the standard factor load was above 0.6. The average variance extracted (AVE) values of each variable ranged from 0.518 to 0.802 [42], indicating that this study had good aggregation validity.

Table 3. Reliability and validity analysis of variables.

Condition Variables	Target Item	Load Factor	Alpha Value	CR	KMO	AVE
Government policy	GP1 GP2 GP3 GP4	0.923 0.842 0.761 0.873	0.912	0.913	0.816	0.726
Supply chain partnership	SCP1 SCP2 SCP3 SCP4 SCP5 SCP6 SCP7 SCP8 SCP9	0.920 0.896 0.894 0.888 0.911 0.849 0.889 0.888 0.925	0.973	0.973	0.942	0.802
Information sharing	IS1 IS2 IS3 IS4 IS5	0.902 0.744 0.858 0.699 0.834	0.920	0.905	0.800	0.657
Risk aversion	RA1 RA2 RA3	0.744 0.721 0.692	0.910	0.763	0.749	0.518
Degree of intelligence	DOI1 DOI2 DOI3 DOI4	0.922 0.745 0.806 0.843	0.897	0.899	0.810	0.691
Supply chain integration	SCI1 SCI2 SCI3 SCI4 SCI5 SCI6 SCI7 SCI8 SCI9 SCI10	0.839 0.777 0.755 0.618 0.755 0.633 0.864 0.902 0.877 0.859	0.938	0.944	0.915	0.630

Note: "GP" means "government policy", "SCP" means " supply chain partnership", "IS" means "information sharing", "RA" means "risk aversion", "DOI" means "degree of intelligence ", "SCI" means "supply chain integration".

3.5. Variable Calibration

According to the fs QCA method, a case belongs to a set, and calibration is the process of assigning membership scores to each case set [43]. The calibration of variables is usually based on existing theoretical knowledge or researchers' experience, and there are three critical points of complete membership, intersection, and complete non-membership on punctuality [44]. The process of transforming variables into sets on 0~1 is fuzzy set analysis. Referring to the Likert five-level scale to measure the degree of formalization of organizational structure, "5" is set as complete membership, "3" as intersection, and "1" as complete non-membership, as shown in Table 4:

Table 4. Variable calibration anchor points.

Condition Variables		The Anchor		
		Completely Unaffiliated	Intersection	Full Membership
External factors	Government policy	1	3	5
	Supply chain partnership	1	3	5
Internal factors	Information sharing	1	3	5
	Risk aversion	1	3	5
	Degree of intelligence	1	3	5
The results of variable	Supply chain integration	1	3	5

4. Result Analysis

4.1. Univariate Result Analysis

Before configuration analysis, it is necessary to discuss whether a single driving factor is a sufficient and necessary condition for the supply chain integration of manufacturing enterprises. According to Ragin's research, when the consistency is greater than 0.9, this condition can be identified as a necessary condition for the outcome variable [44]. The analysis results of single factor necessary conditions are shown in Table 5. The results show that all ante-cause conditions of manufacturing enterprises' integration factors are less than 0.9, indicating that government policy support, supply chain partnership, information sharing, risk avoidance, and intelligence degree are not the necessary conditions for manufacturing enterprises' supply chain integration. This result verified the combination of factors of supply chain integration in manufacturing enterprises, that is to say, the supply chain integration of manufacturing enterprises is a complex system, no single factor has a significant effect, and the supply chain integration path of manufacturing enterprises is multiple.

Table 5. Single-factor necessity analysis results of supply chain integration of manufacturing enterprises.

Condition Variables	Supply Chain Integration	
	Consistency	Coverage
Government policy	0.521544	0.832768
~ Government policy	0.736571	0.831089
Supply chain partnership	0.699207	0.889444
~ Supply chain partnership	0.539776	0.743050
Information sharing	0.617448	0.856042
~ Information sharing	0.632900	0.799856
Risk aversion	0.646881	0.955556
~ Risk aversion	0.631510	0.755772
Degree of intelligence	0.807538	0.814060
~ Degree of intelligence	0.422124	0.810900

Note: "~" means "not" in logical operation.

4.2. Configuration Analysis

Through truth table analysis, three kinds of solutions are obtained, which are a complex solution, reduced solution, and intermediate solution. Among them, the core conditions appear in both the simplified and intermediate solutions, while the edge conditions only appear in the intermediate solutions [44]. The configuration analysis results obtained through Boolean algebra operation are shown in Table 6. Fs QCA analysis needs set case frequency thresholds and consistency threshold filtering to explain the variables with a sufficient explanation of the former conditions [44]. Among them, consistency represents the proportion of all cases covering the result, and the case frequency threshold refers to the corresponding degree of the set relationship expressed in the case and the solution.

According to the research of Fiss [45] and Du Yunzhou et al. [46], the consistency threshold is set as 0.8, the case frequency threshold is set as 1, and the PRI threshold is set as 0.75.

Table 6. Configuration of supply chain integration factors.

Condition Variables	Supply Chain Integration				
	Solution 1	Solution 2A	Solution 2B	Solution 3A	Solution 3B
Government policy	•	⊠	⊠	⊠	•
Supply chain partnership	⊠	⬤	⬤	⬤	⬤
Information sharing	⬤		⊠	⬤	⬤
Risk aversion	⊠	•			
Degree of intelligence	⊠		•	⊠	•
Consistency	0.879171	0.992582	0.906883	0.998173	0.998146
Original coverage	0.28436	0.459489	0.515984	0.26008	0.30807
Unique coverage	0.0925514	0.0402256	0.120922	0.0155342	0.0294334
Total consistency			0.883473		
The total coverage			0.717194		

Note: ⬤ represents that the core condition exists; ⊠ represents the absence of a core condition; • represents the existence of marginal conditions; ⊠ represents the absence of marginal conditions; "Blank" indicates that the condition can be present or absent.

The Fs QCA3.0 software shows a group consistency of 5, which is greater than the 0.8 threshold factors combination of manufacturing enterprise supply chain integration, as shown in Table 5. Although the integration factors contained in each set of results are different, each configuration can promote the supply chain integration of manufacturing enterprises to the same extent. Therefore, each combination of integration factors is equivalent. According to the total coverage, it can be concluded that the five integrated configurations explain about 71.72% of the supply chain samples of manufacturing enterprises. The consistency indexes of the five configurations were 0.879171, 0.992582, 0.906883, 0.998173, and 0.998146, respectively, showing a high consistency level. The five configuration results in Table 5 show that the original coverage is 0.2844, 0.4595, 0.5160, 0.2601 and 0.3081 respectively, indicating that solutions 1, 2A, 2B, 3A and 3B, the ratio of 5 configuration cases to all manufacturing supply chain integration cases were 28.44%, 45.95%, 51.60%, 26.01% and 30.81%, respectively. The unique coverage is the ratio of manufacturing enterprise supply chain integration cases explained by a single configuration after excluding common cases of other configurations. The unique coverage of five configurations is 0.0926, 0.0402, 0.1209, 0.0155, and 0.0294, respectively. Comparatively, solution 2B configuration has the highest coverage.

Under the complex action of government policy, supply chain partnership, information sharing, risk avoidance, and intelligence degree, supply chain integration presents five realization configurations, namely, configurations 1, 2A, 2B, 3A, and 3B. According to the existence or absence of different elements in each configuration, and whether they are core conditions or edge conditions, the five configurations obtained by running fs QCA are divided into the following three types of supply chain integration paths of manufacturing enterprises:

Path 1: No matter whether other conditions exist or not, as long as information sharing exists as the core condition and government policies are assisted as the marginal condition, supply chain integration can be effectively promoted and collaborative efficiency among enterprises can be improved (configuration 1).

In configuration 1, the consistency level is 0.8792 and the unique coverage is 0.0926. In this configuration, the manufacturing enterprise supply chain integration configuration 1 shows that, even if supply chain partnership (core condition) does not exist, as long as there is information sharing as core condition and government policy as edge condition, supply chain integration can be promoted even if there is no risk avoidance and intelligence degree. We named this path "sharing". This shows that in the process of promoting supply chain integration, information sharing plays a central role, and government policies provide a complementary role. Good information sharing can effectively reduce the distortion of supply and demand information caused by a bullwhip effect and promote the visualization of supply chain. In addition, according to the study of Flynn et al. [2], the relationship between information sharing and the supply chain integration promotes each other and spirals upward. Effective exchange of information between supply chain partners, on the one hand, to promote the cooperation between enterprises, to enhance the degree of trust between them, on the other hand, supply chain integration is the integration of all enterprise supply chain for the entire supply network, it will inevitably increase the connection between the various enterprises, to promote the degree of information sharing between them, improve supply chain reaction speed. Enterprises with government support can better attract partners and improve their competitive advantages [25]. Information sharing enhances between suppliers, manufacturers, and customers to a understanding of each other's needs, inventory, improves the production efficiency and inventory turns, government support for the more stable development of enterprises laid a good foundation, so they can effectively promote the integration of the supply chain, improve the quality of cooperation between enterprises, enhance the stability of the supply chain.

Path 2: Supply chain partnership exists as a core condition, and risk aversion exists as a marginal condition, which can promote supply chain integration (configuration 2A), or supply chain integration can be promoted as long as supply chain partnership exists as the core condition and intelligence degree exists as the edge condition (configuration 2B).

In configuration 2A and 2B, the consistency level is 0.9926 and 0.9069, and the unique coverage is 0.0402 and 0.1209, respectively. In this path, manufacturing enterprise supply chain integration configuration solution 2A shows that whatever information sharing (core conditions) and risk aversion (edge conditions) exists or not, as long as there is as a condition of a core of supply chain partnership and as a condition of the edge of intelligent degree, even if the lack of government policy, can also promote the integration of the supply chain. Solution 2B shows that whether risk avoidance (edge condition) exists or not, supply chain integration can be promoted even without government policies and information sharing as long as there is supply chain partnership as the core condition and intelligence degree as the edge condition. We named this path "cooperative", this means that enterprises can also promote supply chain integration by relying on supply chain partnership as the core condition and the existence of risk avoidance or intelligence degree as the edge condition. This is because the partnership is the relationship between different enterprises to jointly invest relevant resources to achieve business objectives, to achieve the goals set by each other [47]. Through establishing supply chain partnerships, enterprises can reduce operating costs, disperse operating risks and improve production efficiency. Through the sharing of key resources between supply chain partners, the competitive position of enterprises can also be improved [27]. Meanwhile, risk avoidance can reduce the complexity and uncertainty of the supply chain system and significantly improve the effectiveness of the supply chain system through supply chain integration [19]. The key for manufacturing enterprises to deal with the current complex business environment is to grasp the opportunity of intelligent manufacturing and create an agile and efficient supply chain [4]. The level of intelligence determines the response speed of the supply chain. The agility and efficiency of the supply chain can effectively shorten the response time of upstream and downstream enterprises of the supply chain and improve the collaborative efficiency of the supply chain. Therefore, either the combination of supply chain partnership and risk avoidance or the combination of supply chain partnership and intelligence degree can promote the supply chain integration of manufacturing enterprises, to promote collaborative development among enterprises.

Path 3: Regardless of the presence or absence of government policies and intelligence, enterprises can effectively promote supply chain integration through the existence of supply chain partnerships and information sharing as core conditions (configurations 3A and 3B).

The consistency level of configuration 3A and 3B is high in the three configurations, 0.9982 and 0.9981, respectively, but the unique coverage is low, 0.0155 and 0.0294, respectively. In this path, solution 3A shows that supply chain partnership and information sharing are core conditions, risk avoidance is optional. Solution 3B shows that all other factors exist except risk aversion. This means that regardless of the presence or absence of government policies and intelligence, enterprises can effectively promote supply chain integration through the existence of supply chain partnerships and information sharing as core conditions. We named this path "cooperative-sharing." The study of Li and Lin [48] proved that the level and quality of information sharing among supply chain partners are positively influenced by the degree of trust. As for the information-sharing behavior of manufacturing enterprises, trust is the key factor for information sharing among members when there is no constraint or norm to regulate the behaviors of members. If enterprises can share profits in the process of information-sharing interaction, they will establish a good trust relationship. The establishment of this relationship can effectively promote the information sharing behavior between the supply chain partners and enhance the stability of the supply chain. This means that the formation of supply chain partnership can promote information sharing, improve the quality and level of shared information, and it is complementary to information sharing. Therefore, enterprises with two core conditions of supply chain partnership and information sharing can also effectively promote supply chain integration and improve the collaborative efficiency of the whole supply chain.

4.3. Robustness Analysis

To ensure the robustness of the configuration, the frequency threshold was set to 2, the consistency threshold remained unchanged, and the data was run again. The results only changed slightly, and the configuration path remained unchanged, as shown in Table 7. Therefore, the robustness of the configurational results can be guaranteed.

Table 7. Configuration of supply chain integration factors.

Condition Variables	Supply Chain Integration				
	Solution 1	**Solution 2A**	**Solution 2B**	**Solution 3A**	**Solution 3B**
Government policy	•	⊠	⊠	⊠	•
Supply chain partnership	☒	●	●	●	●
Information sharing	●		☒	●	●
Risk aversion	⊠	•			
Degree of intelligence	⊠		•	⊠	•
Consistency	0.864172	0.932552	0.906883	0.918103	0.926126
Original coverage	0.28536	0.446489	0.501874	0.27208	0.30807
Unique coverage	0.1025214	0.040232	0.127622	0.015642	0.027435
Total consistency			0.884623		
The total coverage			0.720164		

Note: ● represents that the core condition exists; ☒ represents the absence of a core condition; • represents the existence of marginal conditions; ⊠ represents the absence of marginal conditions; "Blank" indicates that the condition can be present or absent.

5. Conclusions and Enlightenment

5.1. Research Conclusions

Combining the 185 valid questionnaire survey data of Tianjin manufacturing companies, from the perspective of configuration, based on further stimulation and enhancement of the efficiency of the overall coordination of supply chain system, building the government policy, supply chain partnership, information sharing, risk aversion and intelligent degree, the relationship between the theoretical model reveals the influence of core supply chain integration and also its complex interactive nature. Using the method of fs QCA analyzing the configuration of the above theoretical model, the conclusions are as follows:

(1) Through the fs QCA3.0 software on the analysis of the questionnaire data processing, it was found that the government policy, supply chain partnership, information

sharing, risk aversion, and intelligent degree of the consistency level is below 0.9, which shows that the five factors are not the necessary factors to promote the integration of the supply chain, and that validation of manufacturing enterprises in the supply chain integration is a complex process, as no single factor played a significant role. This shows that the effect of manufacturing enterprises' supply chain integration factors alone on supply chain integration is far less than after the configuration of these factors. Each supply chain integration configuration requires the existence of at least one of the core conditional supply chain partnerships or information sharing. The operation results of fs QCA give the configuration of five supply chain integration factors, and the five configuration results forms three supply chain integration paths, and each path promotes the supply chain integration of manufacturing enterprises to the same extent.

(2) Although there are many problems to be solved in the supply chain operation and management of manufacturing enterprises in Tianjin, enterprises have been actively exploring various factors effecting supply chain integration, and some enterprises have gradually launched the integration of the supply chain. This article uses the fuzzy qualitative comparative method to research the supply chain integration path, through the interaction between various factors affecting the manufacturing enterprise supply chain integration mechanism and the configuration of the anatomy, define a "sharing" type, "cooperative" type, " cooperative–sharing type" three supply chains integration path, supply chain partnership and information sharing are the core conditions. The formation of the supply chain partnership can improve the quality and level of enterprise information sharing, and the relationship between information sharing and supply chain integration is mutually promoting and spiraling. Therefore, supply chain partnership and information sharing combined with other integration factors can effectively promote the supply chain integration, thus improving the collaborative efficiency of supply chain.

(3) There are a variety of manufacturing enterprises in China, and different types of manufacturing enterprises have different functional requirements for the supply chain. As an important industrial production base in China, Tianjin has complete industrial product chains, which provide abundant supply chain samples of manufacturing enterprises for research. It is worth mentioning that although the operation results of fs QCA show that the degree of intelligence is not the core condition of supply chain integration at present, with the deepening of the new development pattern of information and digitalization, the intelligent development of manufacturing enterprises in the future is inevitable. Therefore, enterprises should continue to pay attention to the changes in the functional requirements of the supply chain due to the degree of intelligence and update the management mode in time to improve the collaborative development of the supply chain.

5.2. Practical Enlightenment

This study provides some practical enlightenment for the supply chain integration of manufacturing enterprises.

(1) Manufacturing enterprises to explore the integration path of the supply chain. Supply chain integration itself is a management method of higher-level cooperation between supply chain partners to provide customers with higher value and improve their competitive advantages [2]. It is a magic weapon for enterprises to obtain greater performance and enhance their competitive advantages. At present, China's manufacturing industry accounts for an increasing proportion in the world, and the demand for supply chain functions is also becoming stronger. Therefore, it is necessary for the manufacturing industry to actively explore the integration path of the supply chain, and continuously improve the management level and collaborative efficiency of the existing supply chain. With the integration of Internet information technology into manufacturing enterprises and the continuous application of the whole supply chain system, the supply chain integration of manufacturing enterprises will be affected by more factors in the future. Therefore, manufacturing enterprises should grasp the current development of the times and constantly arm themselves with new technologies to achieve better development.

(2) Each subject in the supply chain should attach importance to establishing cooperative partnerships in the process of supply chain integration. Through the operation of fs QCA software, it can be seen that supply chain partnership is one of the core elements in supply chain integration, and supply chain partnership exists in four of the five configuration combinations, indicating the importance of supply chain partnership in supply chain integration. Between different manufacturing companies, to achieve the same business objectives to invest together related resources, to achieve common targets set by the cooperation with each other, make up the supply chain partnership, stable and close relations of cooperation and can improve the level of trust among supply chain partners, increase the number of transactions of upstream and downstream enterprises of the supply chain, improve the efficiency of cooperation. Partnerships between enterprises can improve each other's production efficiency, enhance the ability of enterprises to deal with risks, and promote collaborative development among enterprises. Therefore, a good partnership is not only an effective way to promote supply chain integration, but also an effective means to improve the performance of enterprises.

(3) Supply chain integration can better improve synergistic efficiency among enterprises. The results show that supply chain integration can effectively reduce information asymmetry in supply chain management. By integrating with suppliers and customers, manufacturing enterprises can establish a stable and close supply chain partnership, which can improve cooperation quality and disperse operational risks. In addition, by sharing key resources between the supply chain partners, non-value-added intermediate links can be also reduced, and operating costs can be reduced. At the same time, the establishment of supply chain partnership between enterprises can enhance mutual trust, improve the quality and level of information sharing, effectively reduce the "long whip" problem caused by information asymmetry, improve the overall collaborative efficiency of the supply chain, and enhance the competitive advantage of upstream and downstream enterprises of the supply chain.

5.3. Research Limitations and Prospects

(1) Increase research samples. This study used manufacturing supply chains in Tianjin as the research object; although the manufacturing system of Tianjin provided reliable support, the different area government policies involved in the process of manufacturing enterprises, supply chain partnership, information sharing, risk aversion, and intelligent degree of the recognition of the role of supply chain integration may create certain differences. In the future, questionnaires can be further conducted in several different regions to expand the number and scope of research samples and further improve and adjust the existing research conclusions.

(2) Mining and integrating details based on specific cases. In this paper, the fuzzy set qualitative comparative analysis method overcomes the defect of traditional linear regression that can only analyze single influencing factors one by one, and better explains how various factors interact to produce complex effects on the integration of supply chain of manufacturing enterprises, but the study of specific integration details is not deep enough. The subsequent research can theoretically explore the corresponding supply chain of manufacturing enterprises according to the three paths obtained in this paper, explore the different characteristics of traditional supply chain integration path and intelligent supply chain integration path, and discover the driving mechanism of various integration factors in different types of supply chains.

(3) Explore a new supply chain integration path. With the emergence of various emerging technologies, consumer demand for products is increasingly personalized and diversified and is in dynamic change. Therefore, an agile supply chain that can respond to dynamic market demand in real-time arises at the right moment. The traditional supply chain management model has many links and emphasizes post-event control, which easily leads to supply and demand dislocation. An agile supply chain management mode emphasizes the enterprise by optimizing the management structure and the implementa-

tion process to establish a highly integrated and shared information system, thus giving companies a quick response and the ability to handle customer orders and to streamline the product circulation. This will have a great influence on production, warehousing, distribution and delivery in the supply chain. The emergence of the new supply chain management model may change the driving factors of the existing supply chain integration, thus changing the existing supply chain integration path. Therefore, the impact or change of emerging technologies on supply chain integration deserves further attention and research.

Author Contributions: The contribution of both authors was balanced in all phases of the development of this study, both in the empirical part (creation and validation of the instrument, data collection, and analysis) and in the writing part of this manuscript and its various parts. All authors have read and agreed to the published version of the manuscript.

Funding: This research received no external funding.

Institutional Review Board Statement: Ethical review and approval were waived for this study, as it did not involve personally identifiable or sensitive data.

Informed Consent Statement: The study does not involve human nor sensitive data.

Data Availability Statement: The data in this paper were obtained through a questionnaire survey.

Conflicts of Interest: The authors declare no conflict of interest.

References

1. Bu, M. Supply chain integration and retail firm performance: The mediating role of supply chain flexibility. *J. Commer. Econ.* **2020**, *22*, 121–124.
2. Flynn, B.B.; Huo, B.; Zhao, X. The impact of supply chain integration on performance: A contingency and configuration approach. *J. Oper. Manag.* **2010**, *28*, 58–71. [CrossRef]
3. Benton, M.J.M.C. Supply chain partnerships: Opportunities for operations research. *Eur. J. Oper. Res.* **1997**, *101*, 419–429.
4. Huo, B.; Ye, Y.; Zhao, X.; Shou, Y. The impact of human capital on supply chain integration and competitive performance. *Int. J. Prod. Econ.* **2016**, *178*, 132–143. [CrossRef]
5. Xu, K.; Wang, R.; Zhang, H.; Chen, G. The impact of technological innovation on sustainable development in China's textile and garment industry: A mediating effect model from the perspective of supply chain. *J. Stat. Inf.* **2015**, *30*, 65–70.
6. Zeng, M.; Zhu, J. The impact of environmental uncertainty and government support on supply chain integration. *Sci. Res. Manag.* **2014**, *35*, 79–86.
7. Pan, A.; Liu, H. An exploratory study on the evolution of relational behavior and organizational knowledge under the culture of rationality. *Nankai Bus. Rev.* **2015**, *18*, 85–94.
8. Zhang, W.; Hua, Z.S. Bullwhip effect in multi-product supply chain and its reduction method. *J. Syst. Eng.* **2010**, *25*, 479–483.
9. Wang, L.; Li, D. Study on supply chain integration and financial performance of manufacturing enterprises. *J. Southeast Univ. (Philos. Soc. Sci.)* **2016**, *18*, 55–57.
10. Huo, Y.; Wang, H.; Qi, E. Building smart logistics and supply chain to help smart manufacturing—Introduction of smart logistics and smart supply chain. *China Mech. Eng.* **2020**, *31*, 2891–2897.
11. Chen, X. A review of supply chain integration research. *China Water Transp. (Acad. Version)* **2007**, *11*, 247–250.
12. Xu, K.; Dong, Y.; Evers, P.T. Towards better coordination of the supply chain. *Transp. Res. E* **2001**, *37*, 35–54. [CrossRef]
13. Zhou, S.; Wan, G. The impact of information technology capabilities on supply chain performance: Based on the perspective of information integration. *J. Syst. Manag.* **2016**, *25*, 90–102.
14. Balsmeier, P.W.; Voisin, W.J. Supply chain management: A time- based strategy. *Ind. Manag.* **1996**, *9*, 24–27.
15. Kang, S.B.; Moon, T. Impact of information exchange and supply chain integration on supply chain performance. *Sci. Technol.* **2016**, *9*, 237–246. [CrossRef]
16. Yu, J.; You, D. Comparative study of supply chain competition and cooperation mode considering the characteristics of risk aversion. *Syst. Eng.—Theory Pract.* **2019**, *39*, 2091–2104.
17. Wu, Q.; Zhang, Y. Performance analysis of commercial circulation enterprises based on supply chain risk management. *J. Commer. Econ.* **2020**, *24*, 114–116.
18. Weingarten, F.; Humphreys, P.; Gimenez, C.; McIvor, R. Risk, risk management practices, and the success of supply chain integration. *Int. J. Prod. Econ.* **2016**, *171*, 361–370. [CrossRef]
19. Roh, J.; Hong, P.; Min, H. Implementation of a responsive supply chain strategy in global complexity: The case of manufacturing firms. *Int. J. Prod. Econ.* **2014**, *147*, 198–210. [CrossRef]
20. Zhou, J.; Yao, X. Advanced manufacturing technology and the new industrial revolution. *Comput. Integr. Manuf. Syst.* **2015**, *21*, 1963–1978.

21. Barton, R.; Thomas, A. Implementation of intelligent systems, enabling integration of SMEs to high-value supply chain networks. *Eng. Appl. Artif. Intell.* **2009**, *22*, 929–938. [CrossRef]
22. Huo, B.; Han, Z.; Zhao, X. The impact of power and relationship commitment on supplier integration. *J. Manag. Sci. China* **2013**, *16*, 33–50.
23. Zeng, M.; Lu, S.; Wu, Q. Research on the relationship between government support, trust and supply chain external integration. *Chin. J. Manag. Sci.* **2014**, *22*, 48–55.
24. Sheng, S.; Zhou, K.Z.; Li, J.J. The effects of business and political ties on firm performance: Evidence from China. *J. Mark.* **2011**, *75*, 1–15. [CrossRef]
25. Dou, C.; Chen, X.; Li, X. Government-backed customers, market perception and investment opportunities: From the perspective of supply chain integration. *Manag. Rev.* **2020**, *32*, 13–28.
26. Liu, H.; Wang, Y.; Li, H. Research on the relationship between partnership, logistics capability and supply chain integration. *Chin. J. Manag. Sci.* **2016**, *24*, 148–157.
27. Ye, F.; Xu, X. Research on the relationship between partner characteristics, partnership, and information sharing in the supply chain. *J. Manag. Sci. China* **2009**, *12*, 115–128.
28. Feng, H.; Shi, Y.; Wei, J. A study on the interaction between social control and supply chain performance: Dependence and supply chain integration as mediators. *Chin. J. Manag. Sci.* **2020**, *28*, 136–145.
29. Fiss, P.C. A set-theoretic approach to organizational configurations. *Acad. Manag. Rev.* **2007**, *32*, 1180–1198. [CrossRef]
30. Cai, S.; Jun, M.; Yang, Z. Implementing supply chain information integration in China: The role of institutional forces and trust. *J. Oper. Manag.* **2010**, *28*, 257–268. [CrossRef]
31. Chen, Y.R.; Yang, C.; Hsu, S.M.; Wang, Y.D. Entry mode choice in China's regional distribution markets: Institution vs. transaction costs perspectives. *Ind. Mark. Manag.* **2009**, *38*, 702–713. [CrossRef]
32. Morgan, R.M.; Hunt, S.D. The commitment-trust theory of relationship marketing. *J. Mark.* **1994**, *58*, 20–38. [CrossRef]
33. Kumar, K.; Dissel, H.G.V. Sustainable collaboration: Managing conflict and cooperation in interorganizational systems. *MIS Q.* **1996**, *20*, 279–300. [CrossRef]
34. Li, M. *The Road to Total Risk Management: Theory and Practice Insight of Risk Management*; China Financial Publishing House: Beijing, China, 2014.
35. Yan, C. *Fundamental Knowledge of Risk Management*; Economy & Management Publishing House: Beijing, China, 2018.
36. Meng, F.; Zhao, G. Study on the influencing factors of the development from traditional manufacturing to intelligent manufacturing. *Sci. Technol. Prog. Policy* **2018**, *35*, 66–72.
37. Guo, Z. *Social Statistical Analysis Method—Application of SPSS Software*; China Renmin University Press: Beijing, China, 1999.
38. Gefen, D.; Straub, D.W.; Boudreau, M. Structural equation modeling and regression: Guidelines for research practice. *Commun. Assoc. Inf. Syst.* **2010**, *4*, 1–78. [CrossRef]
39. Joseph, F.; Hair, J.R.; Anderson, R.E.; Tatham, R.L.; Black, W.C. *Multivariate Date Analysis with Reading*, 4th ed.; Prentice-Hall International: Upper Saddle River, NJ, USA, 1995.
40. Schwarzwald, J.; Weisenberg, M.; Solomon, Z. Factor invariance of SCL-90-R: The case of combat stress reaction. *Psychol. Assess.* **1991**, *3*, 385–390. [CrossRef]
41. Alem, A.; Kebede, D.; Woldesemaiat, G.; Jacobsson, L.; Kullgren, G. The prevalence and socio-demographic correlates of mental distress in Buiajira Ethiopia. *Acta Psychiatr. Scand.* **1999**, *100*, 48–55. [CrossRef]
42. McEachan RR, C.; Conner, M.; Taylor, N.J.; Lawton, R.J. Prospective prediction of health-related behaviours with the theory of planned behaviour: A meta-analysis. *Health Psychol. Rev.* **2011**, *5*, 97–144. [CrossRef]
43. Ragin, C.C. *Research on Redesign of Social Sciences*; Machinery Industry Press: Beijing, China, 2019; p. 48.
44. Ragin, C.C. *Redesigning Social Inquiry: Fuzzy Sets and Beyond*; University of Chicago Press: Chicago, IL, USA, 2008.
45. Fiss, P.C. Building better causal theories: A fuzzy set approach to typologies in organization research. *Acad. Manag. J.* **2011**, *54*, 393–420. [CrossRef]
46. Du, Y.; Jia, L. Configuration perspective and qualitative comparative analysis (QCA): A new approach to management research. *Manag. World* **2017**, 155–167.
47. Rigby, D.K.; Buchanan, W.T. Putting more strategy into strategic alliances. *Dir. Boards* **1994**, *18*, 14–19.
48. Li, S.; Lin, B. Accessing information sharing and information quality in supply chain management. *Decis. Support Syst.* **2007**, *42*, 1641–1656. [CrossRef]

processes

MDPI

Article

Comparison of Performance and Economic Efficiency for Greenhouse Solar versus Hot Air Drying: A Case of Crispy Mango Production

Kanokwan Kingphadung [1], Patsanan Kurdkaew [1], Prasong Siriwongwilaichat [1] and Suphaphat Kwonpongsagoon [2,3,*]

1 Department of Food Technology, Faculty of Engineering and Industrial Technology, Silpakorn University, Nakhon Pathom 73000, Thailand; kingphadung_k@su.ac.th (K.K.); patsanan_maamaymae@hotmail.com (P.K.); deersong1@yahoo.com (P.S.)

2 Department of Sanitary Engineering, Faculty of Public Health, Mahidol University, Bangkok 10400, Thailand

3 Center of Excellence on Environmental Health and Toxicology (EHT), Bangkok 10400, Thailand

* Correspondence: suphaphat.kwo@mahidol.ac.th; Tel./Fax: +66-2354-8540

Citation: Kingphadung, K.; Kurdkaew, P.; Siriwongwilaichat, P.; Kwonpongsagoon, S. Comparison of Performance and Economic Efficiency for Greenhouse Solar versus Hot Air Drying: A Case of Crispy Mango Production. *Processes* **2022**, *10*, 311. https://doi.org/10.3390/pr10020311

Academic Editors: Jun-Wei Lim and Worapon Kiatkittipong

Received: 17 January 2022
Accepted: 2 February 2022
Published: 5 February 2022

Abstract: This study aimed to compare the performance and economic efficiency of greenhouse solar and hot air drying. These are widely used and inexpensive techniques for preserved fruit production in Thailand. Two drying methods were used to dry mango fruit, var. Nam Dok Mai. During the process of both drying methods, the moisture content of mango slices was reduced from an average 85.6% to 12.5% wet basis. To determine the most profitable dryer system for investment, economic analyses of greenhouse solar and hot air drying, involving the main variables affecting their profitability, were investigated, including production and maintenance cost, capital costs, revenue, and drying capacities. The internal rate of return (IRR), net present value (NPV), benefit to cost ratio (BCR) and payback period (PBP) were all evaluated. The results of economic evaluation revealed that high values of IRR, NPV, and BCR, and low value of PBP made the two drying systems cost efficient. However, when the incremental rate of return (ROR) was determined, 66% incremental ROR was much greater than the minimum attractive rate of return (MARR = 4.875%). Consequently, the greenhouse solar drying method proved to be the most suitable method for small-to-medium-scale entrepreneurs to produce crispy mango.

Keywords: crispy mango; drying; economic analysis; greenhouse solar; hot air

1. Introduction

Mango is an important export fruit and grown nationwide in Thailand. Of more than 60 described mango varieties grown in Thailand, 7 are grown for export: 'Nam Dok Mai', 'Khiao Sawoei', 'Chok Anan', 'Nang Klang Wan', 'Pim Sen Dang', 'Maha Chanok' and 'Rad' [1]. The export volume of fresh mangoes in 2020 was 90,245 tons, which was valued at USD 79 million. Nam Dok Mai is the most popular variety for export [2]. Thailand exports mangoes mainly to Japan, Korea, Vietnam, China and Malaysia [3]. Nam Dok Mai mango is preferred in many countries due to favorable characteristics, such as attractive color, good taste and great flavor [1]. The main harvest period is limited to about three months yearly in Thailand, April to June. The lack of adequate storage leads to high postharvest losses of the perishable fruits. In addition, overproduction causes low prices and increases the demand for appropriate preservation methods of fresh mango flesh [4]. Processed mango products are fast gaining markets and commanding better prices than other processed tropical fruits. Drying is a major postharvest process of fruit product in Thailand responding to demands of dried fruits from both domestic and export markets [5].

Two drying methods, namely, greenhouse solar and hot air dryers, are commonly used to dry mangoes to obtain a product that provides a beneficial effect on human health, a high-quality product and a positive trade potential [4–6]. Solar and hot air drying are available to prevent the deterioration of products and to reduce the drying time [7]. The technology, particularly solar air type, can be the most profitable compared with open-air sun drying, due to better quality and quantity [8]. The greenhouse solar dryer is a system that uses solar power, a renewable source of energy, and is applied primarily by many small-to-medium-sized agricultural industries. It can reduce operational costs in terms of fuel [9]. To dry mangoes on a large or commercial scale, many enterprises use electricity-powered, hot air dryers and liquefied petroleum gas burners. It is generally claimed that hot air dryers are not economical due to their high energy cost [7].

However, greenhouse solar drying is known as a capital-intensive technology in terms of initial investment and operating costs, compared to hot air dryers. Thus, it becomes necessary that small-and-medium-sized enterprises (SMEs) using dryers are able to not only offset the higher cost of dried fruits, but also provide adequate profits. Many studies have reported the economic analysis of solar dryers used for different kinds of agricultural products, but rarely for hot air dryers. Janjai [5] showed that using a solar dryer—a parabolic roof structure covered with polycarbonate sheets on a concrete floor—offered payback periods (PBP) of about 0.65 years for drying tomatoes. Boonyasri et al. [10] reported that the PBP of a greenhouse solar dryer used for drying pork was 1.15 years. In India, a hybrid solar–electrical dryer used to dry ginger was reported to gain PBP of 0.5 years with an initial investment of INR 9000 [7]. The net present value (NPV) of such a hybrid solar dryer is INR 20,700, and the benefit-to-cost ratio of a hybrid solar dryer is 2.3. Kondareddy et al. [11] analyzed that the PBP and benefit-to-cost ratio of a modified solar dryer (MSD) with thermal energy storage for drying blood fruit in India were 2.16 years and 2.07, respectively. Another related study carried out an economic evaluation of the solar dryer for drying fruit and vegetables, revealing high values of an internal rate of return (IRR) at 84.4%, and low value of PBP at 1.42 years [8]. ELkhadraoui et al. [12] analyzed the economics of solar drying systems for red pepper, indicating that solar dryers took 17 h to reduce the initial moisture content from 12.15 to 0.17 g water/g dry matter, with 1.77 years of PBP.

This study aimed (a) to investigate the effect of temperature and velocity of air drying on dried mangoes quality; (b) to compare performance and economic efficiency of greenhouse solar and hot air drying; and (c) to determine the most suitable method for drying crispy mangoes in SMEs.

2. Materials and Methods

2.1. Material and Sample Preparation

Thai mango cultivar, 'Nam Dok Mai', samples were purchased during the 2021 harvest season from a local market in Nakhon Pathom, Thailand. Then, they were cleaned and washed. Green mature fruits were allowed to reach a degree of approximately 75% to 80% full ripeness [4]. The moisture contents of the fresh mangoes were 85% wet basis (wb), and the mangoes were left for ripening at room temperature. The mangoes were hand-peeled and cut into slabs of 30 × 70 mm and 2 mm thick, using a specially designed stainless-steel cutter. Slices of approximately 3 g were placed on trays as a single layer for all drying experiments.

2.2. Experimental Study

To compare the two drying methods in the first experiment, mango slices were dried using a hot air electric dryer separate from the greenhouse type solar dryer. A hot air electric dryer contains 24 trays, produced by Kluay Namthai Trading Groups, Bangkok, Thailand. The greenhouse solar dryer consisted of a parabolic roof structure made from polycarbonate sheets on a concrete floor. The small-scale solar system used in this study had a width of 6.0 m, length of 8.2 m, and height of 3.25 m. For hot air drying, three drying temperatures were conducted at 70, 80, and 90 °C, and the air velocity was fixed at 1 m/s. Samples were dried to a constant weight with a final moisture content of 12.5% wb to study the drying kinetics. The same procedures were applied for greenhouse solar drying, in which each experiment was conducted from 9:00 a.m. to 4:00 p.m. on each day until the desirable product moisture content of 12.4% wb was obtained.

In the second experiment, a combination of the same types of greenhouse solar and hot air dryers was used to improve the quality and profitability of production. The mango slices were firstly dried in the greenhouse solar dryer for 3 h at 60 °C, followed by the hot air dryer at 90 °C for 30 min with a fixed air velocity at 1 m/s.

2.3. Drying Rate of Mango Slices

The initial moisture content of mangoes on wb was calculated using the equation below.

$$M_i = \frac{w_i - w_f}{w_i} \quad (1)$$

where M_i is the initial moisture content of mango slices on wb expressed in %, w_i is the initial weight of mango slices in g, and w_f is the final weight of mango slices in g [8].

The samples were weighed every 10 min in the first hour of drying until the weight of the samples was constant, and the initial and final moisture contents of the samples were determined. The recorded moisture contents for each sample were then used to plot the drying curves. The drying rate is expressed as the amount of evaporated moisture over time. The drying rate of mango slices was calculated using the following equation.

$$DR = \frac{MC_1 - MC_2}{t_1 - t_2} \quad (2)$$

where MC_1 and MC_2 are the moisture content of samples (kg water/kg dry matter) at time t_1 and t_2 (m), respectively [13].

The water activity (a_w) of the samples was determined using a water activity meter (AQUALAB 4TE, METER Group, Pullman, WA, USA), measured at 25 °C. All measurements were carried out in triplicate.

2.4. Color Measurement of Mango Slices

Three slices of fresh and dried samples were individually taken for a color measurement in the CIE color coordinates (L*, a*, b*), using a colorimeter (Colorflex E spectrophotometer, Hunter Lab, Reston, VA, USA). L*, a*, and b* values represent black to white (0–100), green to red (−60 to +60) and blue to yellow (−60 to +60), respectively. The total difference of the color parameters was calculated using the following equation [5,13].

$$\Delta E = \sqrt{(L_0^* - L_t^*)^2 + (a_0^* - a_t^*)^2 + (b_0^* - b_t^*)^2} \quad (3)$$

where L_o^*, a_o^*, and b_o^* are the initial values of color parameters at the initial time, zero and L_t^*, a_t^*, and b_t^* are the value at drying time, t.

2.5. Total Phenolics and Antioxidant Analysis

2.5.1. Sample Extraction

Sample extraction was modified following Singleton and Rossi [14]. Fresh and dried slices mango samples were finely chopped. Then 5 g of fresh sliced mango samples and 1 g of dry sliced mango samples were placed in a 50 mL centrifuge tube, followed by adding 20 mL of 100% methanol. Then samples were homogenized for 15 s and extracted by an ultrasonic machine for 30 min at control temperature around 38 to 40 °C. After that, the samples were filtered with Whatman no. 4 filter paper, facilitated using a vacuum pump. Then, 50 mL of the extracted sample was used to analyze total phenolic content and antioxidant capacity by both the DPPH and FRAP methods. All analyses were triplicated.

2.5.2. Total Phenolic Content

A slight modification of the Folin–Ciocalteau assay stated by Hossain et al. [15] was adopted to analyze the total phenolic content of the extracted samples. A 0.2 mL sample was placed in an amber flask, followed by adding 1 mL of 10% Folin–Ciocalteu's phenol reagent (vol/vol). Then, samples were mixed using a shaker for 5 min, and 1.6 mL of 7.5% sodium carbonate solution (weight/volume) was added. Shaking and storing in the dark for 120 min before absorbance was measured at a wavelength of 765 nm. Values were reported in milligram equivalent gallic acid (GAE)/100 g dry weight. The standard equation obtained for total phenolic content is y = 0.062x + 0.0239, where y represents absorbance values at a wavelength of 765 nm and x is the concentration of standard solutions of gallic acid.

2.5.3. DPPH and DPPH Radical Scavenging Activity

The DPPH antioxidant analysis was slightly modified from the method of Siddiq et al. [16]. A 0.1 mL sample was placed in a 20 mL centrifuge tube, then 3.9 mL of DPPH solution was added. The mixture was mixed using a vortex mixer and stored in the dark for 120 min before being measured for absorbance at a wavelength of 517 nm using methanol as a blank. Values were reported in milligrams of Trolox/100 g dry weight. The standard equation obtained for DPPH is y = −0.02x + 0.6265, where y represents absorbance values at a wavelength of 517 nm and x is the concentration of standard solutions of Trolox. The findings are presented as percentage of radical scavenging activity using the following equation.

$$\text{DPPH radical scavenging activity } (\%) = \frac{A_O - A_S}{A_O} \times 100 \quad (4)$$

where A_o is absorbance of control solution and A_s is absorbance of the DPPH solution with sample extracts.

2.5.4. Ferric Reducing Antioxidant Power (FRAP) Assay

The FRAP antioxidant capacity analysis was slightly modified from the method of Vongsak et al. [17]. A 950 μL FRAP solution was transferred to a microcentrifuge tube, followed by 50 μL of sample then mixed well before keeping in the dark for 10 min. Absorbance was measured at a wavelength of 595 nm using methanol as a blank. Values were reported in mg of ferrous sulfate per 100 g of sample. The standard equation obtained for the FRAP antioxidant capacity is y = 0078x + 0.0399, where y represents absorbance values at a wavelength of 595 nm and x is the concentration of standard solutions of ferrous sulfate.

2.6. Sensory Evaluation of Mango Slices

Sensory evaluation of dried mango slices for texture appearance was performed using 100 participants between 20 and 25 years of age (mean = 23 years old) recruited in Silpakorn University on Nakhon Pathom Campus. A 9-point hedonic score was used to evaluate the difference between samples. The score for texture appearance, taste, and overall acceptability was individually assessed: 1 to 4 for strongly disliked to slightly disliked, 5 for indifference, and 6 to 9 for slightly liked to strongly liked

2.7. Economic Analysis of the Dryer

For the success and commercialization of dryer technology, establishing the economic viability of the product is important. The economics of greenhouse solar and hot air dryers was analyzed. The total capital cost (C_c), annualized capital cost (C_{ac}), annualized salvage value (S_a), annualized cost (C_a), and annualized electricity cost of fans (C_{ae}), all expressed in monetary terms, were computed using Equations (5)–(9), respectively [10,18]. In the equations, C_{mf} is the cost of fresh material used, C_l is the labor cost of drying operation, i is the interest rate, n is the dryer life span (yr), S is salvage, C_m is annualized maintenance cost, C_p is the cost of packaging yearly, R is the number of hours the fans operated yearly, W is the rated power consumption of fans, and C_e is the unit charge for electricity.

$$C_c = C_{mf} + C_l \tag{5}$$

$$C_{ac} = C_c \times \frac{i(1+i)^n}{(1+i)^{n-1}} \tag{6}$$

$$S_a = S \times \frac{i}{(1+i)^{n-1}} \tag{7}$$

$$C_a = C_{ac} + C_m + C_p + C_{ae} - S_a \tag{8}$$

$$C_{ae} = R \times W \times C_e \tag{9}$$

In addition, the most commonly used economic viability, namely, NPV, IRR, PBP, and BCR, were all investigated in this study.

NPV is the difference between the present value of cash inflows and the present value of cash outflows over a specific period of time and is used to quantify the expected profitability of an investment. It has to be >0 to be profitable and is calculated using the following formula [8,9].

$$NPV = \sum_{n=0}^{N} \frac{R_n - C_n}{(1+i)^n} \tag{10}$$

where C_n denotes operating cost of dryers; R_n denotes the benefit received or revenue of dryers; i denotes the interest rate; n denotes the specific year of investing and N is the total number of years of investing which was 15 years in this study according to the lifespan of the solar dryer [5].

The IRR is the discount rate making the NPV of all cash flows equal to zero. IRR calculations rely on the same formula as NPV. IRR is used to estimate the profitability of potential investment and is calculated using the formula below [8,9].

$$IRR = i : \sum_{n=0}^{N} \frac{R_n}{(1+i)^n} = \sum_{n=0}^{N} \frac{C_n}{(1+i)^n} \tag{11}$$

PBP refers to the amount of time recover the cost of an investment, and the shortest PBP is an investment before the cost of investment can be recovered. PBP is calculated using the formula below [8,9,18].

$$PBP = \frac{\sum \text{investment costs}}{\sum \text{annual benefit} - \text{annual costs}} \tag{12}$$

The BCR is a ratio used to summarize the overall relationship between the relative costs and benefits and can be expressed in monetary terms. A BCR greater than 1.0 is expected to deliver a positive NPV to investors. BCR was calculated using the formula below [8,18].

$$BCR = \frac{\sum \text{benefits received per year}}{\text{total capital cost of the dryer}} \tag{13}$$

2.8. Statistical Analysis

Statistical analyses were performed using SPSS (Version 18, SPSS Inc., Chicago, IL, USA). Significant differences among the means for each treatment at 5% of the probability level were evaluated using one-way ANOVA and Tukey–Kramer's multiple comparison test.

3. Results and Discussion

3.1. Drying Curves and Drying Rate

Drying curves of mango slices at different temperatures using hot air dryers are presented in Figure 1. Typical drying characteristics were observed, i.e., drying rates as indicated by moisture reduction over time. Drying rates at 80 and 90 °C were higher than that at 70 °C, especially during the first 20 min. The initial moisture content of the mango slices ranged from 85.5 to 85.7% wb, while the final moisture of the dried product ranged from 12.4% to 12.6% wb. The water activity of the final product ranged from 0.2 to 0.3.

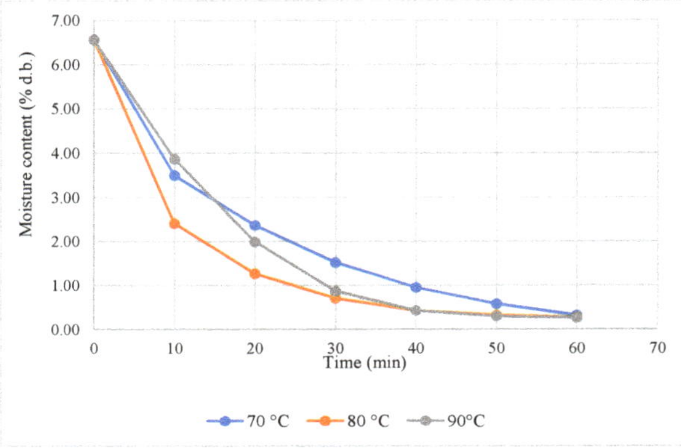

Figure 1. Drying curves of hot air dryers at 70, 80 and 90 °C.

Drying experiments of mango slices in greenhouse dryers were carried out on 26 June 2021. Drying curves and drying rates of mango slices using greenhouse solar dryers are shown in Figures 2 and 3. Regarding the variation of measured moisture content (dry basis) of mango slices each time of drying, it could be observed that the moisture content reduced from 8.68% to 0.14% within 1 day when using greenhouse solar dryers. The drying rate in greenhouse solar dryers increased sharply when the moisture fell below 6.95%. The shape of the drying curve indicates a rapid moisture content removal from the product at the initial stage, which later decreased with the increase in drying time. Thus, the moisture ratio decreased continually with the drying time [7,8].

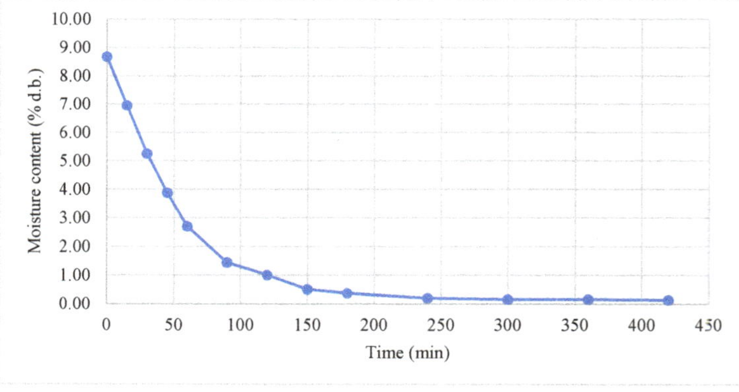

Figure 2. Drying curve of greenhouse solar dryers.

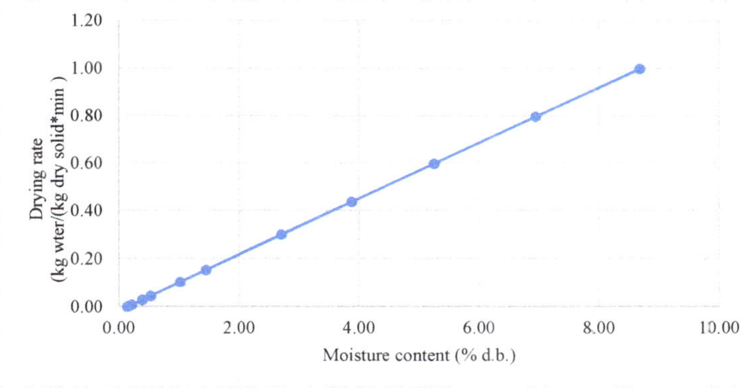

Figure 3. Drying rate of greenhouse solar dryers.

Figure 4 shows the hourly variation of solar radiation during the typical experiment runs of solar drying in the solar greenhouse dryers. When drying mango slices, solar radiation increased sharply from 9:00 a.m. to noon but considerably decreased in the afternoon. The solar radiation on the horizontal surface outside the greenhouse solar dryers reached the highest point at 955 W/m^2.

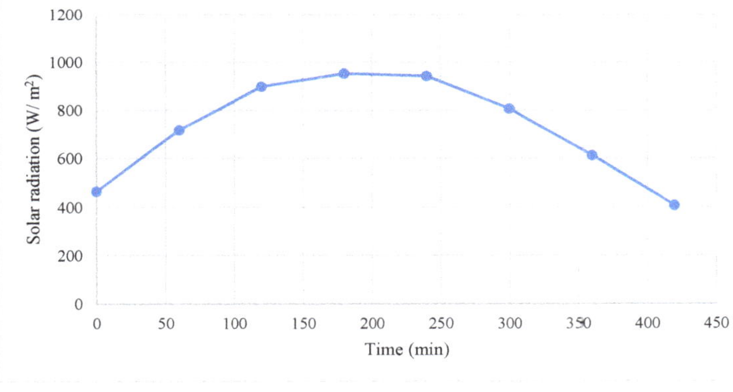

Figure 4. Variation of solar radiation with the time of day when drying mango slices.

Relative air humidity under the greenhouse and drying air temperatures inside greenhouse dryers are shown in Figures 5 and 6. The relative humidity of the air under the greenhouse ranged from 19.8% to 67.2%. The drying air temperature recorded inside the greenhouse ranged from 26.7 to 59.1 °C. During the drying process, the moisture content of the mango slices was reduced from 85.5% wb to 12.4% wb within 1 day (7 h) of exposure in greenhouse solar dryers. The hot air dryer showed a higher drying temperature than that of greenhouse solar dryers. The total time of drying using hot air dryers was lower at 1.5 h (at 70 °C for 1 h, followed by 90 °C for 30 min). In the greenhouse solar drying system, solar radiation was transmitted and absorbed by air and the concrete floor, and then collected in the system rather than reflected to the environment due to the parabolic shape and double layer of the polycarbonate sheet [13]. The temperature in this dryer is directly related to the solar radiation. Thus, increased solar radiation increases air temperature. The relative humidity (%) of the air inside greenhouse solar dryers decreased when the temperature increased. When considering mango slices of this study, one sunny day was reasonably sufficient to finish the drying process using greenhouse solar dryers, similar to the study of Pott et al. [4].

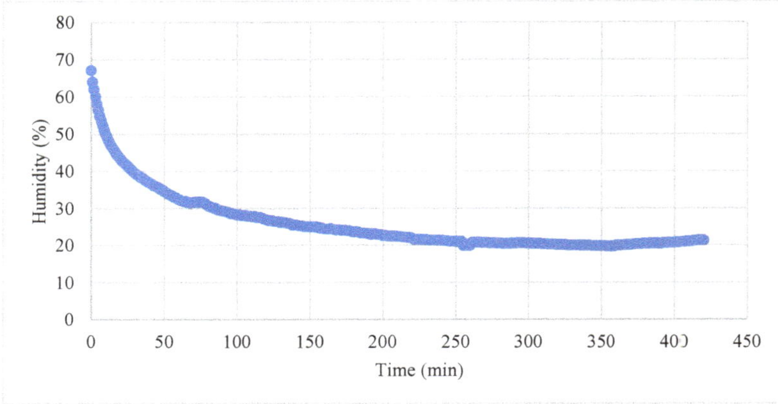

Figure 5. Relative humidity inside greenhouse solar dryers while drying mango slices.

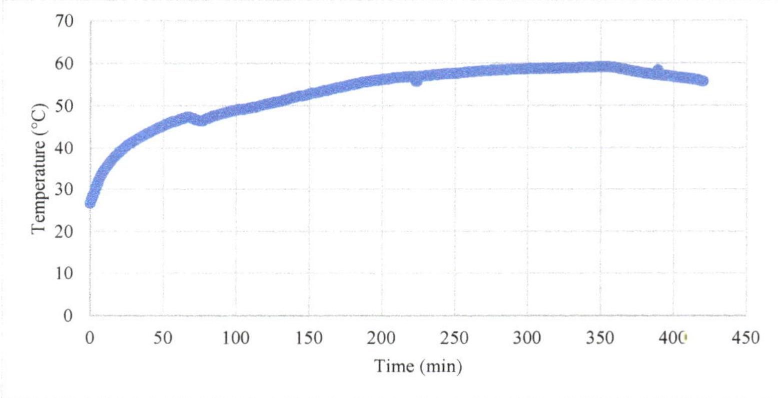

Figure 6. Temperature inside greenhouse solar dryers while drying mango slices.

3.2. Color Parameters

The appearance of fresh and dried mango slices from different drying conditions are illustrated in Figure 7. The color characteristics of fresh and dried mango slices from different drying conditions are shown in Table 1. The color of fresh mango slices was 63.81 for L*, 1.14 for a*, and 27.65 for b*. After drying mango slices using hot air and solar dryers, their L* and b* values tended to be lower than that of fresh mango slices, whereas a* values slightly increased. Notably, the drying systems significantly affected the color parameters. The solar greenhouse dryer gave a significantly higher total color change of dried mango slices compared to that of fresh samples ($p \leq 0.05$).

Table 1. Color characteristics of fresh and dried mango slices from different drying conditions.

Drying Methods	L*	a*	b*	ΔE
Fresh	63.81 ± 0.85 [c]	1.14 ± 0.27 [a]	27.65 ± 0.56 [c]	0.00 [a]
Hot air dryer	37.94 ± 0.27 [b]	3.01 ± 0.29 [c]	25.86 ± 0.55 [b]	26.00 ± 0.23 [b]
Solar greenhouse dryer	34.50 ± 0.21 [a]	2.56 ± 0.32 [b]	18.88 ± 0.52 [a]	30.63 ± 0.74 [c]

Note: Values followed by different letters in the same column indicate significant differences ($p \leq 0.05$).

Fresh Hot air dryer Greenhouse solar dryer

Figure 7. Appearance of fresh and dried mango slices from different drying systems.

3.3. Sensory Evaluation

After the initial comparison of two separate drying systems, only dried mango slices from hot air drying were suitable for sensory evaluation, while those from greenhouse solar drying were not only non-crispy, but also turned into rubberlike pieces. Therefore, the sensory evaluation of mango slices from hot air drying is shown only in Table 2. The 100 participating panelists for dried mango slices tasting included females (70%) and males (30%). The majority of participants fell within the ages ranging from 20 to 25 years old, with only 5% of participants being aged 25 years and above. The mean liking scores for all quality attributes were between 7.0 and 8.0, denoting that the sample was satisfactory at moderately like.

Table 2. Sensory evaluation of mango slices using hot air drying.

Color	Sweetness	Sourness	Crispness	Overall Quality
7.72 ± 1.04	7.55 ± 1.13	7.37 ± 1.19	7.89 ± 0.83	7.89 ± 0.83

3.4. Economic Analysis of the Dryers

Economic analysis of the hot air and greenhouse solar dryers was based on actual cost incurred, and other necessary parameters were taken based on the present economic situation in Thailand as shown in Table 3. The values of the four economic attributes, namely, BCR, NPV, and IRR and PBP, are compared in Table 4.

According to details in Table 3, if crispy mango production was conducted year-round, assuming working 5 days a week, each technology would result in the production of 240 batches annually. Based on the difference in drying capacity for each drying system, solar dryers could perform at 96 kg per batch, while hot air dryers were used twice to obtain the same capacity of solar dryers. The ratio of fresh mangoes to dried mango (final product) weight was 3 to 1, so approximately 7680 kg of final dried mango slices were produced annually.

The BCR for hot air and greenhouse solar dryers were found to be 24.47 and 15.73, respectively indicating a high positive NPV to an investor. The calculated PBPs for hot air and greenhouse solar dryers were 0.43 and 0.67 years, respectively. The PBPs are apparently short, considering the life span of the dryers to be 15 years, suggesting both dryers to be cost effective. Compared with other research, evidently, the PBP of solar dryer in this study was less than one year, similar to the studies of Janjai [5] and Chaudhari et al. [7] but is considered to be better than other related studies, such as Poonia et al. [8], Kondareddy et al. [11], and ELkhadraoui et al. [12] as detailed in the Introduction. The IRRs of hot air and greenhouse solar dryers were about 233% and 150%, respectively, while the NPVs of hot air and greenhouse solar dryers were USD 151,417 and USD 190,050, respectively. In this case, the IRR of a solar dryer was much higher than the minimum attractive rate of return (MARR) (4.875%) (see Table 3), similar to the study by Poonia et al. [8] revealing the high IRR of a solar dryer for drying fruit and vegetables, making the project economically feasible. Initially, hot air drying offered a better return on investment than greenhouse solar drying, according to its higher IRR and BCR values with lower PBP value. However, when considering only NPV, greenhouse solar dryers appeared to be the better choice.

Table 3. Details for computation of economic analysis.

No.	Item	Value		Units
		Hot Air Dryer	Greenhouse Solar Dryer	
1	Capital cost of dryer	6452	12,903	USD
2	Life span of dryer	15	15	Years
3	Capacity of dryer	48	96	Kg/batch
4	Price of fresh mango	1.61	1.61	USD/kg
5	Price of dried mango	7.74	7.74	USD/kg
6	Salvage value	10% of capital cost	10% of capital cost	USD
7	Maintenance cost	10% of capital cost	1% of capital cost	USD
8	Operational labor cost	12.90	6.45	USD/batch
9	Electricity cost	3.68	-	USD/batch
10	Packaging cost	0.06	0.06	USD/kg
11	Interest rate	4.875	4.875	%/year

Note: (1 USD = 31 Baht).

Table 4. Economic analysis of mango slices using hot air and greenhouse solar dryers.

Items	Hot Air Dryer	Greenhouse Solar Dryer	Unit
NPV	151,417	190,050	USD
IRR	233	150	%
PBP	0.43	0.67	Years
BCR	24.47	15.73	-

With the choice controversy arising from evaluating economic viability mentioned above, it became necessary to undertake an incremental cash flow between hot air and greenhouse solar dryers. According to Blank and Tarquin [19], when two mutually exclusive alternatives are evaluated, using an incremental rate of return (ROR) can identify the one alternative that is the best economically. Table 5 summarizes the analysis of an incremental ROR of the two technologies. Because the ROR of 66% on the extra investment is more than 4.875% minimum attractive rate of return (MARR) (see also Table 3), the higher cost of the greenhouse solar dryer was selected. The extra investment of USD 6,451 was economically justified by the higher annual revenue and salvage estimates at the end of year 15.

Table 5. Incremental ROR analysis of mango slices using hot air and greenhouse solar dryers.

Year	Cash Flow in USD		Incremental Cash Flow (GH-HA)
	Hot Air Dryer *	Greenhouse Solar Dryer **	
0	−6452	−12,903	−6451
1–15	15,051	19,328	4277
15	645	1290	645
	Incremental i * (ROR)		66%

Note: * Hot air dryer = HA; ** Greenhouse solar dryer = GH.

In addition, using greenhouse solar dryers can help saving electricity costs up to USD 1766 annually. This agreed with other studies reporting that the use of greenhouse solar dryers considerably reduced postharvest losses in terms of energy consumption and carbon emissions released [5,8,9,12].

3.5. Improving the Quality and Profitability of Crispy Mango Production for a Small-Scale Entrepreneur Using a Combination Hot Air and Greenhouse Solar Dryers

Regarding the first phase results described previously, it remained clear that the disadvantage of producing crispy mango using greenhouse solar dryers was the poor product quality. The advantage, however, was higher production volumes and profits. When using hot air dryers, the disadvantage was high energy consumption with high cost, while the advantage was that the product quality was more satisfactory. After discussing with the manager of one of the small community enterprises for processing fruit and agricultural products, located in Sam Phran District, Nakhon Pathom Province, the critical success factors of such a small-scale entrepreneur are customer satisfaction and entrepreneur profitability. It was suggested to combine the two drying methods of greenhouse solar and hot air dryers to improve product quality, increase production capability and cost saving.

Based on the later request of the main target customers, one of the main quality criteria of crispy mangos was their functional properties in terms of total phenolic content and antioxidant capacity properties. Therefore, profitability, sensory evaluation, total phenolics, and antioxidant capacity properties were subsequently assessed to choose the most suitable method for drying crispy mango production in SMEs.

The total time of crispy mango slices using a combination of greenhouse solar and hot air dryer (GH+HA) was 3.5 h at 60 °C for 3 h in greenhouse solar dryers, followed by 90 °C for 30 min in hot air dryers. With this combination process, the entrepreneur can increase the production capability of two batches daily (from 96 to 192 kg/day) when using greenhouse solar dryers. Hot air dryers reduced drying time from 1.5 to 0.5 h for fresh sliced mango (48 kg/batch) saving electricity cost around 67% (from USD 3.38/batch to USD 1.22/batch). Thus, further economic analysis was carried out.

The NPV is a method to analyze the profitability of a project investment. Therefore, NPV was undertaken for the combined greenhouse and hot air drying as shown in Table 6. The cash inflow of USD 34,204 at the end of the first year to the fifteenth year was discounted at the rate of 4.875%, and the present value was calculated as USD 358,044. The initial investment was USD 19,355. Hence, the NPV was USD 338,689. Such cash flow points out the advantage of a combined drying method over a single method, either hot air or greenhouse drying (see also Table 4).

Table 6. Net present value analysis of dried mango slices using a combination of greenhouse solar and hot air dryers.

Year	Cash Flow in USD			Compound Interest Factor	Present Value (PV, USD)
	HA	GH	GH+HA	(P/A, 4.875%, 15)	
0	−6452	−12,903	−19,355	-	−19,355
1–15	11,205	22,999	34,204	10.4679	358,044
		NPV			338,689

Table 7 presents the results of sensory evaluation of mango pieces obtained by using greenhouse solar combined with hot air dryers. The 100 participating panelists for dried mango slices tasting included females (80%) and males (20%). The majority of participants fell within the ages ranging from 20 to 25 years old, with only 10% of participants being aged 25 years and above. The mean preference scores for all quality attributes were slightly higher than the previous result presented in Table 2. However, the result was found to be very satisfactory based on the 9-point hedonic scale.

Table 7. Sensory evaluation of mango slices using a combination of greenhouse solar and hot air drying.

Color	Sweetness	Sourness	Crispness	Overall Quality
8.28 ± 0.97	8.05 ± 1.38	8.00 ± 1.21	8.04 ± 1.03	8.02 ± 0.89

As mentioned above, as per customer request, the impact of different drying processes on antioxidant activity was an additional objective in this study phase. DPPH, percentage of radical scavenging and FRAP were the parameters to evaluate the properties using Trolox as a standard. The antioxidant capacity of the mango studied (value of DPPH and FRAP) related to free radical scavenging in the samples. The value of DPPH and percentage of free radical scavenging are summarized in Table 8. The highest free radical scavenging was found in fresh samples, which was 72.28 ± 2.64 mgTrolox/100 g db, while dried mango slices (crispy mango) had significantly lower contents of 62.98 ± 1.09 and 64.65 ± 0.71 mgTrolox/100 g db in HA, and GH+HA, respectively. The percentages of radical scavenging also showed a similar trend as the DPPH value of fresh and dried mango slices. Nevertheless, a significant difference was found in percentage of radical scavenging between the HA and GH+HA samples, which were 34.19 ± 0.61 and 37.64 ± 0.43, respectively. Based on FRAP in Table 8, fresh mango showed the highest values, while the results of HA and GH+HA did not significantly differ.

Table 8. Total phenolic and antioxidant activity of fresh and dried mango slices from different drying methods.

Drying Method	DPPH (mgTrolox/100 g db)	%Radical Scavenging	FRAP Value (μmoL/100 g db)	Total Phenolic (mg GAE/100 g db)
Fresh	72.28 ± 2.64 ^{b}	39.26 ± 1.48 ^{b}	40.62 ± 0.11 ^{b}	34.41 ± 4.91 ^{b}
HA	62.98 ± 1.09 ^{a}	34.19 ± 0.61 ^{a}	29.07 ± 2.32 ^{a}	20.85 ± 3.18 ^{a}
GH+HA	64.65 ± 0.71 ^{a}	37.64 ± 0.43 ^{b}	30.23 ± 2.31 ^{a}	29.55 ± 4.59 ^{b}

Note: All the values in the table are shown as mean ± SD (n = 3). Different letters for the same measurement from the same location (leaf shoots and mature leaves) and standard compounds indicate $p \leq 0.05$ using one-way ANOVA.

The total phenolic content of the mango products was reduced after drying as shown in Table 8, which was associated with reduced DPPH and FRAP. The results were similar to those of the studies of Fratianni et al. [20] and Rumanum et al. [21]. The results indicated an impact of the drying treatment on antioxidant activity. In the drying process, temperature of food increases, causing plant damage and decreased phenolic contents. Nevertheless, new antioxidant compounds are formed due to the Maillard reaction. Hence, the antioxidant potential is raised after drying [15]. Önal et al. [22] also pointed out that evaporation can decrease phenolic compounds due to high temperature [23]. Nonetheless, the products decompose rapidly under high temperature. The antioxidant activity of dried sample did not significantly differ from other studies. The antioxidant capacity (DPPH and FRAP) of crispy mango using combined greenhouse solar and hot air drying did not significantly differ ($p \leq 0.05$). However, the GH and HA combination provided dried mango slices of %radical scavenging and total phenolic not significantly different from those of fresh mango ($p \leq 0.05$), but significantly higher than those of HA.

4. Conclusions

The drying process of mango slices using hot air dryers was better than that of greenhouse solar dryers in terms of product quality and drying time. The use of hot air dryers considerably reduced the drying time and improved the dried product quality. The drying time to achieve a moisture content of 12.4% wb for the dried product using the hot air dryers was reduced to 1.5 h compared with 7 h when using greenhouse solar dryers. Nevertheless, the economic evaluation of the greenhouse solar and hot air dryers indicated high IRR values (150%, 233%) and low PBP values (0.67 years, 0.43 years), making these two separate drying systems very cost efficient. The BCR of both drying systems was high (>1), indicating high positive NPV to an investor.

Comparing both technologies, the incremental ROR evaluation of greenhouse solar and hot air dryers equaling 66% was greater than MARR, so the incremental investment in greenhouse solar over hot air dryers was desirable, thus making greenhouse solar dryers cost effective, economically viable and environmentally sustainable. However, more studies to improve drying processes and methods when using only the greenhouse solar drying system are still needed to maintain a satisfactory quality of dried mangoes. Nonetheless, after the experiment of using a greenhouse solar dryer together with a hot air dryer improved the quality of the crispy mango product for a small-scale entrepreneur, crispy mango ultimately provided higher customer satisfaction, profitability, total phenolic content and antioxidant capacity.

Author Contributions: Conceptualization, creation, investigation, data analysis, writing—original draft and preparation, K.K.; data collection and experiment, P.K.; validation, writing—review and editing, P.S. and S.K.; supervision, S.K. All authors have read and agreed to the published version of the manuscript.

Funding: This research received no external funding.

Institutional Review Board Statement: Ethic review and approval were waived for this study, as it did not involve personally identifiable or sensitive data.

Informed Consent Statement: The study did not involve human or sensitive data.

Acknowledgments: The authors would like to thank the Department of Physics, Faculty of Science, Silpakorn University for giving us permission to use the greenhouse solar dryer throughout this study.

Conflicts of Interest: The authors declare no conflict of interest.

References

1. Yasunaga, E.; Fukuda, S.; Takata, D.; Spreer, W.; Sardsud, V.; Nakano, K. Quality changes in fresh mango fruits (*Mangifera indica* L. 'Nam Dok Mai') under actual distribution temperature profile from Thailand to Japan. *Environ. Control Biol.* **2018**, *56*, 45–49. [CrossRef]
2. Office of Agriculture Economics (OAE). Foreign Agricultural Trade Statistics of Thailand in 2020. Available online: http://impexp.oae.go.th/service/export.php (accessed on 11 June 2021).
3. Matulaprungsan, B.; Wongs-Aree, C.; Penchaiya, P.; Boonyaritthongchai, P.; Srisurapanon, V.; Kanlayanarat, S. Analysis of critical control points of post-harvest diseases in the material flow of Nam Dok Mai mango exported to Japan. *Agriculture* **2019**, *9*, 200. [CrossRef]
4. Pott, I.; Konard, S.; Scherer, R.; Wiriyacharee, P.; Mühlbauer, W. Quality of five Thai mango cultivars (*Mangifera indica* L.) using a solar drying system. *CMU J.* **2004**, *3*, 189–198.
5. Janjai, S. A greenhouse type solar dryer for small-scale dried food industries: Development and dissemination. *Int. J. Energy Environ. Eng.* **2012**, *3*, 383–398.
6. Guarte, R.C.; Pott, I.; Mühlbauer, W. Influence of drying parameters on β-carotene retention on mango leather. *Fruits* **2005**, *60*, 255–265. [CrossRef]
7. Chaudhari, R.H.; Gora, A.; Modi, V.M.; Chaudhari, H. Economic analysis of hybrid solar dryer for ginger drying. *Int. J. Curr. Microbiol. App. Sci.* **2018**, *7*, 2725–2731. [CrossRef]
8. Poonia, S.; Singh, A.K.; Santra, P.; Jain, D. Economic analysis of inclined solar dryer for drying of fruit and vegetables. *Int. J. Agric. Sci.* **2019**, *11*, 9154–9159.
9. Krungkaew, S.; Kingphadung, K.; Kwonpongsagoon, S.; Mahayothee, B. Costs and benefits of using parabolic greenhouse solar dryers for dried herb products in Thailand. *Int. J. GEOMATE* **2020**, *18*, 96–101. [CrossRef]
10. Boonyasri, M.; Lertsatitthanakorn, C.; Wiset, L.; Poomsa-ad, N. Performance analysis and economic evaluation of a greenhouse dryer for pork drying. *KKU Eng. J.* **2011**, *38*, 433–442.
11. Kondareddy, R.; Sivakumaran, N.; Radha Krishnan, K.; Nayak, P.K.; Sahu, F.M.; Singha, S. Performance evaluation and economic analysis of modified solar dryer with thermal energy storage for drying of blood fruit (*Haematocarpus validus*). *J. Food Process. Preserv.* **2021**, *45*, e15653. [CrossRef]
12. ELkhadraoui, A.; Kooli, S.; Farhat, A. Study on effective of mixed mode solar greenhouse dryer for drying of red pepper. *Int. J. Sci. Res. Eng. Technol.* **2015**, *3*, 143–146.

13. Mahayothee, B.; Thamsala, T.; Khuwijitjaru, P.; Janjai, S. Effect of drying temperature and drying method on drying rate and bioactive compounds in cassumunar ginger (*Zingiber montanum*). *J. Appl. Res. Med. Aromat. Plants* **2020**, *18*, 100262. [CrossRef]
14. Singleton, V.L.; Rossi, J.A. Colorimetry of total phenolics with phosphomolybdic phosphotungstic acid reagents. *Am. J. Enol. Vitic.* **1965**, *16*, 144–158.
15. Hossain, M.A.; Dey, P.; Joy, R.I. Effect of osmotic pretreatment and drying temperature on drying kinetics, antioxidant activity, and overall quality of taikor (*Garcinia pedunculata* Roxb.) slices. *Saudi J. Biol. Sci.* **2021**, *28*, 7269–7280. [CrossRef]
16. Siddiq, M.; Sogi, D.S.; Dolan, K.D. Antioxidant properties, total phenolics, and quality of fresh-cut 'Tommy Atkins' mangoes as affected by different pre-treatments. *LWT-Food Sci. Technol.* **2013**, *53*, 156–162. [CrossRef]
17. Vongsak, B.; Kongkiatpaiboon, S.; Jaisamut, S.; Konsap, K. Comparison of active constituents, antioxidant capacity, and α-glucosidase inhibition in Pluchea indica leaf extracts at different maturity stages. *Food Biosci.* **2018**, *25*, 68–73. [CrossRef]
18. Kiburi, F.G.; Kanali, C.L.; Kituu, G.M.; Ajwang, P.O.; Ronoh, E.K. Performance evaluation and economic feasibility of a solar-biomass hybrid greenhouse dryer for drying banana slices. *Renew. Energy Focus* **2020**, *34*, 60–68. [CrossRef]
19. Blank, L.; Tarquin, A. *Engineering Economy*, 7th ed.; McGraw-Hill: New York, NY, USA, 2012; pp. 278–300.
20. Fratianni, A.; Adiletta, G.; Di Matteo, M.; Panfili, G.; Niro, S.; Gentile, C.; Farina, V.; Cinquanta, L.; Corona, O. Evolution of carotenoid content, antioxidant activity and volatiles compounds in dried mango fruits (*Mangifera indica* L.). *Foods* **2020**, *9*, 1424. [CrossRef] [PubMed]
21. Rumainum, I.M.; Worarad, K.; Srilaong, V.; Yamane, K. Fruit quality and antioxidant capacity of six Thai mango cultivars. *Agric. Nat. Resour.* **2018**, *52*, 208–214. [CrossRef]
22. Önal, B.; Adiletta, G.; Crescitelli, A.; Di Matteo, M.; Russo, P. Optimization of hot air drying temperature combined with pre-treatment to improve physico-chemical and nutritional quality of 'Annurca' apple. *Food Bioprod. Process.* **2019**, *115*, 87–99. [CrossRef]
23. AlFarsi, M.; Alasalvar, C.; Morris, A.; Baron, M.; Shahidi, F. Compositional and sensory characteristics of three native sun-dried date (*Phoenix Dactylifera* L.) varieties grown in Oman. *Agric. Food Chem.* **2005**, *53*, 7586–7591. [CrossRef] [PubMed]

processes

MDPI

Article

Utilization of Aerobic Compression Composting Technology on Raw Mushroom Waste for Bioenergy Pellets Production

Wen Yi Chia [1], Kit Wayne Chew [2,3,*], Cheng Foh Le [4], Chelsea Siew Chyi Chee [5], Mae See Luan Ooi [5] and Pau Loke Show [1,*]

1 Department of Chemical and Environmental Engineering, Faculty of Science and Engineering, University of Nottingham Malaysia, Jalan Broga, Semenyih 43500, Selangor Darul Ehsan, Malaysia; wenyichia@gmail.com
2 School of Energy and Chemical Engineering, Xiamen University Malaysia, Jalan Sunsuria, Bandar Sunsuria, Sepang 43900, Selangor Darul Ehsan, Malaysia
3 College of Chemistry and Chemical Engineering, Xiamen University, Xiamen 361005, China
4 School of Biosciences, Faculty of Science and Engineering, University of Nottingham Malaysia, Semenyih 43500, Selangor Darul Ehsan, Malaysia; chengfoh.le@nottingham.edu.my
5 Mentari Alam EKO (M) Sdn Bhd (MAEKO), 192, Jalan LP 7/4, Kinrara Uptown, Taman Lestari Perdana, Seri Kembangan 43300, Selangor Darul Ehsan, Malaysia; chelsea@maeko.com.my (C.S.C.C.); mae@maeko.com.my (M.S.L.O.)
* Correspondence: kitwayne.chew@xmu.edu.my (K.W.C.); pauloke.show@nottingham.edu.my (P.L.S.)

Abstract: Raw mushroom waste has been an enormous solid waste, not only causing a huge cut on profit margin of mushroom industries but also leading to environmental pollution. Unfortunately, the current utilization methods, such as pharmaceutical extractions, are unable to keep up with the waste generation rate due to the large-scale mushroom production. Yet, the utilization of raw mushroom waste to produce biomass pellets for energetic purposes and the role of an electric composter on shortening the processing time remain unexplored. This is important because conventional composting, which takes a relatively long period (e.g., weeks to months), is less practical when it comes to commercial use of the biomass pellets. To explore this issue, an industrial composter with initial compost was utilized to process the raw mushroom waste, followed by pelletization. Extraction of the material inside the composter at different timing was carried out to determine the optimal processing time for optimal texture to form pellets. It was found that prolonged composting hour affected the pelletization process since moisture, which acts as a natural binder, reduced when the composting hour increased. The gross calorific value increased from 14.07 MJ/kg to 18.76 MJ/kg for raw mushroom waste and compost pellets at the fifth hour, respectively. This study revealed that the raw mushroom waste compost could serve as a valuable renewable energy source and that the production of energy-rich biomass compost fuel pellets without using any binder within a short composting duration is achievable with the aid of an in-vessel composter.

Keywords: biomass; energy utilization; pelletization; calorific value; composter

Citation: Chia, W.Y.; Chew, K.W.; Le, C.F.; Chee, C.S.C.; Ooi, M.S.L.; Show, P.L. Utilization of Aerobic Compression Composting Technology on Raw Mushroom Waste for Bioenergy Pellets Production. *Processes* **2022**, *10*, 463. https://doi.org/10.3390/pr10030463

Academic Editors: Worapon Kiatkittipong, Alok Kumar Patel and Jun-Wei Lim

Received: 25 October 2021
Accepted: 9 January 2022
Published: 24 February 2022

1. Introduction

Increased depletion of non-renewable energy resources has become a serious concern with the increased energy demand from population growth; there has been a focus on the need for novel alternative energy sources that are renewable and sustainable [1]. Environmental-friendliness has been consciously considered, since several global environmental issues, such as air pollution and global warming, have been triggered by the heavy consumption of fossil fuels [2]. The waste-to-energy approach has been a main research focus, as bioenergy development is reported to be sustainable in nature [3]. Generation of bioenergy is favourable, with minimal cost of operational expenses, reduction in waste volume and minimisation of waste disposal problems, as well as contribution to the economy, as the energy produced can be sold [4,5].

Generally, composting is considered as one of the prospective methods to recycle the organic portion of municipal solid waste, such as food waste, papers, and garden waste, through aerobic or anaerobic decomposition [6]. Compost produced is generally used to add total organic carbon to the soil, acting as soil amendment or fertiliser. However, beneficially applying compost on soil is not always possible, since there are variations in composting quality (i.e., the composts do not fulfil the local legal requirements for agricultural use) and market demand (i.e., the composts are overly produced or farmland area faces steady decline) [7,8]. In such cases, new uses for compost are being developed, including its valorisation as an energy source via combustion, pyrolysis or gasification [9].

The energetic use of compost has been investigated, including compost made from cattle feedlot manure [10], pig manure [11], forestry waste [12,13], shredded maize straw [14], food waste [15] and municipal solid waste [16]. It is important to determine the energy potential of the composts, since compost treatments result in different water content, as well as fuel characteristics [17]. The suitability of the composts as biofuel with or without additional pre-treatment and modification is also crucial. For instance, mixed compost made using passive heap technology studied by Malaťák, et al. possessed a low calorific value (8.51 MJ/kg) and high ash content (61.7%), while combustion with the addition of wood chips increased the calorific value to 13.14 MJ/kg [18].

Nevertheless, compost of raw mushroom waste has not been investigated for the energetic purpose. The undesirable parts of mushrooms are disposed of by mushroom and food industries; the volume drastically increases every year due to the exponential growth in mushroom consumption [19]. This underutilized cut-off waste makes up around 25 to 33% of fresh mushroom weight because of their poor palatability and tough texture [20,21]. Besides, irregular shape and dimensions of mushroom during cultivation and harvesting are also discarded [22]. This enormous solid waste not only causes a huge cut on the profit margin of mushroom industries but also leads to environmental pollution burdening mushroom producers and waste management authority [23]. According to Zou, et al., the current methods of utilizing raw mushroom waste such as pharmaceutical extractions and absorption material development are unable to keep up with the waste generation rate of large-scale mushroom production [24].

In addition, most of the studies in the literature used composts provided by composting plants or made via composting processes, which take a relatively long period (e.g., weeks to months). This is less practical when it comes to decentralized energy generation to solve the waste problems at its origin. The idea proposed in this study was to produce energetic material at the site of waste production for energy generation, which at least replaces a part of coal usage. With this approach, transportation and time could be saved without involving external composting plants, leading to cost reduction. Hence, organic waste from different industries and more advanced composting technology should be examined, but these are lacking in the reported literature. In this study, raw mushroom waste from the mushroom industry was used to generate compost pellets using an aerobic electric composter, where the processing time needed was investigated to greatly reduce the composting time. The aim of this study was to utilize this raw mushroom waste by measuring its energetic and pelletising potential with different short composting ages in order to achieve the most suitable and profitable operating conditions for the production of raw mushroom compost energy pellets.

2. Materials and Methods

2.1. Materials, Composting and Pelletization

Raw mushroom (Shiitake) waste from the food industry of mushroom soup powder production was kindly provided by Mentari Alam Eko (M) Sdn Bhd, together with initial compost with enzymes as well as a composter (MW30). The in-vessel composter with dimensions of 900 × 680 × 1300 mm was equipped with rotary blades and an air blower to provide good aeration (54 L), rotation and agitation of the compost (8 rpm), producing uniform end compost product. This composter's function is to compost all type and sizes of food waste using microbial aerobic compression composting technology with an agitator system (Patent No.: P12015000732). Proper rotation with the good mixing, aeration and agitation of the compost was the efficient technique for the uniform product formation. The temperature inside the composting chamber was controlled at 55 ± 2 °C using the sensors and heating functions equipped, regardless of the ambient temperatures. This is because a suitable thermophilic stage is essential for the inactivation of pathogens; temperature above 55 °C was demanded in the directives or guidelines [25].

In each processing or composting cycle, 30 L of raw mushroom waste, which was in dry pieces, as provided, was added into the composter containing the initial compost at shaft level (approximately 36 L; 30% of the total chamber size of 120 L), followed by 9 L of water, which was required due to the low moisture content (<10%) of the raw mushroom waste. Material inside the composter, or compost, was extracted at different timing by collecting it at 1 h intervals, up to 6 h. Short processing or composting time was possible, as the composter comes with patented aerobic microbial compression technology. The collected samples were processed into pellets by compacting and cutting into small pellets. The moisture content (MC) of the pellets was determined following ASTM E1756-08 [26] standard procedure by the loss of mass when heated in a drying oven at 105 °C for 24 h. These dried pellets were stored in a desiccator at room temperature. The remaining compost inside the composter was composted for 24 h and the product was discharged from the composter until the shaft level. The product that remained inside the composter was recycled in the next composting batch.

2.2. *Physicochemical Analysis*

2.2.1. Pellet Density and FESEM Analysis

The pellet weight was measured using a digital measuring balance with an accuracy of 0.01 g (Mettler Toledo, Greifensee, Switzerland). The height and diameter of the pellets in cylindrical shape were determined with a Vernier caliper (Mitutoyo, Kawasaki, Japan). The density of pellets was calculated by dividing the weight over the pellet volume and the average was taken from at least 10 sets of repetition. This method follows the EN ISO 17225 standard method. The surface morphology of the pellets, including microstructure and bonding behavior within pellets, was studied using a field emission scanning electron microscope (FESEM, FEI Quanta 400F, Oregon, USA).

2.2.2. CHNS and FTIR Analysis

The element compositions, carbon (C), hydrogen (H), nitrogen (N) and sulphur (S), were determined by a CHNS analyzer (Elementar, Langenselbold, Germany). The determination of these elements is important for the study of the carbon cycle, combustion and incineration technology, as well as soil organic materials. The functional group changes were analyzed by recording Fourier transform infrared (FT-IR) spectra of the pellet surface using a FT-IR spectrophotometer (Frontier, Perkin Elmer, Massachusetts, USA) with an average of 32 scans over 4000–400 cm^{-1}.

2.2.3. Proximate Analysis, Thermal Gravimetric and Calorific Value Analysis

Combustion characterization of the pellets was conducted using a thermogravimetric analyzer (Perkin-Elmer STA6000, Massachusetts, USA). Approximately 10 mg of each sample was heated from 30 °C to 800 °C at a heating rate of 20 °C min^{-1} using an air flow of 100 mL min^{-1}. Proximate analysis was carried out using the thermogravimetric analysis method [27]. The gross heating values (GCV) of the pellets were determined using a bomb calorimeter (Parr, Illinois, USA) from three replicates, following BS EN 14918. A sample of approximately 1 g was placed in a crucible linked to the fused wire for combustion to take place. Volumetric energy density of the pellets was calculated by multiplying the GCV with the mass density of the pellets.

2.2.4. Compression Strength, Mechanical Durability and Wettability Index

Compression strength of the pellets was measured by a universal materials testing machine (Ametek Company, Leicester, UK). Before the compression, height and diameter of the pellets were measured. Using a 5 kN load cell moving with a constant velocity of 1 mm/min, compression was carried out in axial direction individually to the pellet, which was laid vertically on a plate until collapsed. The maximum force recorded before breakage of the compressed pellet was taken as the compressive strength (N) of the pellet, following the ASTM D695 method [28]. The tensile strength (MPa) of the pellets was calculated using the equation below.

$$T_P = \frac{2f}{\pi l d} \quad (1)$$

where f is the maximum force, d is the pellet diameter, and l is the pellet length.

Mechanical durability of the pellets was measured through a drop test to determine the ratio of mass retained to the initial weight in percentage. Pellet samples after two weeks of storage were dropped from a height of 1.85 m on a metal plate [29].

Moisture resistance of the pellets was expressed as the wettability index (WI). Preweighed pellet was submerged in water for 30 s and then reweighed. The percentage of the difference in pellet weight before and after submerging from the initial weight was taken as the WI [30].

$$WI = \frac{m_a - m_b}{m_b} \times 100 \quad (2)$$

where m_b and m_a are the pellet weight (g) before and after submerging, respectively.

3. Results and Discussion

3.1. Characterization of Raw Materials and Pellets

Mushrooms generally possess a high nutritional value and are rich in protein, including an important content of fiber and essential amino acids. Shiitake consists of dietary fiber with soluble and insoluble structures, the former of which contains β-glucans and proteins; the latter contains polyuronide (acidic polysaccharide), hemicellulose, chitin, lignin and β-glucan chains with heterosaccharide [31]. The chemical properties of raw mushroom waste and the composted mushroom waste are shown in Table 1. Raw mushroom waste contains a high content of organic matter; the compost pellets at all time periods showed a higher carbon content, which is desirable for energy and heating application.

Table 1. Properties of raw mushroom waste and compost extracted at 1 h (B1), 2 h (B2), 3 h (B3), 4 h (B4) and 5 h (B5).

Ultimate Analysis (%)	Raw Mushroom Waste	B1	B2	B3	B4	B5
Carbon	36.44	42.91	41.46	42.30	41.67	42.09
Hydrogen	6.83	6.46	6.53	6.54	6.54	6.54
Nitrogen	5.36	3.96	3.94	4.04	4.02	4.09
Sulphur	0.37	0.26	0.27	0.26	0.26	0.25
C/N ratio	6.80	10.84	10.52	10.47	10.37	10.29

The composting process involves the reduction of organic substances in the compost feed into decomposed materials [32]. The higher carbon content in compost as compared to the raw mushroom waste might be due to the addition of the initial compost with enzymes. It can be seen that there was reduction in carbon with the increase composting hour through the rapid composting process performed by the electric composter. On the other hand, there was a slight drop in hydrogen and sulphur content, which were likely consumed during the composting process. The nitrogen content decreased by 23.7–26.5% for the raw mushroom waste, as nitrogen is consumed during the composting process.

To be specific, there was a loss of NH_4^+ during the initial stage of the composting process by NH_3 volatilization or with nitrification, followed by a subsequent loss of NO_3^+ via denitrification [33]. The relatively low content of sulphur indicates that the combustion of the samples is environmentally compatible [24]. On the other hand, the H/C ratio decreased from 0.19 to 0.15 after composting, where lower H/C showed better fuel quality and higher thermal efficiency [34].

There was no significant difference between any CHNS compositions for the raw mushroom waste compost of 1–5 h. However, it can be observed that the carbon-to-nitrogen ratio (C:N ratio) gradually decreased due to the release of the organic matter content [35]. The apparent difference was in the observation of the compost texture becoming drier with each subsequent hour of composting. The in-vessel composter was able to complete the composting process in a well-mixed condition and short duration. The study on the effect of time showed that the results were comparable for all durations; extending the composting period mainly affected the moisture content within the compost.

The FTIR spectra of raw mushroom waste and raw mushroom waste compost pellets of different hours are shown in Figure 1. The characteristics of the compost are mainly derived from the raw mushroom residue, which contains carbohydrates, nitrogen compounds that include crude protein and decomposition products, and aliphatic compounds that are mainly crude fat. These compounds are mainly decomposed to carbon dioxide, ammonium ion, amide nitrogen, nitrate ion, protein and its decomposition products [36]. The FTIR spectrum of raw mushroom waste showed peaks corresponding to different functional groups, including carbohydrates such as cellulose, hemicellulose and lignin, proteins, silicate minerals and amide compounds, where the peak at 3253 cm^{-1} corresponds to ≡C-H stretch; 2927 cm^{-1} and 2853 cm^{-1} corresponds to aliphatic C-H stretch; 1635 cm^{-1} corresponds to C=O stretch; 1148 cm^{-1} and 1021 cm^{-1} corresponds to C-O stretch [37].

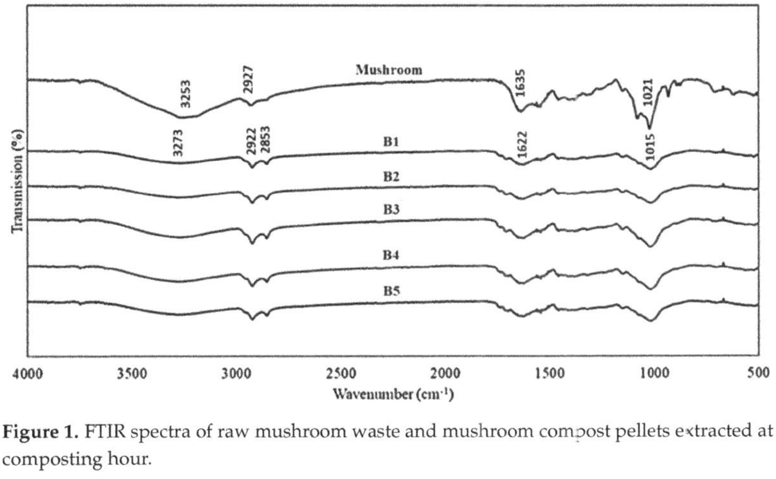

Figure 1. FTIR spectra of raw mushroom waste and mushroom compost pellets extracted at different composting hour.

There was an obvious reduction in the absorption peaks when comparing the FTIR spectra of raw material and compost pellets, while there was just a gradual reduction among the compost pellets at different composting hour. Firstly, there is a strong band detection between 930 and 1148 cm^{-1} which corresponds to the carbohydrate or polysaccharides substance. Since polysaccharides are an important component in raw mushroom caused by the starchy fibers of organic waste plant materials, there exists a sharp peak [36]. The composted pellets still retain raw mushroom as its main material and, hence, the peak is observed for all durations. The intensity of peaks at 1021 cm^{-1} decreased greatly with maturation due to the progressive consumption of polysaccharides and stabilization [38].

Another band at 1635 cm^{-1} in the raw mushroom is likely the presence of carboxylates and C=C from aromatics and alkenes. Mushrooms contain many functional properties, and the composition of carboxylic acid has contributed to the peak. Composted mushroom still showed a slight peak at the respective band. The band at 1635 cm^{-1}, which was also related to lignin, experienced a relative decrease after composting [36]. The presence of a band around 2927 cm^{-1} in the mushrooms is attributed to the humic acid component. Humic acid is widely used for the cultivation of mushroom, as it helps to promote its growth; the raw mushroom waste was likely to contain traces of the humic acid from cultivation steps. While the range of 2940 to 2850 cm^{-1} indicated the saturation of humic acid, which occurs during the composting process, the peaks were observed for composted fractions at all durations. The development of the band at 2927 cm^{-1} was also used as an indicator of organic matter decomposition [39]. The strength at the peak of 2927 cm^{-1} had weakened compared to the raw mushroom waste, which could have been caused by losing specific constituents in cell walls, such as pectin [40].

The range of O-H and N-H stretching vibration band was obtained around 3200–3300 cm^{-1}, representing the intermolecular hydrogen bonding that corresponds to phenolic groups. Phenolic compounds are the main components that support the antioxidant activity from the raw mushroom extract, hence, the peak is observed for raw mushroom waste; however, the peak declines for composted fractions, as phenolic compounds tend to degrade at temperatures higher than 50 °C, which occurs during the composting step. Additionally, the peak at 3253 cm^{-1}, which could refer to the presence of a hydrogen band in cellulose, changed from a relatively sharp peak to a gentle peak, indicating the degradation of lignocellulose [41,42].

3.2. Combustion Characteristics

Proximate analysis was carried out for both raw mushroom waste and compost pellets produced. The results of raw mushroom waste are similar to results by Zou, et al., who found that the mushroom stipe contained 8.19% moisture, 73.27% volatile matter, 13.83% fixed carbon and 4.72% ash [24]. The samples in the reported study [24] were dried before analysis, while, in the present study, raw mushroom waste was analyzed in its original form; hence, there are slight differences in terms of moisture content and higher heating value. According to Ahmad, et al., moisture content of less than 10% in lignocellulosic biomass is considered as suitable for combustion [43]. It is known that high volatile matter content makes biomass a highly reactive fuel with a faster combustion rate, while fixed carbon—which is the remaining solid fuel after volatile matter—acts as a main energy generation for combustion [44]. The volatile matter (Table 2) was found to increase and decrease, respectively, after the composting process. Some of the substrate and compounds were broken down during the decomposition; hence, the amount of volatile matter was lower during the early stages of the composting process. The highest volatile matter in B5 indicated that it would burn vigorously with high reactivity and show low combustion enthalpies during direct combustion [45].

Table 2. Proximate analysis, gross heating value, mass and energy densities of raw mushroom waste compost pellets prepared at different composting hour.

Sample	Moisture (%)	Volatile Matter (%)	Fixed Carbon (%)	Ash (%)	GCV (MJ/kg)	Mass Density (kg/m^3)	Energy Density (GJ/m^3)
Raw mushroom waste	9.81	74.02	13.83	2.10	14.07	-	-
B1	5.47	64.52	21.78	8.18	18.85	1055.46	19.90
B2	6.06	73.92	15.18	4.94	19.08	1076.70	20.54
B3	2.64	65.01	23.82	8.58	18.90	1099.71	20.78
B4	5.06	62.61	25.76	6.43	18.83	1107.67	20.86
B5	3.33	75.42	13.50	7.17	18.76	1130.37	21.21

The same trend was observed for the fixed carbon content, where durations of 1–4 h showed a good increase in fixed carbon; however, the carbon content dropped to a lower value than the control at a duration of 5 h. Continuous processing of compost within the in-vessel system may lead to exposure of high temperatures that are a result of the continuous operation and decomposition. These high temperatures will eventually decompose the organic properties of compost; hence, there was a loss of carbon content. The ash content of all the compost pellets was classified as a low ash coal based on the Chinese National Standard (Classification for quality of coal. Part 1: Ash content, GB/T 15224.2-2010) and International Standard (Classification of coals, ISO 11760-2005) [46]. The ash contents for the compost pellets fall in the range of 4.936–8.582%. The ash content increased from the initial ash content in raw mushroom waste. This is due to the decomposition process in composting, which produces ash compounds.

The mass density was increased after the pelletization process and increased for every hour. According to Pua, et al., high moisture content in biomass tends to cause reduction in pellet density and raw fibers of low density will cause an issue in product storage and transportation [44,47]. Hence, fuel pellets that have high bulk and energy density are favorable to save space and minimize expenses. According to European guidelines, the accepted standard density range of a single pellet is between 1000 to 1400 kgm^{-3} [48], making pellet density in this study, with a density range of 1055.46–1130.37 kgm^{-3}, acceptable. The mass density of the compost pellets showed an increasing trend with longer composting duration. During the composting process, the raw mushroom waste was broken down into smaller fractions by the substrates and bacteria; hence, a longer composting period will result in more parts of the fine compost material. The pelletization of finer compounds might lead to higher density, as it is easier to compact fine powdery materials, although it is not significant [49].

Calorific value can indicate the combustion properties of a pellet. Lower composting hours were selected as lower composting degradation, which means that a higher content of organic matter gives higher heating values and that the heating value must be reduced after the active composting process [16]. The gross calorific value (GCV) of low-rank coals according to the Chinese National Standard (Classification for quality of coal. Part 3: Calorific value, GB/T 15224.32010) is below 16.30 MJ/kg [46]; the results in Table 2 showed that all compost pellets possessed higher GCV. Once densified, these pellets can show a high GCV value, making them suitable as heating sources. The GCV for all time durations are comparable, and are significantly higher than the pellets made from other composts reviewed by Chia, et al. [6]. The increase in density and similar range of GCV also resulted in increasing energy density of the compost pellets, in line with the composting time. The energy density values were similar to torrefied biomass pellets and coal, which reaches 18 GJ/m^3 and exceeds 20 GJ/m^3, respectively, while raw biomass typically has an energy density of 10 to 11 GJ/m^3 [50,51].

The combustion process and thermal stability of the resulting pellets from compost extracted at the fifth hour were investigated with a thermogravimetric analyzer, as shown in Figure 2. The pellet faced three stages of mass loss process, which were between temperatures of 100 to 200 °C, 200 to 400 °C, and 400 to 600 °C, respectively. Mass loss in the first stage, which is about 3.87%, may be due to evaporation of moisture and loss of a small number of volatile substances. It can be seen that the moisture content was low as the pellets were dried after the pelletization. There might also be slight moisture loss during the pelletization. Mass loss, about 49.37%, in the second stage, called the volatile component combustion stage, is the main stage of weight loss because the main chemical components, such as cellulose, hemicellulose and other organic functional groups in the sample faced gradual degradation. To be specific, hemicellulose, which has low polymerization degree, should break first, while cellulose faced a partly depolymerizing reaction [24]. The maximum weight loss was estimated at 9.97%/min at 289.46 °C. Lastly, degradation of stable lignin is the main cause of the mass loss in the third stage (36.56%), which is the oxidation stage of fixed carbon. The TG curve tends to be flat after 600 °C as the combustion was basically completed with remaining residual ash and other non-decomposable substances [52].

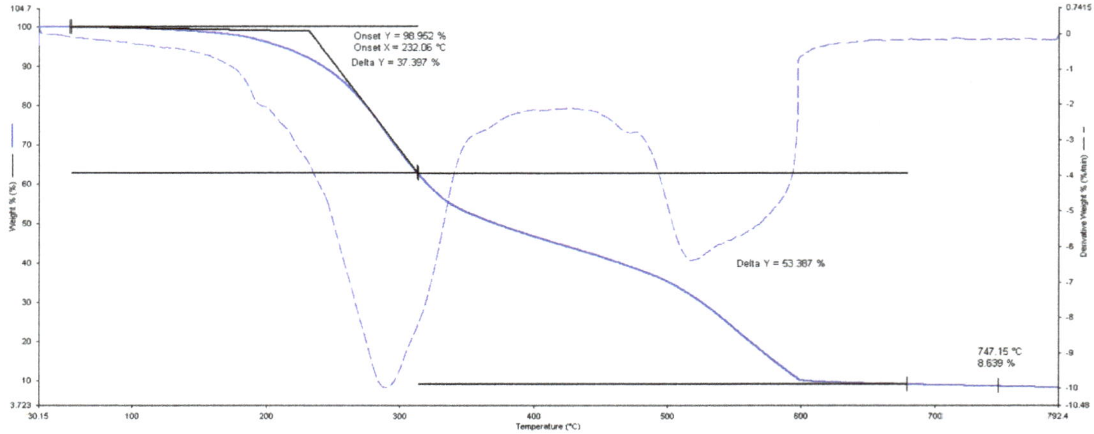

Figure 2. Thermal degradation (TGA) and differential mass loss (DTG) curve of the pellets prepared from compost extracted at 5 h.

3.3. Mechanical Properties

The composts were pelletized using a small extruder pelletizer into pellets of around 5.5 ± 0.1 mm diameter and 20 ± 0.15 mm length. Pellets that were below this range were sieved and removed. The pellets produced were able to form into the size fitting to ISO/DIS 17225-6, which biofuel pellet refers to densified biofuel in a cylindrical form, with a diameter up to 25 mm and length between 5 mm to 40 mm, produced by biomass compression. The physico-chemical properties of the compost pellet at different durations are presented in Table 3. The moisture content of the compost was found to decrease proportionally with the composting time. A longer duration of composting would subject the materials to more heat and decomposition, which will result in lower final moisture content [53]. The resultant pellets showed a similar trend of decreasing moisture content with an increasing composting time interval. A minimum amount of moisture is required to enhance the particle adhesion during pelletization. Generally, organic fibrous material can contain a moisture content of 16–28%; however, when the moisture content drops to 16–18%, the pellet mill dies tend to choke, which is not desirable [54]. Meanwhile, Zajonc, et al. reported that the optimum moisture value is within 25 to 30% and that higher values are not suitable in the production of pellets, although this might change depending on the composition of compost [30].

Table 3. Moisture content, compressive and tensile strength, durability and wettability index of raw mushroom waste compost pellets prepared at 1 h (B1), 2 h (B2), 3 h (B3), 4 h (B4) and 5 h (B5).

Sample	Moisture Content before Pelletization (%wt)	Impact Resistance (%)	Maximum Compressive Force (N)	Tensile Strength (MPa)	Wettability Index (%)
B1	29.62 ± 1.06	99.90 ± 0.09	266.27	1.56	7.93
B2	27.88 ± 0.89	99.91 ± 0.08	255.73	1.50	8.65
B3	26.46 ± 0.59	99.84 ± 0.11	258.68	1.52	8.94
B4	25.22 ± 0.70	99.90 ± 0.03	302.62	1.77	10.77
B5	24.34 ± 0.73	99.94 ± 0.04	280.84	1.65	13.70

The pellets with lower moisture content also showed signs of surface crack in the surface morphology study, as shown in Figure 3. Compost pellets made from 5 h of composting show large defects on the surface of the pellets, while, from 2–3 h of composting, minor defects were observed. The presence of a gap in Figure 3e showed poor adhesion, which may cause reduction of stiffness and deformation, leading to particle movement within the pellet matrix and making the pellets less durable mechanically [55]. The moisture content plays a significant role in pellet production; a small decrease in percentage (up to 5.82%) can affect the overall mechanical stability of pellets [54]. Hence, further extraction of compost at the sixth hour was avoided to produce pellets with good mechanical durability without the addition of any binder. This is because pellets with an ideal combination of mechanical shock resistance and calorific value are the most suitable for commercialization.

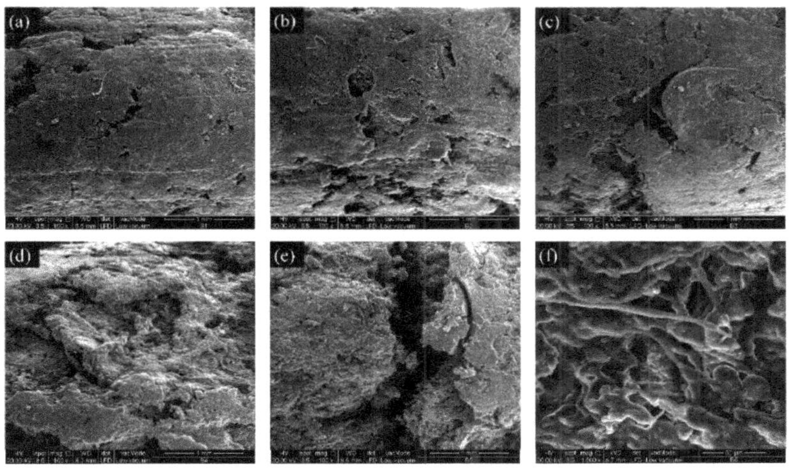

Figure 3. FESEM images on the compost pellet surface at 1 mm for compost extracted at 1 h (**a**), 2 h (**b**), 3 h (**c**), 4 h (**d**), 5 h (**e**) and at 50 µm for compost extracted at 5 h (**f**).

The impact resistance of the pellets was found to be above 99.8% for compost pellets from all durations of composting. This indicates that the compost pellet exhibited high resistance to external force and would be able to withstand the load during transportation and handling [56]. Most pellets were able to show good impact resistance as the pellets were densified and the structure of the pellet showed a closely packed interconnected structure, which held its strength when subjected to load [55]. During the biomass densification or pelletization, the formation of a solid bridge, mechanical interlocking bonds and attraction forces between solid particles, is the most important binding mechanism, because the solid bridges, which were shown in Figure 3f, mainly created the bonding between particles [57].

The analysis on the compressive force and tensile strength of the compost pellets showed that samples at 4 h and 5 h had better strength. At 1 h duration, it is likely that the composting process has not fully begun; hence, the raw mushroom waste compost extracted was at the beginning stage of composting and retained a major part of the initial raw mushroom waste properties. While, for the duration of 2–3 h, the composting process was in progress and its extraction led to slightly incomplete composting. Raw mushroom contains phenolic compounds, which require more composting time and more effort by bacteria to break down, as their chemical structure is complex [58]. Nevertheless, the raw material in this study, which is the raw mushroom waste, was in small pieces that would accelerate the composting process, as there was increased surface area with more susceptibility to degradation [6]. While, for 4 and 5 h of composting, the pellets produced showed an increase of 13.7% and 13.5% for compressive strength and tensile strength, respectively. At this duration, the compost extracted had been processed for a more sufficient duration and the compost exhibited suitable properties for pellet making. Overall, the compressive strength of the raw mushroom waste compost pellets was higher than compost pellets produced with municipal solid wastes with 44 to 113 N [57], and comparable to compost pellets made from food waste and green waste with additives, which exhibited the highest strength up to 297.1 N [28].

For the wettability index (WI), the increasing duration of composting produced pellets of increasing WI. This trend is in line with the initial moisture content of the pellets. Each pellet has a saturation point for moisture and, hence, the pellets with lower moisture were able to uptake more liquid before achieving saturation. Composts from longer duration are also more porous due to the decomposition step, which transforms the compost into more porous-like structures. The availability of pores allows more fluid to be absorbed within the pellet, which is seen from the results. A work by Zajonc, et al. showed that compost that were drier exhibited higher WI as well. Comparing to average 18% WI and the individual WI results with <30% moisture in the study by Zajonc, et al., the compost pellets in the present study generally exhibited lower WI [30]. The WI between 7.93 to 13.70% indicates that the pellets can absorb up to 13.7% moisture content with humid situations. Besides, the WI of compost pellets reported in a study by Chew, et al. was between 18.1% and 19.7% which were higher as well. The low WI presented in this study is desirable to ensure that the compost pellets could be stored in appropriate conditions within an adequate period without worry on deterioration [28].

Based on the overall observation, the suitable timing for the composting is most favorable at 4 h. Compost from 4 h of processing showed the highest compressive strength and tensile strength, with high impact resistance and feasible wettability index. For the pelletized compost, at 4 h of processing, the compost pellets showed reasonable moisture content, high fixed carbon and calorific value, with high mass density and energy density, falling slightly below that of 5 h. The duration of 1–2 h of composting did not show sufficient decomposition of the raw mushroom waste, while 3 h of composting exhibited qualities of compost that were much lower than 4–5 h. The duration of 4 h is considered as the most satisfactory, as the design of the electric composter is supposed to function as a rapid composter to reduce the overall composting time, which, on average, could take less than 24 h for a closed composting system [59,60]. This reduction in time needed for composting and the evaluation of the subsequent pellets characteristics, which are suitable as biomass pellets, indicate the effectiveness of the rapid electric composter design.

Nevertheless, the compost investigated in this study was not regarded as matured compost since fertilizing potential was not the focus to be examined and the curing process was not carried out. The main objective of this study was to generate pellets from the abundantly available raw mushroom waste using an electric composter and to analyze their energetic potential with the variable of different composting hours. Even though composting using this MW30 composter was said to be mature towards 24 h, short composting hours were experimented to reduce the processing time, retain the carbon content that reduces over composting and reduce the moisture loss due to continuous heat exposure in the composter. The aerobic electric composter was able to process and decompose the raw organic waste into sludgy or powdery material that can be pelletized within the short composting hours. Comparing the results of raw mushroom waste and compost pellets prepared, it can be observed that compost pellets exhibited higher carbon content, lower H/C ratio, lower sulphur content, lower moisture content, higher gross calorific value and higher density, all of which indicate better fuel quality. Even though the ash contents were higher after composting, they were still within the standards for low ash coal. The insignificant results between composting hours might be because the difference in composting hours was too small, in that only five hours were examined. However, moisture content at 6 h was too low to form pellets with good form and shape, hence it was not included. Besides, the insignificant results of trends with time might be due to the large amount of heterogenous samples despite the number of replicates of at least three sets. However, from the view of much higher tensile strength, wettability index and energy density, compost pellets of 4 h and 5 h are deemed more suitable than the ones of 1–3 h.

4. Conclusions

Raw mushroom waste was processed using an aerobic composting machine to a compost-like material, which was then extracted at a different processing hour to generate pellets for energetic purpose. Raw mushroom waste was used as it is abundantly available from the food factory; it would be beneficial if factories could possess equipment to process the raw mushroom waste to an energy fuel material to replace the depleting energy sources. The results presented that the mushroom compost pellets contained a low sulphur content and a low H/C ratio which indicated better environmental compatibility and thermal efficiency, respectively. Besides, thermal and combustion properties were potential for all the pellets due to the high volatile matter content, low ash content and high calorific value (GCV > 18.76 MJ/kg). Additionally, the pellets exhibited excellent mechanical properties, especially the impact resistance, as high as 99.94%, and the wettability index, between 7.93 to 13.70%, which were much lower than other pellets in the literature. This study concluded that a short processing period (<5 h) was sufficient to generate material feed for biomass pelletization that can be utilized as a solid fuel source. This study also suggests the implementation of on-site waste processing to generate valuable material with energy purposes. Nevertheless, combustion test analysis would be beneficial to study boiler efficiency using the end-product as a biofuel source to improve the overall economics.

Author Contributions: Conceptualization, W.Y.C., C.S.C.C. and M.S.L.O.; methodology, W.Y.C. and M.S.L.O.; software, W.Y.C.; validation, W.Y.C. and K.W.C.; formal analysis, W.Y.C. and K.W.C.; investigation, W.Y.C.; resources, C.S.C.C. and M.S.L.O.; data curation, W.Y.C. and K.W.C.; writing—original draft preparation, W.Y.C. and K.W.C.; writing—review and editing, W.Y.C., K.W.C. and C.F.L.; visualization, W.Y.C.; supervision, K.W.C., C.F.L. and P.L.S.; project administration, P.L.S.; funding acquisition, K.W.C. and P.L.S. All authors have read and agreed to the published version of the manuscript.

Funding: This research was funded by Mentari Alam EKO (M) Sdn Bhd (MAEKO), Fundamental Research Grant Scheme, Malaysia [FRGS/1/2019/STG05/UNIM/02/2] and MyPAIR-PHC-Hibiscus Grant [MyPAIR/1/2020/STG05/UNIM/1]. This work was also supported by Xiamen University Malaysia (XMUM) under the XMUM Research Fund (Grant number: XMUMRF/2021-C7/IENG/0033) and Hengyuan International Sdn. Bhd.

Acknowledgments: We would like to express appreciation to Mentari Alam EKO (M) Sdn Bhd (MAEKO) for providing technical support, composting machine and materials used.

Conflicts of Interest: The authors declare no conflict of interest.

References

1. Khalil, M.; Berawi, M.A.; Heryanto, R.; Rizalie, A. Waste to energy technology: The potential of sustainable biogas production from animal waste in Indonesia. *Renew. Sustain. Energy Rev.* **2019**, *105*, 323–331. [CrossRef]
2. Abdeshahian, P.; Lim, J.S.; Ho, W.S.; Hashim, H.; Lee, C.T. Potential of biogas production from farm animal waste in Malaysia. *Renew. Sustain. Energy Rev.* **2016**, *60*, 714–723. [CrossRef]
3. Srivastava, R.K.; Shetti, N.P.; Reddy, K.R.; Aminabhavi, T.M. Sustainable energy from waste organic matters via efficient microbial processes. *ScTEn* **2020**, *722*, 137927. [CrossRef]
4. Dhanya, B.S.; Mishra, A.; Chandel, A.K.; Verma, M.L. Development of sustainable approaches for converting the organic waste to bioenergy. *ScTEn* **2020**, *723*, 138109. [CrossRef]
5. Iqbal, M.W.; Kang, Y. Waste-to-energy supply chain management with energy feasibility condition. *J. Clean. Prod.* **2021**, *291*, 125231. [CrossRef]
6. Chia, W.Y.; Chew, K.W.; Le, C.F.; Lam, S.S.; Chee, C.S.C.; Ooi, M.S.L.; Show, P.L. Sustainable utilization of biowaste compost for renewable energy and soil amendments. *Environ. Pollut.* **2020**, *267*, 115662. [CrossRef]
7. Jeong, K.-H.; Lee, D.-j.; Lee, D.-H.; Lee, S.-H. Combustion characteristics of cow manure pellet as a solid fuel source. *J. Korea Org. Resour. Recycl. Assoc.* **2019**, *27*, 31–40. [CrossRef]
8. Sardarmehni, M.; Levis, J.W.; Barlaz, M.A. What Is the Best End Use for Compost Derived from the Organic Fraction of Municipal Solid Waste? *Environ. Sci. Technol.* **2021**, *55*, 73–81. [CrossRef] [PubMed]

9. Díaz, M.J.; Ruiz-Montoya, M.; Palma, A.; de-Paz, M.-V. Thermogravimetry Applicability in Compost and Composting Research: A Review. *Appl. Sci.* **2021**, *11*, 1692. [CrossRef]
10. Sweeten, J.M.; Heflin, K.; Annamalai, K.; Auvermann, B.W.; McCollum, F.T.; Parker, D.B. Combustion-Fuel Properties of Manure or Compost from Paved vs. Un-Paved Cattle Feedlots. In Proceedings of the 2006 ASAE Annual Meeting, Portland, OR, USA, 9–12 July 2006; p. 1.
11. Troy, S.M.; Nolan, T.; Leahy, J.J.; Lawlor, P.G.; Healy, M.G.; Kwapinski, W. Effect of sawdust addition and composting of feedstock on renewable energy and biochar production from pyrolysis of anaerobically digested pig manure. *Biomass Bioenergy* **2013**, *49*, 1–9. [CrossRef]
12. Skanderova, K.; Malat'ák, J.; Bradna, J.J. Energy use of compost pellets for small combustion plants. *Agron. Res.* **2015**, *13*, 413–419.
13. Malaťák, J.; Bradna, J.; Velebil, J. Combustion of briquettes from oversize fraction of compost from wood waste and other biomass residues. *Agron. Res.* **2016**, *14*, 525–532.
14. Kopeć, M.; Gondek, K.; Mierzwa-Hersztek, M.; Zaleski, T. Effect of the composting process on physical and energetic changes in compost. *Acta Agrophys.* **2016**, *23*, 607–619.
15. Yeo, J.; Oh, J.-I.; Cheung, H.H.L.; Lee, P.K.H.; An, A.K. Smart Food Waste Recycling Bin (S-FRB) to turn food waste into green energy resources. *J. Environ. Manag.* **2019**, *234*, 290–296. [CrossRef] [PubMed]
16. Doña-Grimaldi, V.M.; Palma, A.; Ruiz-Montoya, M.; Morales, E.; Díaz, M.J. Energetic valorization of MSW compost valorization by selecting the maturity conditions. *J. Environ. Manag.* **2019**, *238*, 153–158. [CrossRef]
17. Vandecasteele, B.; Boogaerts, C.; Vandaele, E. Combining woody biomass for combustion with green waste composting: Effect of removal of woody biomass on compost quality. *Waste Manag.* **2016**, *58*, 169–180. [CrossRef]
18. Malaťák, J.; Bradna, J.; Velebil, J.; Gendek, A.; Ivanova, T. Evaluation of dried compost for energy use via co-combustion with wood. *Agron. Res.* **2018**, *16*, 157–166. [CrossRef]
19. Wang, C.-Y. A Review on the Potential Reuse of Functional Polysaccharides Extracted from the By-Products of Mushroom Processing. *Food Bioprocess Technol.* **2020**, *13*, 217–228. [CrossRef]
20. Chou, W.-T.; Sheih, I.-C.; Fang, T.J. The Applications of Polysaccharides from Various Mushroom Wastes as Prebiotics in Different Systems. *J. Food Sci.* **2013**, *78*, M1041–M1048. [CrossRef]
21. Li, S.; Wang, A.; Liu, L.; Tian, G.; Wei, S.; Xu, F. Evaluation of nutritional values of shiitake mushroom (Lentinus edodes) stipes. *J. Food Meas. Charact.* **2018**, *12*, 2012–2019. [CrossRef]
22. Heleno, S.A.; Prieto, M.A.; Barros, L.; Rodrigues, A.; Barreiro, M.F.; Ferreira, I.C. Optimization of microwave-assisted extraction of ergosterol from Agaricus bisporus L. by-products using response surface methodology. *Food Bioprod. Process.* **2016**, *100*, 25–35. [CrossRef]
23. Ahmed, M.; Abdullah, N.; Shuib, A.S.; Abdul Razak, S. Influence of raw polysaccharide extract from mushroom stalk waste on growth and pH perturbation induced-stress in Nile tilapia, Oreochromis niloticus. *Aquaculture* **2017**, *468*, 60–70. [CrossRef]
24. Zou, H.; Evrendilek, F.; Liu, J.; Buyukada, M. Combustion behaviors of pileus and stipe parts of Lentinus edodes using thermogravimetric-mass spectrometry and Fourier transform infrared spectroscopy analyses: Thermal conversion, kinetic, thermodynamic, gas emission and optimization analyses. *Bioresour. Technol.* **2019**, *288*, 121481. [CrossRef] [PubMed]
25. Bernal, M.P.; Sommer, S.G.; Chadwick, D.; Qing, C.; Guoxue, L.; Michel, F.C. Chapter Three—Current Approaches and Future Trends in Compost Quality Criteria for Agronomic, Environmental, and Human Health Benefits. In *Advances in Agronomy*; Sparks, D.L., Ed.; Academic Press: Cambridge, MA, USA, 2017; Volume 144, pp. 143–233.
26. *ASTM E1756-08*; Standard Test Method for Determination of Total Solids in Biomass. ASTM International: West Conshohocken, PA, USA, 2015.
27. Cai, J.; He, Y.; Yu, X.; Banks, S.W.; Yang, Y.; Zhang, X.; Yu, Y.; Liu, R.; Bridgwater, A.V. Review of physicochemical properties and analytical characterization of lignocellulosic biomass. *Renew. Sustain. Energy Rev.* **2017**, *76*, 309–322. [CrossRef]
28. Chew, K.W.; Chia, S.R.; Yap, Y.J.; Ling, T.C.; Tao, Y.; Show, P.L. Densification of food waste compost: Effects of moisture content and dairy powder waste additives on pellet quality. *Process Saf. Environ. Prot.* **2018**, *116*, 780–786. [CrossRef]
29. Azargohar, R.; Soleimani, M.; Nosran, S.; Bond, T.; Karunakaran, C.; Dalai, A.K.; Tabil, L.G. Thermo-physical characterization of torrefied fuel pellet from co-pelletization of canola hulls and meal. *Ind. Crops Prod.* **2019**, *128*, 424–435. [CrossRef]
30. Zajonc, O.; Frydrych, J.; Jezerska, L. Pelletization of Compost for Energy Utilization. *IERI Procedia* **2014**, *8*, 2–10. [CrossRef]
31. Finimundy, T.C.; Dillon, A.J.P.; Henriques, J.A.P.; Ely, M.R. A review on general nutritional compounds and pharmacological properties of the Lentinula edodes mushroom. *Food Nutr. Sci.* **2014**, *5*, 1095–1105. [CrossRef]
32. Yu, H.; Xie, B.; Khan, R.; Shen, G. The changes in carbon, nitrogen components and humic substances during organic-inorganic aerobic co-composting. *Bioresour. Technol.* **2019**, *271*, 228–235. [CrossRef]
33. Chalk, P.M.; Inácio, C.T.; Chen, D. An overview of contemporary advances in the usage of 15N natural abundance (δ15N) as a tracer of agro-ecosystem N cycle processes that impact the environment. *Agric. Ecosyst. Environ.* **2019**, *283*, 106570. [CrossRef]
34. Wang, X.; Zhai, M.; Wang, Z.; Dong, P.; Lv, W.; Liu, R. Carbonization and combustion characteristics of palm fiber. *Fuel* **2018**, *227*, 21–26. [CrossRef]
35. Akratos, C.S.; Tekerlekopoulou, A.G.; Vasiliadou, I.A.; Vayenas, D.V. Chapter 8—Cocomposting of olive mill waste for the production of soil amendments. In *Olive Mill Waste*; Galanakis, C.M., Ed.; Academic Press: Cambridge, MA, USA, 2017; pp. 161–182.

36. Xiang, Y.; Jun, Q.; Zhen, L.; Yaning, L.; Anlong, Z.; Yingkun, S. Research on the infrared spectroscopy of spent mushroom compost. *Nat. Envir. Pollut. Technol.* **2016**, *15*, 701.
37. Parmar, R.; Kumar, D. Studyof chemical composition in wild edible mushroom Pleurotus cornucopiae (Paulet) from Himachal Pradesh, India by using Fourier transforms infrared spectrometry (FTIR), Gas chromatography-mass spectrometry (GCMS) and X-ray fluorescence (XRF). *Biol. Forum Int. J.* **2015**, *7*, 1057–1066.
38. Li, X.; Pang, Y.; Zhang, R. Compositional changes of cottonseed hull substrate during P. ostreatus growth and the effects on the feeding value of the spent substrate. *Bioresour. Technol.* **2001**, *80*, 157–161. [CrossRef]
39. Grube, M.; Lin, J.G.; Lee, P.H.; Kokorevicha, S. Evaluation of sewage sludge-based compost by FT-IR spectroscopy. *Geoderma* **2006**, *130*, 324–333. [CrossRef]
40. Singh, B.; Avci, U.; Eichler Inwood, S.E.; Grimson, M.J.; Landgraf, J.; Mohnen, D.; Sørensen, I.; Wilkerson, C.G.; Willats, W.G.T.; Haigler, C.H. A specialized outer layer of the primary cell wall joins elongating cotton fibers into tissue-like bundles. *Plant Physiol.* **2009**, *150*, 684–699. [CrossRef]
41. Abidi, N.; Cabrales, L.; Haigler, C.H. Changes in the cell wall and cellulose content of developing cotton fibers investigated by FTIR spectroscopy. *Carbohydr. Polym.* **2014**, *100*, 9–16. [CrossRef] [PubMed]
42. Lou, Z.; Sun, Y.; Zhou, X.; Baig, S.A.; Hu, B.; Xu, X. Composition variability of spent mushroom substrates during continuous cultivation, composting process and their effects on mineral nitrogen transformation in soil. *Geoderma* **2017**, *307*, 30–37. [CrossRef]
43. Ahmad, M.S.; Mehmood, M.A.; Liu, C.-G.; Tawab, A.; Bai, F.-W.; Sakdaronnarong, C.; Xu, J.; Rahimuddin, S.A.; Gull, M. Bioenergy potential of Wolffia arrhiza appraised through pyrolysis, kinetics, thermodynamics parameters and TG-FTIR-MS study of the evolved gases. *Bioresour. Technol.* **2018**, *253*, 297–303. [CrossRef]
44. Pua, F.-L.; Subari, M.S.; Ean, L.-W.; Krishnan, S.G. Characterization of biomass fuel pellets made from Malaysia tea waste and oil palm empty fruit bunch. *Mater. Today. Proc.* **2020**, *31*, 187–190. [CrossRef]
45. Gao, L.; Volpe, M.; Lucian, M.; Fiori, L.; Goldfarb, J.L. Does hydrothermal carbonization as a biomass pretreatment reduce fuel segregation of coal-biomass blends during oxidation? *Energy Convers. Manag.* **2019**, *181*, 93–104. [CrossRef]
46. Wang, T.; Li, Y.; Zhang, J.; Zhao, J.; Liu, Y.; Sun, L.; Liu, B.; Mao, H.; Lin, Y.; Li, W.; et al. Evaluation of the potential of pelletized biomass from different municipal solid wastes for use as solid fuel. *Waste Manag.* **2018**, *74*, 260–266. [CrossRef] [PubMed]
47. Tumuluru, J.S. Effect of process variables on the density and durability of the pellets made from high moisture corn stover. *Biosys. Eng.* **2014**, *119*, 44–57. [CrossRef]
48. Emadi, B.; Iroba, K.L.; Tabil, L.G. Effect of polymer plastic binder on mechanical, storage and combustion characteristics of torrefied and pelletized herbaceous biomass. *ApEn* **2017**, *198*, 312–319. [CrossRef]
49. Pradhan, P.; Mahajani, S.M.; Arora, A. Pilot scale production of fuel pellets from waste biomass leaves: Effect of milling size on pelletization process and pellet quality. *Fuel* **2021**, *285*, 119145. [CrossRef]
50. Qin, Y.; Zhao, Z.; Wiltowski, T.; Aloqaili, M.; Liang, Y. Investigation of Co-Gasification Reactivity of Torrefied Jatropha Seed Cake with Illinois# 6 Coal Char. *BioResources* **2016**, *11*, 7624–7636.
51. Nobre, C.; Gonçalves, M.; Mendes, B.; Vilarinho, C.; Teixeira, J. Torrefaction Effects on Composition and Quality of Biomass Wastes Pellets. In Proceedings of the 3rd International Conference on Wastes: Solutions, Treatments and Opportunities, Viana Do Castelo, Portugal, 14–16 September 2015; pp. 171–176.
52. Jia, G. Combustion Characteristics and Kinetic Analysis of Biomass Pellet Fuel Using Thermogravimetric Analysis. *Processes* **2021**, *9*, 868. [CrossRef]
53. Waqas, M.; Almeelbi, T.; Nizami, A.-S. Resource recovery of food waste through continuous thermophilic in-vessel composting. *Environ. Sci. Pollut. Res.* **2018**, *25*, 5212–5222. [CrossRef]
54. Ungureanu, N.; Vladut, V.; Voicu, G.; Dinca, M.-N.; Zăbavă, B.-Ș. Influence of biomass moisture content on pellet properties—Review. *Eng. Rural. Dev.* **2018**, *17*, 1876–1883. [CrossRef]
55. Sharma, H.B.; Dubey, B.K. Binderless fuel pellets from hydrothermal carbonization of municipal yard waste: Effect of severity factor on the hydrochar pellets properties. *J. Clean. Prod.* **2020**, *277*, 124295. [CrossRef]
56. Liu, Z.; Liu, X.e.; Fei, B.; Jiang, Z.; Cai, Z.; Yu, Y. The properties of pellets from mixing bamboo and rice straw. *Renew. Energy* **2013**, *55*, 1–5. [CrossRef]
57. Zafari, A.; Kianmehr, M.H. Factors affecting mechanical properties of biomass pellet from compost. *Environ. Technol.* **2014**, *35*, 478–486. [CrossRef] [PubMed]
58. Ayilara, M.S.; Olanrewaju, O.S.; Babalola, O.O.; Odeyemi, O. Waste Management through Composting: Challenges and Potentials. *Sustainability* **2020**, *12*, 4456. [CrossRef]
59. Malakahmad, A.; Idrus, N.B.; Abualqumboz, M.S.; Yavari, S.; Kutty, S.R.M. In-vessel co-composting of yard waste and food waste: An approach for sustainable waste management in Cameron Highlands, Malaysia. *Int. J. Recyl. Org. Waste Agric.* **2017**, *6*, 149–157. [CrossRef]
60. Abasalizadeh, H.; Fathi Gerdelidani, A.; Khalaj, S.; Jalili Aramesh, S. Evaluation of environmental performance of the Starclean composter machine in converting the wet wastes into compost. *J. Environ. Sci. Technol.* **2017**.

processes

MDPI

Article

Impact of Various Visible Spectra on Attached Microalgal Growth on Palm Decanter Cake in Triggering Protein, Carbohydrate, and Lipid to Biodiesel Production

Zhi Wei Tiong [1], Hemamalini Rawindran [1], Wai Hong Leong [1], Chin Seng Liew [1], Yi Ying Wong [2], Worapon Kiatkittipong [3,*], Eman Alaaeldin Abdelfattah [4], Pau Loke Show [5], Anisa Ur Rahmah [6], Woei Yenn Tong [7] and Jun Wei Lim [1,8,*]

1 HICoE-Centre for Biofuel and Biochemical Research, Institute of Self-Sustainable Building, Department of Fundamental and Applied Sciences, Universiti Teknologi PETRONAS, Seri Iskandar 32610, Perak Darul Ridzuan, Malaysia
2 HICoE-Centre for Biofuel and Biochemical Research, Institute of Self-Sustainable Building, Department of Chemical Engineering, Universiti Teknologi PETRONAS, Seri Iskandar 32610, Perak Darul Ridzuan, Malaysia
3 Department of Chemical Engineering, Faculty of Engineering and Industrial Technology, Silpakorn University, Nakhon Pathom 73000, Thailand
4 Lecturer of Biochemistry and Molecular Science, Entomology Department, Faculty of Science, Cairo University, Giza 12613, Egypt
5 Department of Chemical and Environmental Engineering, Faculty of Science and Engineering, University of Nottingham Malaysia, Semenyih 43500, Selangor Darul Ehsan, Malaysia
6 Department of Chemical Engineering, Universitas Muhammadiyah Surakarta, Surakarta 57162, Indonesia
7 Branch Campus Institute of Medical Science Technology, Universiti Kuala Lumpur, Kajang 43000, Selangor, Malaysia
8 Department of Biotechnology, Saveetha School of Engineering, Saveetha Institute of Medical and Technical Sciences, Chennai 602105, India
* Correspondence: kiatkittipong_w@su.ac.th (W.K.); junwei.lim@utp.edu.my or jun_wei_85@yahoo.com (J.W.L.)

Citation: Tiong, Z.W.; Rawindran, H.; Leong, W.H.; Liew, C.S.; Wong, Y.Y.; Kiatkittipong, W.; Abdelfattah, E.A.; Show, P.L.; Rahmah, A.U.; Tong, W.Y.; et al. Impact of Various Visible Spectra on Attached Microalgal Growth on Palm Decanter Cake in Triggering Protein, Carbohydrate, and Lipid to Biodiesel Production. *Processes* **2022**, *10*, 1583. https://doi.org/10.3390/pr10081583

Academic Editor: Juan Luis Gomez Pinchetti

Received: 30 June 2022
Accepted: 9 August 2022
Published: 12 August 2022

Abstract: Attached microalgal growth of *Chlorella vulgaris* on palm decanter cake (PDC) under irradiation with various visible monochromatic and polychromatic spectra to produce biodiesel was studied in this work. The results demonstrated that the white spectrum cultivation exhibited the highest microalgal density of 1.13 g/g along with 1.213 g/L day of microalgal productivity. Correspondingly, the biodiesel obtained was comprised mainly of C16 and C18 fatty acids, possessing a high cetane number and oxidation stability from the high saturated fatty acid content (70.38%), which was appealing in terms of most biodiesel production requirements. Nevertheless, the highest lipid content (14.341%) and lipid productivity (93.428 mg/L per day) were discovered with green spectrum cultivation. Blue and white spectra led to similar protein contents (34%) as well as carbohydrate contents (61%), corroborating PDC as a feasible carbon and nutrient source for growing microalgae. Lastly, the energy feasibilities of growing the attached microalgae under visible spectra were investigated, with the highest net energy ratio (NER) of 0.302 found for the yellow spectrum. This value outweighed that in many other works which have used suspended growth systems to produce microalgal fuel feedstock. The microalgal growth attached to PDC is deemed to be a suitable alternative cultivation mode for producing sustainable microalgal feedstock for the biofuel industry.

Keywords: microalgae; visible spectrum; biodiesel; attached growth; palm decanter cake; net energy ratio

1. Introduction

There has been an increase in population and urbanization over the last few decades as a result of global industrialization. Consequently, the need for fossil fuels has increased significantly. This has inevitably depleted the stores of fossil fuels since they are non-renewable resources. According to BP's statistical analysis report, fossil fuels will be completely exhausted within the next 50 years if they are exploited continuously at the current rate [1]. Besides, fossil fuels also create environmental disadvantages, mainly contributing to the greenhouse effect due to the emission of carbon dioxide (CO_2) from fossil fuel combustion. Thus, microalga have been found to be a sustainable feedstock for producing biofuels with astounding benefits [2,3]. Microalgae have a high growth rate and lipid content, and do not compete for agricultural land. Notably, they could sequester CO_2 for assuaging global warming. Furthermore, to date, biodiesel from microalgal cultivation has been projected to be the front runner for replacing diesel fuel because of the high cetane number and low sulfur content [1].

Generally, there are four types of microalgal cultivation modes: photoautotrophic, heterotrophic, mixotrophic and photoheterotrophic, with the high biomass productivity usually being reported under mixotrophic cultivation [4]. Moreover, compared with the suspended cultivation system, numerous researchers have indicated that the attached cultivation system exhibits better harvesting performance and higher biomass yield as well as lower water and energy consumption [5]. Hence, attached microalgal cultivation has become the predominant choice for growing microalgae. In the attached cultivation system, the microalgae are grown on a solid support to form a biofilm. A number of support materials had been investigated, such as filter paper, glass fiber, fabric–hydrogel composites, polytetrafluoroethylene, etc. Nonetheless, these substrates are not ideal for long-term cultivation, as their performances would deteriorate after multiple re-culturing [6]. In fact, they lack the essential amounts of nutrients for microalgal growth, which incurs the cost of supplementary nutrients in the culture medium. In spite of that, attached microalgal cultivation has been widely explored. With the aim of reducing operational costs, agricultural wastes such as palm decanter cake (PDC) have been proposed for use as support media for attached microalgal cultivation, which was initially postulated to proffer both nutrients and a platform for biofilm formation. Interestingly, PDC has also been reported as a source of nutrients for plantations [7]. Furthermore, PDC can be easily collected from palm oil milling plants as a waste by-product of oil palm plantations that is prevalent in Southeast Asia.

The effect of the visible spectrum stands among the least studied factors for stimulating microalgal growth. Recent breakthroughs in microalgal cultivation have revealed the plausible potential influence of monochromatic and polychromatic LED wavelengths on enhanced microalgal growth [8]. This has also been supported by various studies on which pigments produced by the microalgae absorb light energy. When a light quantum is absorbed, these pigment molecules are stimulated and transformed to a state of high energy. When they return to their original condition, they release energy, which drives photochemical reactions. The energy hub of microalgae, known as the photosystem, contains a distinctive set of light-harvesting pigments that provide a unique absorption spectrum. Therefore, different spectral compositions of light are known to be effectively captured by distinct chlorophyll pigments, resulting in enhanced microalgal growth [9].

Therefore, this study highlighted the efficacy of attached microalga (i.e., *Chlorella vulgaris*) on PDC under irradiation by polychromatic and various monochromatic visible spectra to enhance microalgal growth and lipid production. In this regard, a novel use of solid nutrients was adopted to culture the microalgal cells on the PDC's surface. Additionally, the quality and composition of fatty acid methyl esters (FAME) in the biodiesel was examined, as well as an energy feasibility assessment to confirm the viability of the microalgal biodiesel production process.

2. Materials and Methods

2.1. Cultivation of Microalgal Stock

A strain of bacteria-free *Chlorella vulgaris* sp. USMAC 24 was used as the inoculum. The microalgal culture was acquired from the Centre for Biofuel and Biochemical Research, Universiti Teknologi PETRONAS, and was carried out in synthetic Bold's Basal medium (BBM) until the required amount for carrying out the batch experiments had been obtained. The media had the following chemical compositions: (1) 10 mL/L of NaCl (2.5 g/L), $MgSO_4{\bullet}7H_2O$ (7.5 g/L), $CaCl_2{\bullet}2H_2O$ (2.5 g/L), K_2HPO_4 (7.5 g/L), KH_2PO_4 (17.5 g/L), and $NaNO_3$ (25 g/L) and (2) 1 mL/L of $Co(NO_3)_2{\bullet}6H_2O$ (0.49 g/L), H_3BO_3 (11.4 g/L), $CuSO_4{\bullet}5H_2O$ (1.57 g/L), $ZnSO_4{\bullet}7H_2O$ (8.82 g/L), $FeSO_4{\bullet}7H_2O$ (4.98 g/L), $MnCl_2{\bullet}4H_2O$ (1.44 g/L), MoO_3 (0.71 g/L), anhydrous EDTA (50 g/L), KOH (31 g/L), and H_2SO_4 (1 mL/L) [3]. A microalgal seed of 500 mL was first introduced in a 5 L bottle containing 4.5 L of the medium. Compressed air at the flow rate of 6.5 L/min was aerated into the cultured medium at room temperature (25 °C ± 1 °C). Simultaneously, the culture was exposed to continuous illumination under a cool-white, fluorescent light at an intensity of 60–70 μmol/m^{2}s. The pH of the medium was kept constant at 3.0 ± 0.1 throughout the cultivation period. The cultivation process was ended once the microalgal growth reached a stationary growth phase prior to experimental use [10].

2.2. Characterization of the Palm Decanter Cake

The palm decanter cake (PDC) was collected from the Nutrition Technologies Sdn Bhd. The PDC was then stored at temperatures below 4 °C in a chiller before being subjected to analysis. The CHNS analyzer was used to determine the elemental content (C, H, N) of the PDC. Oxygen gas was required to oxidize the subjected elements into their oxide molecules at 1000 °C. An inert carrier gas, namely helium, was used to sweep the combusted products and pass them over to heated high-purity copper at 600 °C, in which the copper was responsible for reduction of the molecules [10]. Determination of the moisture content and dry matter of the PDC were carried out using a standard oven-drying method: IS 2720-2 (1973).

2.3. Experimental Setup for Attached Microalgal Growth

Attached microalgal cultivation was carried out in 1 L Erlenmeyer flasks that served as the photobioreactors. Initially, each 1 L Erlenmeyer flask contained 900 mL of tap water to which 100 mL of the microalgal stock culture was added, with the pH being kept constant at 3.0 ± 0.1 throughout the cultivation period. Next, PDC at a concentration of 10 g/L was added into each cultivation flask to act as a nutritional support substratum. All setups were aerated with compressed air at a flow rate of 1.3 L/min at room temperature (25 °C ± 1 °C). Different colors of lights were used to separately illuminate every photobioreactor containing the attached microalgal growth on PDC. In this regard, five lightbulbs with different irradiant spectrum colors, namely, red, blue, white, yellow, and green, were used at an intensity of 100 μmol/m^{2}s each. The cultivation period was scheduled for 9 days of exposure to each irradiant spectrum before measuring the microalgal biomass and lipid production. The flasks were then sealed with aluminum foil to prevent air contamination. Each setup was covered with a black box to minimize random light radiation from the surroundings. To ensure the consistency of results, each cultivation flask was at least duplicated [10,11].

2.4. Analytical Methods

2.4.1. Attached Microalgal Biomass Productivity

Gravitational sedimentation was adopted to harvest the mature attached microalgae from the PDC. The aeration was stopped and the attached microalgae with residual PDC was allowed to settle at the bottom of the culture medium. The supernatant was then decanted carefully without agitating the attached microalgae on the PDC layer. The remaining biomass was then subjected to centrifugation at 6000 rpm for 6 min for further dewatering. The PDC with attached microalgal biomass was subsequently dried in an oven at 60 °C until a constant weight was achieved (W_1). After that, the residual PDC was separated from the microalgal biomass by introducing mixed solvents at the ratio of 2:1 of methanol and chloroform, respectively. The mixture in the solution was gently stirred until a suspension of the residual PDC appeared, which was isolated physically. The solution was then rested, and the separated residual PDC was dried at 60 °C before being weighed as W_2. Thereafter, the attached microalgal biomass productivity was calculated by using Equation (1) [12]:

$$\text{Attached microalgal biomass productivity}\left(\frac{\text{g}}{\text{L day}}\right) = \frac{\text{Dry attached microalgae (g)}}{\text{Cultivation volume (L)} \times \text{Cultivation period (day)}} \quad (1)$$

where dry attached microalgae were calculated by subtracting W_1 from W_2 [10].

2.4.2. Attached Microalgal Biochemical Productivity

The microalgal residue in the chloroform and methanol mixture was transferred into a capped sampling bottle and placed on an orbital shaker for 24 h at 250 rpm. Thereafter, the solvent mixture was filtered by using Whatman Grade 1 filter paper to retain the solid microalgal biomass. The solvent filtrate was evaporated by purging with dry inert gas to obtain the extracted microalgal lipids. Finally, the lipid content Equation (2) and lipid productivity Equation (3) were calculated from weighing the residue in the glass vial [10].

$$\text{Lipid content (\%)} = \frac{\text{Lipid yield (g)}}{\text{Dry attached microalgae (g)}} \times 100\% \quad (2)$$

$$\text{Lipid productivity}\left(\frac{\text{mg}}{\text{L day}}\right) = \frac{\text{Lipid yield (mg)}}{\text{Cultivation volume (L)} \times \text{Cultivation period (day)}} \quad (3)$$

Subsequently, the protein content was determined on the basis of the nitrogen content in the microalgae by Equation (4). The carbohydrate content was calculated by subtracting the lipid and protein contents from 100%.

$$\text{Protein content (\%)} = \frac{\text{Nitrogen content in attached microalgae(g)} \times 5.3}{\text{Dry attached microalgae (g)}} \times 100\% \quad (4)$$

The multiplication factor of 5.3 was proposed to correlate the nitrogen and protein content in palm decanter cake [13]. The nitrogen content in the attached microalgae was verified by analyzing the nitrogen content of the microalgal stock culture in a CHNS analyzer and multiplied by the weight of the microalgae after lipid extraction (Equation (5)) [11].

$$\text{Microalgae weight after lipid extraction (g)} = [100\% - \text{Lipid content (\%)}] \times \text{Dry attached microalgae (g)} \quad (5)$$

2.4.3. Profile of Fatty Acid Methyl Esters

Sample mixing was enhanced by introducing 1 mL of tetrahydrofuran to the extracted lipid. After that, methanol and lipid at a 15:1 ratio in 3 wt.% of concentrated sulfuric acid catalyst were mixed thoroughly before initiating the transesterification process in an incubator shaker for 3 h at 60 °C. After that, 2 mL of methanol, 10 mL of hexane, 4 mL of 10% sodium chloride, and 4 mL of distilled water were added and mixed thoroughly. The solvent mixture was later transferred into a centrifugation tube and centrifuged at 5000 rpm for 5 min. Thereafter, two immiscible layers were formed, in which the upper layer consisted of mixed hexane with dissolved fatty acid methyl esters (FAME). The upper layer was extracted and dried in an oven at 105 °C to a constant weight. Lastly, 1 mL of the internal standard C17:0 of 0.8 mg/mL in hexane was introduced to the dried FAME and 1 µL of the FAME mixture was then subjected to gas chromatography (Shimadzu GC-2010 plus), with the operating conditions referred to in the literature [3]. Equation (6) was used to calculate the FAME percentages from each attached microalgal cultivation [14].

$$\text{FAME (\%)} = \frac{A_{\text{FAME}}}{A_{\text{ISTD}}} \times \frac{C_{\text{ISTD}} \times V_{\text{ISTD}}}{m} \times 100\% \quad (6)$$

where A_{FAME} is the FAME peak area, A_{ISTD} is the C17:0 internal standard's peak area, C_{ISTD} is the concentration of the C17:0 internal standard (mg/L), V_{ISTD} is the volume of the C17:0 internal standard (L), and m is the mass of the crude biodiesel sample before mixing with the C17:0 internal standard (g)

2.4.4. Analysis of the Net Energy Ratio (NER)

Energy input was an important parameter for assessing the feasibility of the current study for potential use in the industrial sector. The amount of water consumed to produce 1 g of lipid could be formulated using a tailor-made formula, as shown in Equation (7):

$$\text{Water usage per g of lipid produced (g)} = \frac{\text{Amount of water used in cultivation (g)}}{\text{Total lipid content }\left(\frac{g}{L}\right)} \times \text{Density of water}\left(\frac{g}{L}\right) \quad (7)$$

where the density of water is equivalent to 1 g/mL.

Consequently, the net energy ratio (NER) of the entire process was calculated to be 0.302 by using Equation (8) [15].

$$\text{NER} = \frac{\text{Primary Energy Output}}{\text{Non-renewable Energy Input}} \quad (8)$$

3. Results and Discussion

3.1. Characterization of Palm Decanter Cake

The viability of using palm decanter cake (PDC) as a solid support was closely associated with the elemental content. The PDC was characterized using elemental analysis as being composed of 42.76 ± 2.82% carbon, 6.98 ± 0.67% hydrogen, and 2.78 ± 0.18% nitrogen. The abundance of carbon as a nutrient source proved to be significant for microalgal biomass and lipid production. Generally, the carbon in PDC mainly comprised cellulose, hemicellulose, and lignin [16]. These polysaccharides are external carbon sources to enhance microalgal growth by promoting the uptake of organic materials and microbial photosynthesis. Therefore, microalgal biomass production was strongly escalated [17]. Additionally, proximate analysis was used in this study, which included dry matter and moisture content. The PDC was composed of 98.57 ± 0.22% dry matter and 1.43 ± 0.22% moisture content, vindicating it a ideal for subsequent application as a substrate and carbon source for the production of attached microalgal biomass. In contrast, integrating organic substances with a high moisture contents with the cultivation medium may manifest in bacterial contamination, which would eventually impede the microalgae's growth [12]. The correlation between protein content (14.50 ± 1.80%) and nitrogen content was evaluated by a multiplication factor of 5.3. The C:N ratio in this study was assessed to be 15.38, satisfying the optimum value for outstanding biofilm formation. A C/N ratio of 18 or below is often necessary for appropriate growth and treatment efficiency. On the other hand, values greater than 22 have a negative impact on performance and induce the growth of filamentous organisms [18]. A recent study reported that an optimal protein content of 43% amino-N could be utilized as a building block for microalgal cells to promote proliferation. Overall, it could be claimed that PDC is beneficial as a nutritious carbon source in microalgal cultivation [12].

3.2. Attached Microalgal Growth on Palm Decanter Cake under Various Visible Spectra of Irradiation

The potential of attached microalgal growth on palm decanter cake was studied under various monochromatic and polychromatic visible spectra and continuous irradiation (a photoperiod regime with a light:dark cycle of 24:0 h) (Figure 1). The results could be related to the activity of chlorophyll, as it plays a role in light harvesting during photosynthesis, promoting the growth of microalgal biomass and accumulation of biochemicals [19,20]. In this regard, the maximum microalgal density and microalgal productivity were achieved under a white light spectrum, namely 1.130 ± 0.06 g/g and 1.213 ± 0.06 g/L per day, respectively. These cultivation parameters were found to outperform other light spectra by at least 5.6%. The white light spectrum resulted in the highest absorption percentage compared with the others, yielding the highest amount of microalgal biomass along with the fastest microalgal growth rate. This implied that the microalgae absorbed most of the photosynthetic active radiation (PAR) from the wavelengths of 400 nm (blue), 500 nm (green), 580 nm (yellow) to 700 nm (red). The amalgamation of the aforementioned colors is responsible for producing the white spectrums, defining the reason why the white spectrum is known as a polychromatic spectrum [21]. However, attached microalgal cultivation under green light produced the least microalgal density (0.623 ± 0.02 g/g) and microalgal productivity (0.655 ± 0.02 g/L per day), due to the fact that *Chlorella vulgaris* is a species of green microalgae in which the pigments or chlorophylls inside the cells are green in color. In this case, the chlorophylls would have reflected most of the green light to make the plant appear green, rather than absorbing it for photosynthesis and cell development. [17]. On the other hand, a previous study proved that the main constituents of *Chlorella vulgaris* are chlorophyll *a* and chlorophyll *b*, of which chlorophyll *a* showed selective absorption at peaks of 440 and 682 nm, while chlorophyll *b* had absorption peaks at 473 and 655 nm [22]. These results indicated that both chlorophylls exhibited blue and red absorption spectrums. Nevertheless, the current research unveiled that the red and blue light spectra contributed to sloughing of the attached microalgal biomasses. The detachment of microalgal biofilms impoverished the attached microalgal density and their ability to thrive on the palm decanter cake's surface. Conversely, the highest biomass production of *Chlorella vulgaris* growing in suspension was measured under red and blue light conditions [9]. These observations could be further rationalized by the use of attached growth versus suspended growth. Accordingly, the longer wavelength of red light was occluded by the presence of the bulk suspended palm decanter cake, preventing it from reaching the microalgal cells due to insufficient photon energy, resulting in stunted growth of the attached microalgae, whereas the blue light with a shorter wavelength had higher energy and could damage the microalgal cells or even lead to cell death.

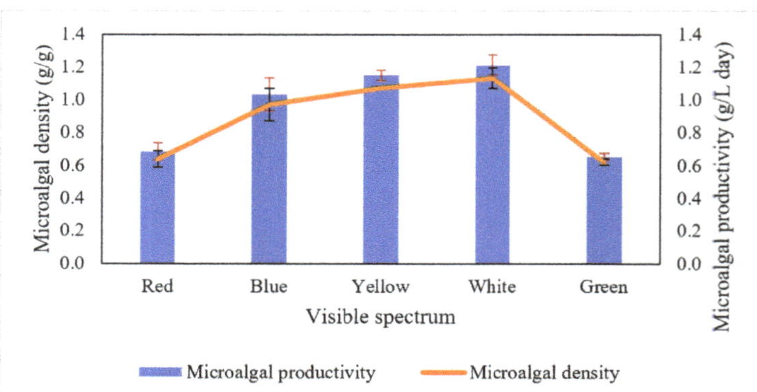

Figure 1. Productivity and density of attached microalgae on palm decanter cake at various visible spectra of irradiation.

According to Table 1, various species of microalgae have tailored optimum spectrums of light colors for cultivation, since each species has its own pigment composition [20]. According to Raqiba and Sibi (2019), red and blue light cultivation could enhance the biomass of *Chlorella vulgaris* compared with other light spectra, which appears to conflict with the findings of this study. However, most of the studies obtained results in line with the current findings, where white light was the best at stimulating the growth of microalgae due to the combined ratio of several light spectra. Although the use of the 24:0 h cycle in this study may have enabled the attached microalgae to experience a longer light duration in which to perform photosynthesis, the 12:12 h cycle offered by mixotrophic cultivation of attached microalgae permits organic carbon assimilation from the palm decanter cake during the light period, while losing inorganic carbon as carbon dioxide in the absence of light. Consequently, the light would have insufficient time to damage the microalgal cells after penetrating the culture under a 12:12 h cycle.

Table 1. Microalgal growth responses to various visible spectra of irradiation.

Visible Spectrum	Intensity (μmol/m^2 s)	Microalgal Species	Stimulating Effect in Decreasing Order	Reference
White, red, blue, green, yellow	100	*Chlorella vulgaris*	White > yellow > blue > red > green	Current study
White, red, blue, green	60	*Botryococcus braunii*	Red > white > blue > green	[23]
White, red, blue, green, yellow	100	*Chlorella vulgaris* *Diacronema lutheri* *Porphyridium purpureum*	White > red > blue > yellow > green Blue > red > green > yellow > white Green > blue > yellow > red > white	[22]
White, red, blue, green, yellow	- - -	*Chlorella vulgaris* *Tetradesmus obliquus* *Arthospira platensis*	White > red > blue > yellow > green Red > blue > white > yellow > green Red > white > blue > yellow > green	[23]
White, red, blue,	-	*Auxenochlorella pyrenoidosa, Scenedesmus quadricaud, Tetradesmus obliquus*	Blue > red > white	[24]
White, red, yellow	-	*Phaeodactylum tricornutum*	Red > white > yellow	[24]
White, red, blue	52	*Isochrysis galbana*	Blue > red > white	[25]
White, red, blue, green	100	*Chlorella sp., Nannochloris oculata*	Blue > red > white > green	[5]

3.3. Lipid Accumulation from Attached Microalgae Grown on Palm Decanter Cake under Various Visible Spectra of Irradiation

The impact of various visible spectra on the attached microalgal biomass could be further comprehended from the perspective of lipid content and productivity, since lipids are the most important biochemical for the microalgae's downstream processes (Figure 2). Among the five spectra, attached microalgal cultivation under green light had the highest lipid content (14.341 ± 0.14 %) and lipid productivity (93.428 ± 0.18 mg/L per day). In this context, the microalgae favored lipid production and accumulation over growth, as shown by the lowest microalgal density and productivity being found under green light irradiation (Figure 1). This primarily stemmed from the green light being reflected the most, which subjected the microalgal cells to light limitation stress [26]. Accordingly, the microalgae received an inadequate level of photon fluxes for cell growth, thus, inducing the lipid to be stored instead of using them as energy for microalgal cell proliferation. Similar trends were observed for *Chlorella vulgaris* and Ettlia pseudoalveolaris, which both had higher lipid contents when cultivated at intensities of 50 and 150 μE/m^2 per s, respectively, as opposed to a higher light intensity [27]. Moreover, Chen et al. [28] confirmed that Chromochloris zongfingienis under continuous dark cycle cultivation had yielded more lipids than continuous light cycle growth. In fact, any stress that inhibits microalgal growth could convert additional carbon and energy into lipid storage by the cells, predominantly in the form of triacyclglycerols, which are effectively packed in the cells and provide energy for oxidation, thus forming the best reserves for cell reconstruction after the stress condition was removed [28]. On the contrary, red spectrum cultivation accumulated the least lipid productivity due to rapid detachment of the microalgal biomass from the palm decanter cake, which was thus lost to the suspended growth form [29]. However, the lipid content of the attached microalgae under red light cultivation was slightly higher than that of blue, yellow, and white light. The rationale for this was a deficiency in the red light's intensity, as it was unable to penetrate into the culture medium and thus triggered light limitation stress, improving the lipid accumulation of the remaining microalgal cells growing on palm decanter cake.

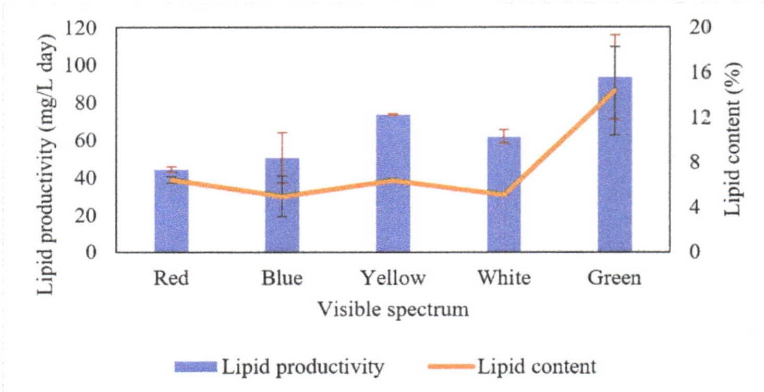

Figure 2. Lipid productivity and contents of attached microalgae growing on palm decanter cake at various visible spectra of irradiation.

Comparing the ability of microalgae to accumulate lipids under various visible spectra of irradiation, Table 2 accentuates that the *Chlorella vulgaris* cultivated under blue light conditions produce the highest amount of lipid [23]. This can be explained by the enzymes within the microalgal cells such as carbonic anhydrase and ribulose biphosphate carboxylase/oxygenase being solely controlled by blue irradiation. Therefore, the presence of blue light enhanced the enzyme activities, triggering triglyceride accumulation in the carbon cycle [30,31]. This earlier finding was contrary to the present study, which showed lower lipid accumulation under blue and white light conditions, considering that the light intensities used here were higher than those in the preceding research. As a result, the photoinhibition process transpired, prompting mutilation of the microalgal cells while degrading the lipid yield.

Table 2. Microalgal lipid accumulation responses to various visible spectra of irradiation.

Visible Spectrum	Intensity (μmol/m^2 s)	Microalgal Species	Stimulating Effect in Decreasing Order	Reference
White, red, blue, green, yellow	100	*Chlorella vulgaris*	Green > red > yellow > white > blue	Current study
White, red, blue, green	60	*Botryococcus braunii*	Red > white > blue > green	[23]
White, red, blue, green, yellow	- - -	*Chlorella vulgaris* *Tetradesmus obliquus* *Arthospira platensis*	Blue > red > white > green > yellow Blue > red > white > green > yellow Blue > red > white > green > yellow	[23]
White, red, yellow	-	*Phaeodactylum tricornutum*	Red > white	[24]
White, red, blue, green	100	*Chlorella sp.* *Nannochloris oculata*	Blue > red > white > green Blue > white > red > green	[5]
White Red Blue Blue-green Pink	100 75 15 20 85	*Nannochloropsis* sp.	Blue > white > blue-green > pink > red	[31]
White, red, blue	70	*Streptomyces. bacillaris*	Red > white > blue	[32]
White, red, blue	50	*Chlorella vulgaris*	Blue > white > red	[30]
White, red, blue	104	*Isochrysis galbana*	Blue > white > red	[25]

3.4. Protein and Carbohydrate Contents of Attached Microalgae Grown on Palm Decanter Cake under Various Visible Spectra of Irradiation

Apart from the attached microalgal lipids, other cellular biochemicals, including carbohydrate and protein, were also affected by changes in the visible spectra. Microalgal carbohydrates can serve as a source of bio-alcohols, such as bioethanol and biobutanol [33]. On the other hand, protein from microalgal feedstocks can be extracted for bio-oil, animal feed, medicine, and pigment production [34]. Figure 3 depicts the protein and carbohydrate contents of attached microalgae upon having been exposed to various visible spectra of irradiation. The blue and white light imparted the greatest protein content enhancement (0.341 g/L), and microalgae under white light accumulated the highest carbohydrate content of 0.919 g/L. This trend was inversely proportional to the lipid accumulation (Figure 2), implying that less of the photosynthetic flow of carbon and energy from protein and carbohydrate was converted into the biosynthesis of lipids. Moreover, this could also engender the highest nitrogen uptake from palm decanter cake for conversion into protein, and the highest carbon dioxide assimilation for protein as well as carbohydrate production within microalgal cells. Conversely, green light cultivation accumulated the least protein and carbohydrate content, amounting to only 0.307 and 0.806 g/L, respectively. In this case, the lipids were a better energy reserve than carbohydrate under green light irradiation, i.e., approximately 2.25 times higher in energy value than carbohydrate. Thus, it was shown that green light could trigger the microalgae to build an efficient energy and carbon storage system as opposed to other visible irradiation spectra, offering the greatest sink for available energy accumulation [35].

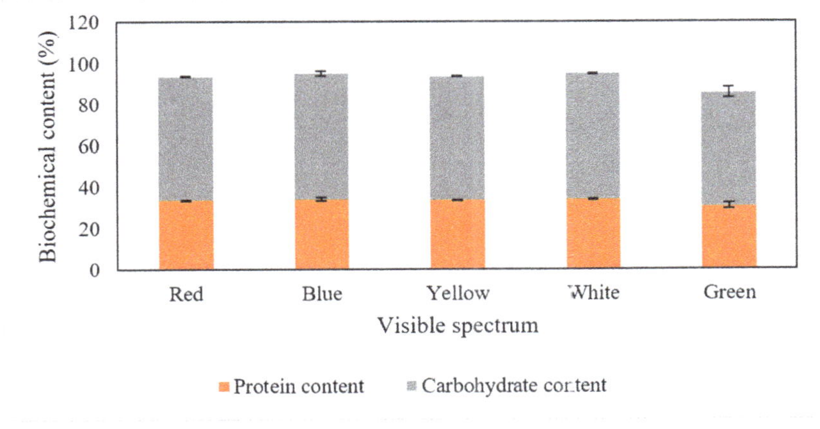

Figure 3. Protein and carbohydrate content of attached microalgae growing on palm decanter cake at various visible spectra of irradiation.

Table 3 shows that the carbohydrate content recorded in the present research was the highest, 0.91 % higher than that found in a study using a glass slide as the surface for attachment. This could be justified by the earlier analysis, which revealed that a bulk material with irregular particle morphologies and tiny particles on the surface of the palm decanter cake was more desirable for microalgae forming an attachment. This was because the rough surface improved the microalgal biomass retention as opposed to the glass slide, which only had a hydrophilic surface [36,37]. As a result, the attached microalgae would thrive by assimilating nutrients derived only from palm decanter cake. Nonetheless, the microalgal attachment on the palm decanter cake experienced protein deficiencies. This was primarily accentuated by the presence of only 2.78% nitrogen content in the palm decanter cake biomass, leading to limited nitrogen uptake. Nevertheless, the fact that the biochemicals derived from the microalgae populating the palm decanter cake could be improved by optimizing other cultivation variables must not be overlocked.

Table 3. Protein and carbohydrate accumulation derived from attached microalgal biomasses growing on various supporting materials.

Supporting Material	Microalgal Species	Protein Content (%)	Carbohydrate Content (%)	Reference
Palm decanter cake	*Chlorella vulgaris*	34.10 ± 0.65	60.91 ± 1.15	Current study
Spent coffee grounds	*Chlorella pyremoidosa*	20.36 ± 1.13	20.69 ± 0.89	[38]
Cottom, linen, mohair	*Scenedesmus.* sp	50.10 ± 10.1	37.60 ± 10.5	[39]
Walnut shells	*Chlorella vulgaris*	27.70 ± 1.25	34.50 ± 1.31	[40]
	Scenedesmus obliquus	34.80 ± 0.55	32.50 ± 0.59	
Cellulose acetate membrane	*Chlorella vulgaris*	52.84 ± 2.50	37.93 ± 1.15	[41]
Glass slide	Consortia	12.00 ± 1.13	60.00 ± 2.07	[42]
Cellulose acetate–nitrate membrane	*Nannochloris oculata*	19.20 ± 2.59	9.10 ± 0.49	[43]
	Chlorella sp.	19.20 ± 2.57	3.70 ± 0.24	
	Chlorella pyrenoidosa	31.00 ± 2.01	9.00 ± 0.50	

3.5. Biodiesel Profile Derived from Attached Microalgal Grown on Palm Decanter Cake under Various Visible Spectra of Irradiation

Biodiesel, one of the major biofuels produced from microalgal lipid feedstock as a substitute for non-renewable fossil fuel, comprises numerous categories and varieties of fatty acid methyl esters (FAME), which will dictate the quality of the biodiesel. As demonstrated in Figure 4, each visible spectrum for cultivating microalgal biomasses had the highest composition of monounsaturated fatty acids (MUFA), except for white light cultivation, which had the highest content of saturated fatty acids (SFAs) (70.38%). This was possibly a result of the high consumption of MUFAs for microalgal metabolism and growth (Figure 1). MUFAs have a single double bond that is more easily broken down than SFAs, which do not have any double bonds in releasing the energy. On the other hand, PUFAs have been reported to be susceptible to oxidative damage from the presence of reactive oxidative species [9]. The increases in SFAs and PUFAs were proportional to the wavelength of the visible spectrum, implying that the photo-enzymes in the microalgal species were activated by the longer light wavelengths and synthesized more fatty acids, particularly SFAs [30,32]. Less of the excitation energy would be absorbed by the cells to impede photodamage on the acceptors of Photosystem II. Nevertheless, the highest percentage of PUFAs was observed under blue light cultivation as a response to the microalgal enzymes in the cells. It was also discovered that white light produced the lowest biomass (Figure 1) but produced highest SFA and lowest MUFA + PUFA grade biodiesel (Figure 4). This could be because irradiation under white light triggered the stimulation of the photoreceptors, boosting the chlorophyll density and resulting in higher growth. It has also been mentioned in the literature that photochromic stress may sometimes lead to changes in the photosystems, thereby affecting the ultrastructure of the thylakoid organelles in microalgal cells and thus affecting the biodiesel profile [44].

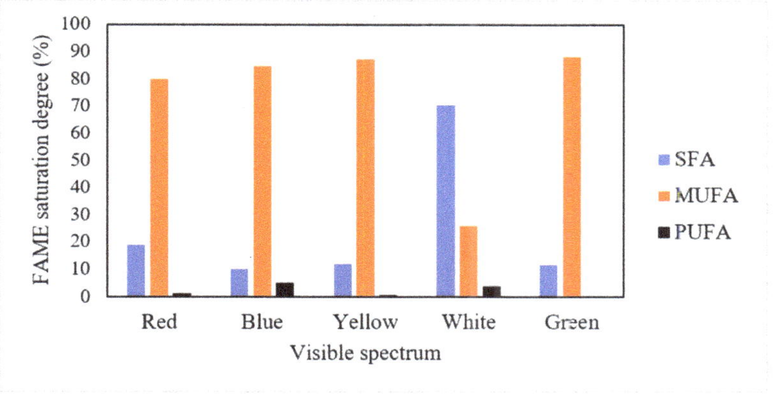

Figure 4. FAME saturation degrees of attached microalgae growing on palm decanter cake at various visible spectra of irradiation.

A study revealed that the general components of biodiesel are C10:0 (capric), C12:0 (lauric), C14:0 (myristic), C16:0 (palmitic), C16:1 (palmitoleic), C18:0 (stearic), C18:1 (oleic), C18:2 (linoleic), C18:3 (linolenic), C20:0 (arachidic), C20:1 (paulinnic) and C22:1 (erucic) [45]. The FAME species of C16 and C18, namely, palmitic (C16:0), stearic (C18:0), oleic (C18:1), linoleic (C18:2) and linolenic (C18:3) acid are best suited for biodiesel production [2]. Chhandama et al. [1] affirmed that the highest cetane number was discovered in C16:0 and C18:0 fatty acids. This refined ignition quality is needed to ensure cold start performance and reduce the development of white smoke. The longer the carbon chain of the fatty acids and the more saturated the molecules, the higher the cetane number will be [46]. Moreover, biodiesel that was enriched with SFAs would have high oxidative stability against reactive oxidation species, despite having low-temperature properties that could potentially clog the biodiesel fuel filters [2,46]. On that note, the MUFAs such as C16:1, C18:1, C20:1, and C22:1 would produce a superior cold flow. Even though the PUFA-loaded biodiesel exhibited excellent cold flow performance, it was vulnerable to the oxidation process [3]. C18:2 and C18:3 have low melting points which barely form crystals at cold temperatures. This avoids the blockage of fuel filters that is commonly due to the formation of cloudy waxlike structures, and which cause poor engine operation [47]. In the present study, the total SFA content of 70.38%, including C16:0, C18:0 and C24:0, indicates the high cetane number and oxidation stability for preventing the formation of gums, deposits, and sediment (Table 4). The cultivation of attached *Chlorella vulgaris* on palm decanter cake yielded 3.65% total PUFAs, which barely affected the oxidation stability of the biodiesel. Aside from that, this biodiesel also showed moderate cold flow characteristics, as reflected by the low fraction of MUFAs (25.97%) and PUFAs. Since the biodiesel did not contain any C18:2 and C18:3, it was proposed to be blended with other microalgal lipids with a preponderance of PUFAs to enhance the FAME composition and thus better biodiesel quality derived from attached microalgal biomass [48]. A comparative analysis was carried out to evaluate the FAME compositions of various studies that used different microalgal strains cultivated in distinct media, as tabulated in Table 5. It can be noted that the highest degree of SFA was achieved with the FAME profile in the present study, authenticating the exceptional oxidative stability of biodiesel for longer storage.

Table 4. Overall FAME profile derived from attached microalgal biomass grown on palm decanter cake.

FAME Species	FAME Content (%)	Saturation Degree
Methyl palmitate (C 16:0)	51.36	SFA
Methyl stearate (C18:0)	5.36	SFA
Methyl lingnocerate (C24:0)	13.66	SFA
cis-Methyl oleate and trans-Methyl 9-octadecenoate (C18:1)	25.97	MUFA
Methyl cis-11,14,17-eicosatrienoate (C20:3)	1.88	PUFA
Methyl cis-5,8,11,14,17-eicosapentaenoate (C20:5)	1.77	PUFA

Table 5. FAME contents derived from various microalgal strains grown on various supporting materials and growth media.

Microalgal Species	Supporting Material/ Culture Medium	Total C16:0, C18:0, C18:1, C18:2, C18:3 (%)	SFA (%)	MUFA (%)	PUFA (%)	Reference
Chlorella vulgaris	Palm decanter cake	82.69 ± 1.40	70.38 ± 2.30	25.97 ± 1.12	3.65 ± 0.28	Current study
Chlorella vulgaris, Oscillatoria tenus, Scenedesmus obliquus	Wastewater	78.67	61.57	20.23	11.92	[49]
Nannochloropsis oculata	-	44.00 ± 4.30	43.03 ± 2.20	37.41 ± 3.40	15.12 ± 1.30	[50]
Thalassiosira weissflogii		29.69 ± 3.55	26.52 ± 4.65	34.25 ± 5.00	36.95 ± 4.50	
Chlorella vulgaris	Polyurethane foam	68.70	35.60	43.10	19.70	[3]
Chlorella vulgaris,	Walnut shells	77.85 ± 5.05	32.60 ± 3.96	2.85 ± 0.11	58.66 ± 5.67	[40]
Scenedesmus obliquus		78.67 ± 5.80	25.99 ± 2.53	15.34 ± 1.60	53.66 ± 4.81	
Chlorella vulgaris	Cotton	92.89 ± 0.93	32.37 ± 0.39	28.04 ± 0.40	39.57 ± 0.34	[51]
Dunaliella tertiolecta	-	97.8	32.17	19.45	47.25	[52]
Phaeodactylum tricornutum	-	18.69 ± 3.62	25.77 ± 3.04	31.65 ± 1.94	29.85 ± 5.18	[9]
Ettlia sp.	Porous membrane	90.10	23.60	29.20	43.50	[53]
Dunaliella salina		53.33	34.76	32.18	33.06	[54]
Graesiella emersonni	Vermicompost	82.70 ± 5.49	47.10 ± 2.43	15.30 ± 1.01	37.60 ± 1.66	[55]

3.6. Energy Feasibility of Lipid Production from Attached Microalgal Grown on Palm Decanter Cake under Various Visible Spectra of Irradiation

The evaluation of the net energy by life cycle analysis (LCA) encompassed the microalgal lipid production process, with the proposed flow diagram depicted in Figure 5. The functional unit was assigned as 1 g of lipid extracted from the attached microalgal biomass. Simapro® 8.4.0 was chosen to compute the energy requirements of each process. In microalgal cultivation, water, nutrient, light, and energy inputs were the primary starting concerns for biodiesel production from lab scale to pilot and industrial setups [32]. Figure 6 demonstrates that yellow and white light required the least electricity (0.324 kWh), while green light demanded the smallest amount of water (859.6 g) to produce 1 g of microalgal lipid. This was also supported by the highest lipid content being recorded under green light cultivation (Figure 2). The reduced water usage during cultivation would lead to a lower cost of microalgal dewatering, while minimizing the subsequent wastewater production. Compared with yellow light, which required more water to cultivate the attached microalgae (902.663 g), green light saved an additional 43.064 g of water per 1 g of lipid produced. Since yellow light cultivation led to the highest lipid yield per energy input, the list of energy demands of various processes to yield 1 g of lipid from attached microalgal cultivation is shown in Table 6. In this regard, a net energy input of 115.72 MJ was needed, of which 110.47 MJ (95.46%) stemmed from non-renewable fossil fuels due to the intensive electricity usage. However, the use of palm decanter cake was not incorporated in this assessment, because it was retrieved as a waste by-product to serve a dual role, namely, a platform and nutritional source for the attached microalgal. Leong et al. [56] claimed that the energy equivalent of 1 g of lipid is 35 MJ.

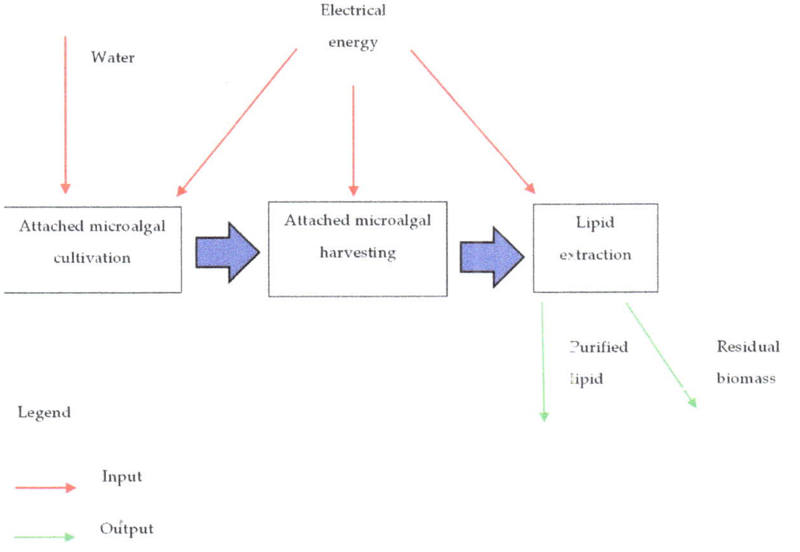

Figure 5. Process flow diagram for microalgal lipid production from attached microalgal growth on a palm decanter cake substrate.

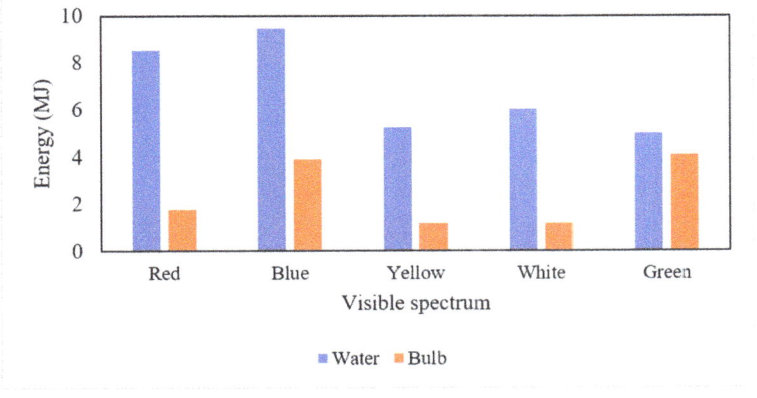

Figure 6. Energy inputs of water and light for growing attached microalgae on palm decanter cake at various visible spectra of irradiation.

Although the system was undesirable because the NER is lower than 1.0, it showed a slight improvement compared with previous research (Table 7) that utilized cool white fluorescent light to irradiate *Chlorella vulgaris* (NER = 0.27) [47]. In this study, light emitting diodes were used, which saved 30% of the power compared with the fluorescent counterpart [57]. Other studies also had NER values lower than 1.0, with the highest being 0.73 for using a mild hydrothermal treatment for lipid extraction that required less electricity (1.78×10^{-4} kWh). This process pumped the microalgal slurry into a reactor at 260 °C and 5 MPa to disintegrate the microalgal cell walls for lipid hydrolysis and solvent-free extraction in batch reactors for acquiring biocrude for biodiesel production. In a raceway pond, optimistic assumptions were presumed during the cultivation of *Haematococcus pluvialis* and *Nannochloropsis* sp., where CO_2 was supposed to be viable in an external carbon capture system coupled to the biodiesel production process. If the CO_2 from biomass combustion emissions was captured, greenhouse gas emissions would ensue. Even so, this study had the merit of highlighting the generation of microalgal residual biomass enriched with various biochemical compounds such as carbohydrate and protein; hence, this could be further exploited as a feedstock for various industries.

According to the author's knowledge, no comprehensive work has been dedicated to the attachment of microalgae onto solid PDC thus far, other than non-agricultural based inert supports. The utilization of low-cost PDC could concomitantly reduce waste from being generated, with downstream benefits for biodiversity and consumers. The utilization of various visible spectra could overcome the limitation of using natural sunlight to cultivate microalgae by modifying the duration of irradiation contingent on maximum microalgal lipid productivity and lipid yield. In the near future, the photoperiod could be adjusted to compare the microalgal lipid productivity and lipid yield while reducing the net energy ratio. Moreover, the current limitation of this study is caused by the maximal energy utilization of various visible spectra operating 24 h. In real applications, energy consumption can be plausibly reduced by utilizing solar energy during the daytime and visible light during the dark period. This would aid in the operational economic aspects during the scale-up process, materializing the exploitation of microalgal feedstock.

Table 6. Direct processes producing 1 g of lipid from attached microalgal biomass grown on palm decanter cake under irradiation with yellow light.

Microalgae Species	Cultivation System	NER	Reference
Chlorella vulgaris	Attached	0.30	Current study
Chlorella vulgaris	Suspended	0.27	[56]
Consortium	Suspended	0.73	[58]
Haematococcus pluvialis	Suspended	0.40	[59]
Nannochloropsis sp.		0.12	

Table 7. The NER values of different microalgal strains in various cultivation systems.

Process	Amount	Energy Equivalent (MJ)
Cultivation:		
Water	0.9 kg	5.25
Electricity	0.89 kWh	3.20
Harvesting:		
Electricity	26.36 kWh	94.90
Lipid extraction:		
Electricity	3.435 kWh	12.37
Total cumulative energy demand	-	115.72

4. Conclusions

Palm decanter cake (PDC) was found to be a suitable substratum that provided nutrients for microalgal growth and a platform for microalgal attachment. In the assessment of various visible spectrums used to grow attached microalgal biomass, white light was found to be the optimum spectrum for cultivating attached microalgae at a light intensity of 100 μmol/m^{2}s under continuous irradiation conditions. In this context, a microalgal density of 1.13 g/g ± 0.06 and microalgal productivity of 1.213 ± 0.06 g/L per day were achieve, along with a high cetane number and oxidation stability in the produced biodiesel. The feasibility of using PDC was further verified by the high protein and carbohydrate contents (34.061 ± 0.14% and 60.838 ± 0.15%) derived from mature microalgal biomass regardless of the spectrum. Conversely, the trade-off of utilizing the white spectrum was the low lipid content and lipid productivity, which could be offset by using green light. Despite that, the improved net energy ratio (0.3) was an indication that non-renewable fossil fuels could be substituted by microalgal fuel to deal with the global warming phenomenon.

Author Contributions: Conceptualization, Z.W.T., H.R., W.K. and J.W.L.; methodology, Z.W.T., H.R., W.K. and J.W.L.; validation, W.K. and J.W.L., formal analysis, Z.W.T. H.R. and W.H.L.; investigation, Z.W.T. and H.R.; resources, W.K. and J.W.L., data curation, Z.W.T. and H.R.; writing—original draft preparation, Z.W.T. and H.R.; writing—review and editing, W.H.L., C.S.L., Y.Y.W., E.A.A., W.K., A.U.R., W.Y.T. and J.W.L.; visualization, P.L.S. and Z.W.T.; supervision, J.W.L. and W.K.; project administration, J.W.L. and W.K.; funding acquisition, J.W.L. and W.K. All authors have read and agreed to the published version of the manuscript.

Funding: This work was financially supported by the Joint Research Project among UTP-UMT-UMP-UCTS with the cost center of 015MD0-019 and The Murata Science Foundation with the cost center of 015ME0-299.

Conflicts of Interest: The authors declare no conflict of interest.

Abbreviations

Palm decanter cake	PDC
Bold's Basal medium	BBM
Net energy ratio	NER
Light-emitting diode	LED
Photosynthetic active radiation	PAR
Fatty acid methyl ester	FAME
Monounsaturated fatty acid	MUFA
Polyunsaturated fatty acid	PUFA
Saturated fatty acid	SFA
Life cycle analysis	LCA

References

1. Chhandama, M.V.L.; Satyan, K.B.; Changmai, B.; Vanlalveni, C.; Rokhum, S.L. Microalgae as a Feedstock for the Production of Biodiesel: A Review. *Bioresour. Technol. Rep.* **2021**, *15*, 100771. [CrossRef]
2. Zhu, L.; Nugroho, Y.K.; Shakeel, S.R.; Li, Z.; Martinkauppi, B.; Hiltunen, E. Using Microalgae to Produce Liquid Transportation Biodiesel: What Is Next? *Renew. Sustain. Energy Rev.* **2017**, *78*, 391–400. [CrossRef]
3. Mohd-Sahib, A.A.; Lim, J.W.; Lam, M.K.; Uemura, Y.; Isa, M.H.; Ho, C.D.; Kutty, S.R.M.; Wong, C.Y.; Rosli, S.S Lipid for Biodiesel Production from Attached Growth Chlorella Vulgaris Biomass Cultivating in Fluidized Bed Bioreactor Packed with Polyurethane Foam Material. *Bioresour. Technol.* **2017**, *239*, 127–136. [CrossRef]
4. Daneshvar, E.; Sik Ok, Y.; Tavakoli, S.; Sarkar, B.; Shaheen, S.M.; Hong, H.; Luo, Y.; Rinklebe, J.; Song, H.; Bhatnagar, A. Insights into Upstream Processing of Microalgae: A Review. *Bioresour. Technol.* **2021**, *329*, 124870. [CrossRef]
5. Yuan, H.; Zhang, X.; Jiang, Z.; Wang, X.; Wang, Y.; Cao, L.; Zhang, X. Effect of Light Spectra on Microalgal Biofilm: Cell Growth, Photosynthetic Property, and Main Organic Composition. *Renew. Energy* **2020**, *157*, 83–89. [CrossRef]
6. Dalirian, N.; Abedini Najafabadi, H.; Movahedirad, S. Surface Attached Cultivation and Filtration of Microalgal Biofilm in a Ceramic Substrate Photobioreactaor. *Algal Res.* **2021**, *55*, 102239. [CrossRef]
7. Embrandiri, A.; Fatemeh Rupani, P.; Ahmed Ismail, S.; Pratap Singh, R.; Hakimi Ibrahim, M.; Omar Abd Kadir, M. The Effect of Oil Palm Decanter Cake on the Accumulation of Nutrients and the Stomatal Opening of Solanum Melongena (Brinjal) Plants. *Int. J. Recycl. Org. Waste Agric.* **2016**, *5*, 141–147. [CrossRef]
8. Abomohra, A.E.F.; Shang, H.; El-Sheekh, M.; Eladel, H.; Ebaid, R.; Wang, S.; Wang, Q. Night Illumination Using Monochromatic Light-Emitting Diodes for Enhanced Microalgal Growth and Biodiesel Production. *Bioresour. Technol.* **2019**, *288*, 121514. [CrossRef]
9. Maltsev, Y.; Maltseva, K.; Kulikovskiy, M.; Maltseva, S. Influence of Light Conditions on Microalgae Growth and Content of Lipids, Carotenoids, and Fatty Acid Composition. *Biology* **2021**, *10*, 1060. [CrossRef]
10. Rawindran, H.; Leong, W.H.; Suparmaniam, U.; Liew, C.S.; Raksasat, R.; Kiatkittipong, W.; Mohamad, M.; Ghani, N.A.; Abdelfattah, E.A.; Lam, M.K.; et al. Residual Palm Kernel Expeller as the Support Material and Alimentation Provider in Enhancing Attached Microalgal Growth for Quality Biodiesel Production. *J. Environ. Manag.* **2022**, *316*, 115225. [CrossRef]
11. Hemamalini, R.; Lim, J.W.; Leong, W.H.; Chew, T.L.; Lam, M.K.; Uganeeswary, S.; Mohamad, M.; Almendrala, M.C.; Shamsuddin, R. Utilization of Solid Palm Kernel Expeller for Attached Growth of *Chlorella vulgaris* Sp. In *Proceedings of the AIP Conference Proceedings*; American Institute of Physics Inc.: College Park, MD, USA, 9 June 2022; Volume 2454. [CrossRef]
12. Zhong, Y.; Jin, P.; Cheng, J.J. A Comprehensive Comparable Study of the Physiological Properties of Four Microalgal Species under Different Light Wavelength Conditions. *Planta* **2018**, *248*, 489–498. [CrossRef] [PubMed]

13. Boni, J.; Aida, S.; Leila, K.; Selmani, N.; Mirghani, M.E.S.; Alam, M.Z. To Cite This Article: Nabila Selmani et Al. *IOP Conf. Ser. Earth Environ. Sci.* **2013**, *16*, 12006. [CrossRef]
14. Leong, W.H.; Lim, J.W.; Lam, M.K.; Lam, S.M.; Sin, J.C.; Samson, A. Novel Sequential Flow Baffled Microalgal-Bacterial Photobioreactor for Enhancing Nitrogen Assimilation into Microalgal Biomass Whilst Bioremediating Nutrient-Rich Wastewater Simultaneously. *J. Hazard. Mater.* **2021**, *409*, 124455. [CrossRef] [PubMed]
15. Gutierrez-Wing, M.T.; Silaban, A.; Barnett, J.; Rusch, K.A. Light Irradiance and Spectral Distribution Effects on Microalgal Bioreactors. *Eng. Life Sci.* **2014**, *14*, 574–580. [CrossRef]
16. Adam, M.A.; Sulaiman, A.; Said, C.M.S.; Som, A.M.; Tabatabaei, M. Enhanced Rigidity of Natural Polymer Composite Developed from Oil Palm Decanter Cake. *BioResources* **2015**, *10*, 932–942. [CrossRef]
17. Choi, H.J. Agricultural Biowaste, Rice Bran, as Carbon Source to Enhance Biomass and Lipid Production: Analysis with Various Growth Rate Models. *Water Sci. Technol.* **2020**, *82*, 1120–1130. [CrossRef]
18. Lessard, P.; Bihan, Y. le Fixed Film Processes. In *Handbook of Water and Wastewater Microbiology*; Academic Press: Cambridge, MA, USA; Elsevier: Cambridge, MA, USA, 2003; pp. 317–336. [CrossRef]
19. Metsoviti, M.N.; Papapolymerou, G.; Karapanagiotidis, I.T.; Katsoulas, N. Effect of Light Intensity and Quality on Growth Rate and Composition of Chlorella Vulgaris. *Plants* **2020**, *9*, 31. [CrossRef]
20. Blair, M.F.; Kokabian, B.; Gude, V.G. Light and Growth Medium Effect on Chlorella Vulgaris Biomass Production. *J. Environ. Chem. Eng.* **2014**, *2*, 665–674. [CrossRef]
21. Yun, Y.S.; Park, J.M. Attenuation of Monochromatic and Polychromatic Lights in Chlorella Vulgaris Suspensions. *Appl. Microbiol. Biotechnol.* **2001**, *55*, 765–770. [CrossRef]
22. Lysenko, V.; Kosolapov, A.; Usova, E.; Tatosyan, M.; Varduny, T.; Dmitriev, P.; Rajput, V.; Krasnov, V.; Kunitsina, A. Chlorophyll Fluorescence Kinetics and Oxygen Evolution in Chlorella Vulgaris Cells: Blue vs. Red Light. *J. Plant Physiol.* **2021**, *258–259*, 153392. [CrossRef]
23. Acuapan-Hernandez, J.; Cañizares-Villanueva, R.O.; Cristiani-Urbina, E. Red Light and Nitrogen Depletion Stimulate the Synthesis of Lipids and N-Alkadienes Susceptible to Be Used as Biofuels in Botryococcus Braunii UTEX 2441 (Race A). *Biotechnol. Ind. J.* **2017**, *13*, 155.
24. Sharma, N.; Fleurent, G.; Awwad, F.; Cheng, M.; Meddeb-Mouelhi, F.; Budge, S.M.; Germain, H.; Desgagné-Penix, I. Red Light Variation an Effective Alternative to Regulate Biomass and Lipid Profiles in Phaeodactylum Tricornutum. *Appl. Sci.* **2020**, *10*, 2531. [CrossRef]
25. Yoshioka, M.; Yago, T.; Yoshie-Stark, Y.; Arakawa, H.; Morinaga, T. Effect of High Frequency of Intermittent Light on the Growth and Fatty Acid Profile of Isochrysis Galbana. *Aquaculture* **2012**, *338–341*, 111–117. [CrossRef]
26. Belaiba, A.; Bouharat, D.; Malvis, A.; Hodaifa, G. Feasibility of the Hybrid Use of Chlorella Vulgaris Culture with the Conventional Biological Treatment in Urban Wastewater Treatment Plants. *Processes* **2021**, *9*, 1640. [CrossRef]
27. Rodolfi, L.; Zittelli, G.C.; Bassi, N.; Padovani, G.; Biondi, N.; Bonini, G.; Tredici, M.R. Microalgae for Oil: Strain Selection, Induction of Lipid Synthesis and Outdoor Mass Cultivation in a Low-Cost Photobioreactor. *Biotechnol. Bioeng.* **2009**, *102*, 100–112. [CrossRef]
28. Chen, T.; Liu, J.; Guo, B.; Ma, X.; Sun, P.; Liu, B.; Chen, F. Light Attenuates Lipid Accumulation While Enhancing Cell Proliferation and Starch Synthesis in the Glucose-Fed Oleaginous Microalga Chlorella Zofingiensis. *Sci. Rep.* **2015**, *5*, 14936. [CrossRef]
29. Muñoz, I.L.; Bernard, O. Modeling the Influence of Temperature, Light Intensity and Oxygen Concentration on Microalgal Growth Rate. *Processes* **2021**, *9*, 496. [CrossRef]
30. Wong, Y. Effect of Different Light Sources on Algal Biomass and Lipid Production in Internal Leds-Illuminated Photobioreactor. *J. Mar. Biol. Aquac.* **2016**, *2*, 8. [CrossRef]
31. Vadiveloo, A.; Moheimani, N.R.; Cosgrove, J.J.; Bahri, P.A.; Parlevliet, D. Effect of Different Light Spectra on the Growth and Productivity of Acclimated Nannochloropsis Sp. (Eustigmatophyceae). *Algal Res.* **2015**, *8*, 121–127. [CrossRef]
32. Mutaf, T.; Oz, Y.; Kose, A.; Elibol, M.; Oncel, S.S. The Effect of Medium and Light Wavelength towards Stichococcus Bacillaris Fatty Acid Production and Composition. *Bioresour. Technol.* **2019**, *289*, 121732. [CrossRef]
33. Sánchez-Zurano, A.; Ciardi, M.; Lafarga, T.; Fernández-Sevilla, J.M.; Bermejo, R.; Molina-Grima, E. Role of Microalgae in the Recovery of Nutrients from Pig Manure. *Processes* **2021**, *9*, 11. [CrossRef]
34. Siddiki, S.Y.A.; Mofijur, M.; Kumar, P.S.; Ahmed, S.F.; Inayat, A.; Kusumo, F.; Badruddin, I.A.; Khan, T.M.Y.; Nghiem, L.D.; Ong, H.C.; et al. Microalgae Biomass as a Sustainable Source for Biofuel, Biochemical and Biobased Value-Added Products: An Integrated Biorefinery Concept. *Fuel* **2022**, *307*, 121782. [CrossRef]
35. He, Q.; Yang, H.; Wu, L.; Hu, C. Effect of Light Intensity on Physiological Changes, Carbon Allocation and Neutral Lipid Accumulation in Oleaginous Microalgae. *Bioresour. Technol.* **2015**, *191*, 219–228. [CrossRef]
36. Wang, J.H.; Zhuang, L.L.; Xu, X.Q.; Deantes-Espinosa, V.M.; Wang, X.X.; Hu, H.Y. Microalgal Attachment and Attached Systems for Biomass Production and Wastewater Treatment. *Renew. Sustain. Energy Rev.* **2018**, *92*, 331–342. [CrossRef]
37. Maniam, G.P.; Hindryawati, N.; Nurfitri, I.; Jose, R.; Mohd, M.H.; Dahalan, F.A.; Yusoff, M.M. Decanter Cake as a Feedstock for Biodiesel Production: A First Report. *Energy Convers. Manag.* **2013**, *76*, 527–532. [CrossRef]
38. Abomohra, A.E.F.; Zheng, X.; Wang, Q.; Huang, J.; Ebaid, R. Enhancement of Biodiesel Yield and Characteristics through In-Situ Solvo-Thermal Co-Transesterification of Wet Microalgae with Spent Coffee Grounds. *Bioresour. Technol.* **2021**, *323*, 124640. [CrossRef] [PubMed]

39. Lin-Lan, Z.; Jing-Han, W.; Hong-Ying, H. Differences between Attached and Suspended Microalgal Cells in SsPBR from the Perspective of Physiological Properties. *J. Photochem. Photobiol. B Biol.* **2018**, *181*, 164–169. [CrossRef]
40. Zou, X.; Xu, K.; Chang, W.; Qu, Y.; Li, Y. A Novel Microalgal Biofilm Reactor Using Walnut Shell as Substratum for Microalgae Biofilm Cultivation and Lipid Accumulation. *Renew. Energy* **2021**, *175*, 676–685. [CrossRef]
41. Yu, H.; Zhuang, L.L.; Zhang, M.; Zhang, J. The Mechanism Study of Attached Microalgae Cultivation Based on Reverse Osmosis Concentrated Water (WROC). *Resour. Conserv. Recycl.* **2021**, *179*, 106066. [CrossRef]
42. Palma, H.; Killoran, E.; Sheehan, M.; Berner, F.; Heimann, K. Assessment of Microalga Biofilms for Simultaneous Remediation and Biofuel Generation in Mine Tailings Water. *Bioresour. Technol.* **2017**, *234*, 327–335. [CrossRef]
43. Zhang, X.; Yuan, H.; Guan, L.; Wang, X.; Wang, Y.; Jiang, Z.; Cao, L.; Zhang, X. Influence of Photoperiods on Microalgae Biofilm: Photosynthetic Performance, Biomass Yield, and Cellular Composition. *Energies* **2019**, *12*, 3724. [CrossRef]
44. Valchev, D.; Ribarova, I. A Review on the Reliability and the Readiness Level of Microalgae-Based Nutrient Recovery Technologies for Secondary Treated Effluent in Municipal Wastewater Treatment Plants. *Processes* **2022**, *10*, 399. [CrossRef]
45. Ramos, M.J.; Fernández, C.M.; Casas, A.; Rodríguez, L.; Pérez, Á. Influence of Fatty Acid Composition of Raw Materials on Biodiesel Properties. *Bioresour. Technol.* **2009**, *100*, 261–268. [CrossRef]
46. Miraboutalebi, S.M.; Kazemi, P.; Bahrami, P. Fatty Acid Methyl Ester (FAME) Composition Used for Estimation of Biodiesel Cetane Number Employing Random Forest and Artificial Neural Networks: A New Approach. *Fuel* **2016**, *166*, 143–151. [CrossRef]
47. Folayan, A.J.; Anawe, P.A.L.; Aladejare, A.E.; Ayeni, A.O. Experimental Investigation of the Effect of Fatty Acids Configuration, Chain Length, Branching and Degree of Unsaturation on Biodiesel Fuel Properties Obtained from Lauric Oils, High-Oleic and High-Linoleic Vegetable Oil Biomass. *Energy Rep.* **2019**, *5*, 793–806. [CrossRef]
48. Rosli, S.S.; Wong, C.Y.; Yunus, N.M.; Lam, M.K.; Show, P.L.; Cheng, C.K.; Wang, D.K.; da Oh, W.; Lim, J.W. Optimum Interaction of Light Intensity and CO2 Concentration in Bioremediating N-Rich Real Wastewater via Assimilation into Attached Microalgal Biomass as the Feedstock for Biodiesel Production. *Process Saf. Environ. Prot.* **2020**, *141*, 355–365. [CrossRef]
49. Zhang, Q.; Li, X.; Guo, D.; Ye, T.; Xiong, M.; Zhu, L.; Liu, C.; Jin, S.; Hu, Z. Operation of a Vertical Algal Biofilm Enhanced Raceway Pond for Nutrient Removal and Microalgae-Based Byproducts Production under Different Wastewater Loadings. *Bioresour. Technol.* **2018**, *253*, 323–332. [CrossRef]
50. Borges, L.; Morón-Villarreyes, J.A.; D'Oca, M.G.M.; Abreu, P.C. Effects of Flocculants on Lipid Extraction and Fatty Acid Composition of the Microalgae Nannochloropsis Oculata and Thalassiosira Weissflogii. *Biomass Bioenergy* **2011**, *35*, 4449–4454. [CrossRef]
51. Gross, M.; Henry, W.; Michael, C.; Wen, Z. Development of a Rotating Algal Biofilm Growth System for Attached Microalgae Growth with in Situ Biomass Harvest. *Bioresour. Technol.* **2013**, *150*, 195–201. [CrossRef] [PubMed]
52. Tizvir, A.; Shojaeefard, M.H.; Zahedi, A.; Molaeimanesh, G.R. Performance and Emission Characteristics of Biodiesel Fuel from Dunaliella Tertiolecta Microalgae. *Renew. Energy* **2022**, *182*, 552–561. [CrossRef]
53. Kim, S.; Moon, M.; Kwak, M.; Lee, B.; Chang, Y.K. Statistical Optimization of Light Intensity and CO_2 Concentration for Lipid Production Derived from Attached Cultivation of Green *Microalga ettlia* Sp. *Sci. Rep.* **2018**, *8*, 15390. [CrossRef] [PubMed]
54. Fakhry, E.M.; Maghraby, D.M. Fatty Acids Composition and Biodiesel Characterization of Dunaliella salina. *J. Water Resour. Prot.* **2013**, *05*, 894–899. [CrossRef]
55. Santhana Kumar, V.; das Sarkar, S.; Das, B.K.; Sarkar, D.J.; Gogoi, P.; Maurye, P.; Mitra, T.; Talukder, A.K.; Ganguly, S.; Nag, S.K.; et al. Sustainable Biodiesel Production from Microalgae Graesiella Emersonii through Valorization of Garden Wastes-Based Vermicompost. *Sci. Total Environ.* **2022**, *807*, 150995. [CrossRef] [PubMed]
56. Leong, W.H.; Azella Zaine, S.N.; Ho, Y.C.; Uemura, Y.; Lam, M.K.; Khoo, K.S.; Kiatkittipong, W.; Cheng, C.K.; Show, P.L.; Lim, J.W. Impact of Various Microalgal-Bacterial Populations on Municipal Wastewater Bioremediation and Its Energy Feasibility for Lipid-Based Biofuel Production. *J. Environ. Manag.* **2019**, *249*, 109384. [CrossRef] [PubMed]
57. Ryckaert, W.R.; Smet, K.A.G.; Roelandts, I.A.A.; van Gils, M.; Hanselaer, P. Linear LED Tubes versus Fluorescent Lamps: An Evaluation. *Energy Build.* **2012**, *49*, 429–436. [CrossRef]
58. Huang, R.; Li, J.; Tang, Y.; Song, W.; Yu, Y.; Yang, W.; Cheng, J. Comparative Life-Cycle Assessment of Microalgal Biodiesel Production via Various Emerging Wet Scenarios: Energy Conversion Characteristics and Environmental Impacts. *Energy Convers. Manag.* **2022**, *257*, 115427. [CrossRef]
59. Razon, L.F.; Tan, R.R. Net Energy Analysis of the Production of Biodiesel and Biogas from the Microalgae: Haematococcus Pluvialis and Nannochloropsis. *Appl. Energy* **2011**, *88*, 3507–3514. [CrossRef]

processes MDPI

Article

Dechlorination of Polyvinyl Chloride via Solvothermal Treatment with Glycerol

Katsuki Kusakabe [1,*], Anna Nagai [2], Wai Hong Leong [3,*], Kouki Yamasaka [1], Takuro Nakaaki [1], Yoshimitsu Uemura [4] and Kazutoshi Ikenaga [1]

[1] Department of Nanoscience, Sojo University, Kumamoto 860-0082, Japan
[2] Division of Materials Science and Chemistry, Faculty of Advanced Science and Technology, Kumamoto University, Kumamoto 860-8555, Japan
[3] Department of Fundamental and Applied Sciences, HICoE-Centre for Biofuel and Biochemical Research, Institute of Self-Sustainable Building, Universiti Teknologi PETRONAS, Seri Iskandar 32610, Perak Darul Ridzuan, Malaysia
[4] Department of Chemical Engineering, Universiti Teknologi PETRONAS, Seri Iskandar 32610, Perak Darul Ridzuan, Malaysia
* Correspondence: kusakabe@nano.sojo-u.ac.jp (K.K.); wai_18003501@utp.edu.my (W.H.L.)

Abstract: Solvothermal treatment using glycerol effectively removes chlorine from polyvinyl chloride (PVC). Additive-free PVC was dechlorinated by treatment with glycerol in an autoclave at 200–240 °C. Liquid glycerol was coexistent with a red powder after the reaction at 200 °C. At reaction temperatures of 220 and 240 °C, the liquid glycerol disappeared, and carbonization proceeded to form a black powder. In this reaction, the glycerol was directly converted into glycidol. Epichlorohydrin was also produced from the glycerol via 1,3-dichloro-2-propanol as an intermediate. The PVC was converted into the polyethylene oxide structure, following the generation of the polyene structure by zip-elimination. Finally, the converted PVC was resinified by an epoxidation reaction with the glycidol and epichlorohydrin. Thus, glycerol acted as a useful reaction medium for this reaction system. The analysis of the decomposition ratio and dechlorination yield showed that the PVC was converted into solid products in high yields via chlorine release, and 30% of the glycerol was solidified by resinification at 240 °C. The effects of the additives on dechlorination were investigated using commercially available PVC sheets. Elemental analysis showed that the calorific value of the solid product dechlorinated at 240 °C was found to be higher than that of standard bituminous coal.

Keywords: waste plastics; dechlorination; solvothermal treatment; glycerol; polyvinyl chloride

Citation: Kusakabe, K.; Nagai, A.; Leong, W.H.; Yamasaka, K.; Nakaaki, T.; Uemura, Y.; Ikenaga, K. Dechlorination of Polyvinyl Chloride via Solvothermal Treatment with Glycerol. *Processes* **2022**, *10*, 2047. https://doi.org/10.3390/pr10102047

Academic Editor: Aneta Magdziarz

Received: 7 September 2022
Accepted: 6 October 2022
Published: 11 October 2022

1. Introduction

Polyvinyl chloride (PVC) has been used in various fields, such as packaging materials, piping materials, agricultural films, and electric wires, because of its low cost and high performance [1]. Therefore, a large amount of waste, including PVC, is generated every year and requires disposal. Thermal recycling (pyrolysis, gasification, hydrogenation, etc.) and incineration are common waste treatment methods for PVC. Owing to the high chlorine content of PVC (56.8%), the incineration method requires an advanced facility for treating the exhaust gas, which contains large amounts of hydrogen chloride and trace amounts of harmful dioxins. When PVC is disposed of in landfills, phthalates and heavy metals contained in the plasticizers and stabilizers are leached, causing environmental problems. Therefore, a highly efficient method for dechlorinating PVC is required. Yu et al. [2] reviewed the recent advances in PVC pyrolysis. The dechlorination of PVC, which is affected by the characteristics of the PVC and its additives, mainly occurred in the temperature range of 250–320 °C [3]. The results of thermogravimetric analysis–mass spectrometry, the thermal decomposition of the PVC was found to be clearly divided into three stages, and dechlorination occurred mainly in the first and second stages [4,5]. Many methods have been used to improve the efficiency of chlorine removal from PVC. Yuan et al. [6] used a gas–liquid fluidized bed to clarify the dechlorination behavior and reported that the chlorine removal ratio was greatly affected by the temperature; this method was able to remove up to 99.5% of the chlorine in 1 min. at 300 °C. Wu et al. [7] investigated the co-pyrolysis behavior of polyethylene, polyvinyl chloride, and polystyrene and reported that the chlorination of the PVC and the cyclization of the conjugated polyenes occurred simultaneously to produce dioxins. Hydrothermal treatments are widely used to generate energy from waste [8–10], in which supercritical water is employed as a useful reaction medium [11]. Endo and Emori [12] reported that chlorine was completely removed by hydrothermal treatment at 19.3 MPa at 300 °C. NaOH is often used to promote the dechlorination of PVC under mild conditions. Shin et al. [13] reported that the addition of NaOH reduced the hydrothermal treatment temperature from 300 °C to 150–250 °C. Dimethyl sulfoxide [14] and ethylene glycol [15] were added to further promote dechlorination with NaOH. However, in these methods, dechlorination was improved by the addition of an organic solvent, but solvent recycling to reduce costs becomes an additional problem.

Recently, it was reported that PVC was effectively dechlorinated by hydrothermal carbonization (HTC) [1]. Furthermore, the hydrothermal co-carbonization (Co-HTC) of biomass and PVC at temperatures of 200–260 °C has been studied, where the OH groups of the lignocellulosic biomass promoted dechlorination [16–20]. The Co-HTC process can provide dechlorinated char, but the organic matter must be recovered from the wastewater, followed by wastewater treatment.

Crude glycerol is a by-product in the production of biodiesel from waste cooking oil, and its production is equivalent to approximately 10% of all biodiesel production. Purification is required to use crude glycerol as the raw material for chemical products. Glycerol purification using distillation is costly because of the high boiling point of glycerol, and most crude glycerol waste is used as fuel or incinerated. There is a demand for effective methods for utilizing crude glycerol without purification. Kusakabe et al. [21] reported the formation of biochar by the solvothermal co-carbonization of PVC and wood chips in glycerol. In this process, the dechlorination of the PVC was promoted by the reaction with the glycerol or wood chips. At a reaction temperature of 200–220 °C, unreacted liquid glycerol was coexistent with the biochar after the reaction, but when the reaction temperature was raised to 240 °C, the liquid substance disappeared, and a resinized biochar was formed. These results showed that glycerol had a significant influence as a reaction medium on dechlorination and carbonization.

In this study, to understand the details of the dechlorination of PVC by glycerol, the dechlorination of PVC powder in glycerol was conducted at a reaction temperature of 200–240 °C. The effects of additives in PVC on the decomposition ratio and dechlorination yield were investigated by using a PVC reagent and commercially available PVC.

2. Materials and Methods

2.1. Dechlorination of PVC

The powdery, additive-free PVC reagent (hereafter referred to as PVC-P) was purchased from FUJIFILM Wako Pure Chemical Corporation, Osaka, Japan. A commercially available PVC resin sheet with a thickness of 0.1 mm (hereafter referred to as PVC-S) was cut into small pieces of several millimeters in size for use. The additive contained in the PVC-S was partially removed by extraction with tetrahydrofuran (THF), and the resulting sample is hereafter referred to as PVC-ST. The THF and glycerol were purchased from FUJIFILM Wako Pure Chemical Corporation, Osaka, Japan. The PVC and glycerol were loaded into the decomposition container (25 mL) of an autoclave reactor (San-ai Kagaku Co. Ltd., Nagoya, Japan) in an N_2-filled glove box. The dechlorination of the PVC was conducted with the PVC/glycerol mass ratios of 4:3, 2:3, and 1:3 at 200, 220, and 240 °C for 30–120 min. The solid and liquid samples were recovered from the autoclave after dechlorination. The solid product was rinsed with ethanol to remove the glycerol solution and dried in a vacuum oven at 105 °C for 1 d. The decomposition ratio was calculated via the following equation:

$$\text{Decomposition ratio} = 1 - (\text{Mass of dried solid sample}/\text{Mass of feedstock (PVC and glycerol)}) \quad (1)$$

2.2. Characterization

The surface morphologies of the dried solid samples were observed using a scanning electron microscope (SEM, VE-9800, Keyence, Osaka, Japan). The presence of functional groups on the solid products was examined by scanning from 4000 to 650 cm^{-1} at a resolution of 1 cm^{-1} using a Fourier-transform infrared spectrometer (FTIR, Spectrum 100, PerkinElmer, Waltham, MA, USA). The chlorine content of the solid products was measured using the flask combustion method. The solid products (50 mg) were wrapped with filter paper and tied with a platinum wire. The tip of the filter paper was ignited externally, and combustion in oxygen was carried out in a closed separable flask with a sodium carbonate solution. The chlorine concentration in the sodium carbonate solution was determined using ion chromatography (Dionex Integrion HPIC system, Thermo Fisher Scientific, Waltham, MA, USA). The dechlorination yield of the solid product was calculated using the following equation:

$$\text{Dechlorination yield} = 1 - (\text{Cl content in solid product})/(\text{Cl content in feedstock}) \quad (2)$$

The elemental contents (C, H, and N) of the solid product were analyzed using an elemental analyzer (UNICUBE, Elementar, Langenselbold, Germany). The oxygen content was determined indirectly by the difference, when considering the Cl content. The higher heating value (HHV) is the amount of heat produced from the complete combustion of the sample, including the heat vaporization of the moisture in the sample and the H_2O produced from the reaction. In this study, the calorific value of the solid product was calculated from a developed correlation using the percentages of carbon, hydrogen, and oxygen from ultimate analysis, using the following equation [22]:

$$\text{HHV (MJ/kg)} = 0.3491\ \text{C} + 1.1783\ \text{H} - 0.1034\ \text{O} \quad (3)$$

3. Results and Discussion

3.1. Dechlorination of PVC-P

Figure 1 shows the physical appearance of the PVC-P pyrolyzed at 200–240 °C for 120 min. The remaining liquid glycerol from the product pyrolyzed at 200 °C was removed with ethanol, resulting in a red powder. Carbonization proceeded with the disappearance of the glycerol at reaction temperatures of 220 °C and 240 °C, forming a black powder. Figure 2 shows the surface morphologies of the pyrolyzed PVC-P. The unreacted PVC-P powder had a diameter of 100–150 μm. The PVC-P pyrolyzed at 200 °C expanded slightly, and surface melting was observed. At 220 °C, the dispersed granules remained in the melt. At 240 °C, the entire sample melted, and the granules disappeared.

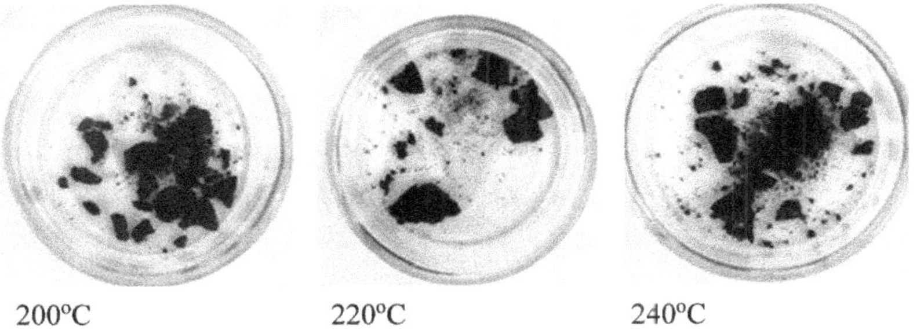

Figure 1. Physical appearance of pyrolyzed PVC-P; PVC/glycerol mass ratio = 2/3 (PVC = 0.4 g, glycerol = 0.6 g), reaction time = 120 min.

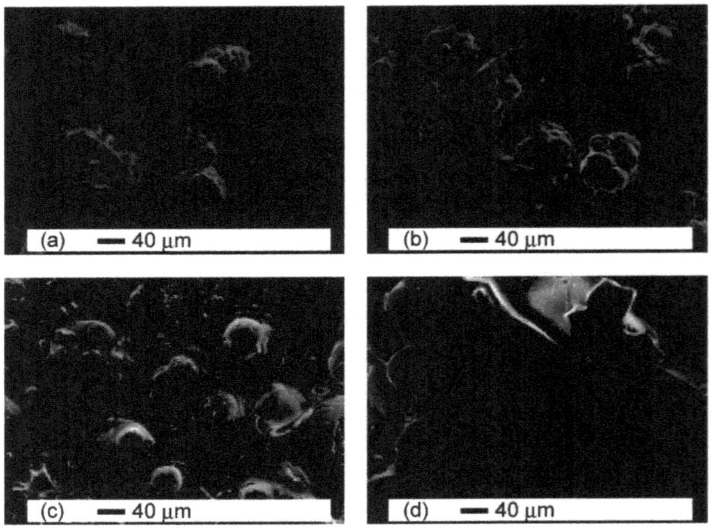

Figure 2. Surface morphology of pyrolyzed PVC-P: (**a**) unreacted PVC-P, (**b**) PVC-P pyrolyzed at 200 °C, (**c**) pyrolyzed at 220 °C, (**d**) pyrolyzed at 240 °C, PVC/glycerol mass ratio = 2/3, reaction time = 120 min.

As indicated in Table 1, the chlorine fraction in the pyrolyzed PVC-P was 47.2, 16.0, and 10.0 wt% for the samples treated at 200, 220, and 240 °C, respectively. Figure 3 shows the mass balance of the pyrolysis products evaluated according to the data in Table 1. During the chemical conversion of PVC, glycerol is thought to be converted into glycidol and epichlorohydrin, according to Equations (4) and (5), respectively, which is in line with reported patents and papers detailing the glycerol chlorination mechanism [23]. These compounds were solidified by an epoxidation reaction with the converted PVC. Therefore, liquid glycerol remained as a reaction product at 200 °C but disappeared at 220 °C. The chlorine content in the solid product decreased sharply from 200 °C to 220 °C, and conversely, the amount of CHO in the solid product increased.

Table 1. Decomposition ratio and dechlorination yield of PVC-P.

Reaction Temperature (°C)	200	220	240
Decomposition ratio (wt%)	61.1	63.3	64.9
Dechlorination yield (wt%)	16.7	71.8	82.4
Chlorine fraction in solid residue (wt%)	47.2	16.0	10.0

PVC/glycerol mass ratio = 2/3; reaction time = 120 min.

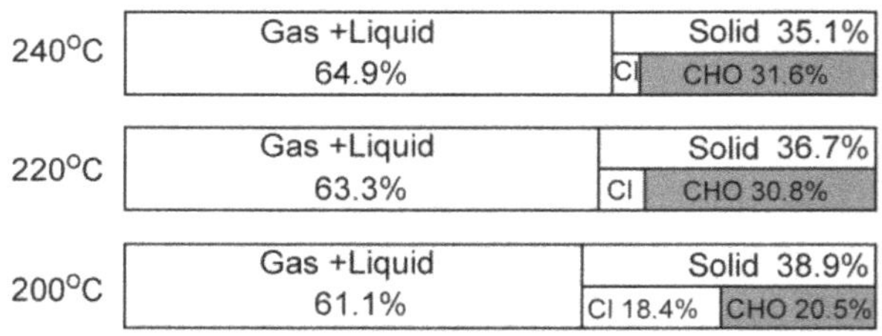

Figure 3. Mass ratio of gas, liquid, and solid phases during PVC-P pyrolysis; PVC/glycerol mass ratio = 2/3, reaction time = 120 min.

$$C_3H_8O_3 \rightarrow C_3H_2O_2 + H_2O \quad (4)$$

$$C_3H_8O_3 + 2HCl \rightarrow HOCH_2CHClCH_2Cl \rightarrow C_3H_5ClO \quad (5)$$

The FT-IR spectra of the unreacted PVC-P and the pyrolyzed PVC-P in glycerol are presented in Figure 4. In the FT-IR spectrum of the unreacted PVC-P, no OH stretching band was observed in the wavenumber range of 3100–3600 cm^{-1}. At a reaction temperature of 200 °C, the OH stretching absorption was observed, because the PVC was partially converted into a polyene structure by zip-elimination (Equation (6)), and further reacted with water to form a polyethylene oxide structure (Equation (7)). PVC containing a polyethylene oxide structure is thought to be resinified by ring-opening polymerization with glycidol and epichlorohydrin at high temperatures. As a result, the OH stretching absorption decreased at 220 °C and disappeared at 240 °C. The absorption peaks of the C=O stretching vibration at 1710 cm^{-1} and the CH out-of-plane bending vibration at 740 cm^{-1} for the sample treated at 220 °C indicated aromatization of the polyene structure.

$$-(CH_2\text{-}CHCl)_n- \rightarrow -(CH{=}CH)_n- + nHCl \quad (6)$$

$$-(CH_2\text{-}CHCl)_n- + [OH] \rightarrow -(CH_2\text{-}CHOH)_n- + [Cl] \quad (7)$$

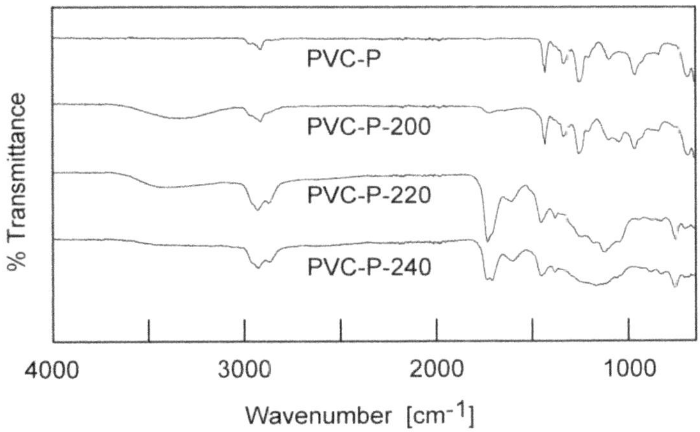

Figure 4. FT-IR spectrum of PVC−P and that pyrolyzed at 200 °C (PVC−P−200), 220 °C (PVC−P−220), and 240 °C (PVC−P−240); PVC/glycerol mass ratio = 2/3, reaction time = 120 min.

Figure 5 shows the decomposition ratio and dechlorination yield of the PVC-P pyrolyzed at 200–240 °C, respectively. The PVC/glycerol mass ratio was 2:3. According to Equation (1), the initial decomposition ratio was 60% because the feedstock contained 60% liquid glycerol. At a reaction temperature of 200 °C, the decomposition ratio initially decreased and then increased to 61.1% at 120 min, which corresponded to the increase in the dechlorination of the PVC (decomposition) (Figure 5). This indicated that glycerol solidified due to the presence of glycidol at the beginning of the reaction (Equation (4)). However, the dechlorination hardly progressed, as shown in Figure 5b. When the reaction temperature was raised to 220–240 °C, the decomposition ratio increased slightly with time and reached 63–65% at 120 min. The initial processes in the PVC thermal degradation were the zip-elimination of HCl and the simultaneous formation of a conjugated double-bond (Equation (6)), which generated a polyene structure in the polymer backbone [24]. Assuming the complete dechlorination of the PVC, the decomposition ratio was calculated to be 71.3%. The decomposition ratio was almost the same at 120 min. Regardless of the reaction temperature, the large differences in the dechlorination yields cannot be explained by the direct dechlorination of the PVC (Equation (6)). To explain the unchanged decomposition ratios and the high dechlorination yields at the high temperatures, solidification with dechlorination must be considered. Hydrogen chloride produced by decomposition reacts with glycerol to produce 1,3-dichloro-2-propanol. The dechlorination yields increased with temperature because of the formation of epichlorohydrin by the dechlorination of 1,3-dichloro-2-propanol (Equation (7)). Finally, glycidol and epichlorohydrin were thought to be resinified by an epoxidation reaction with the converted PVC. The reaction scheme, surmised on the basis of these results, is shown in Figure 6.

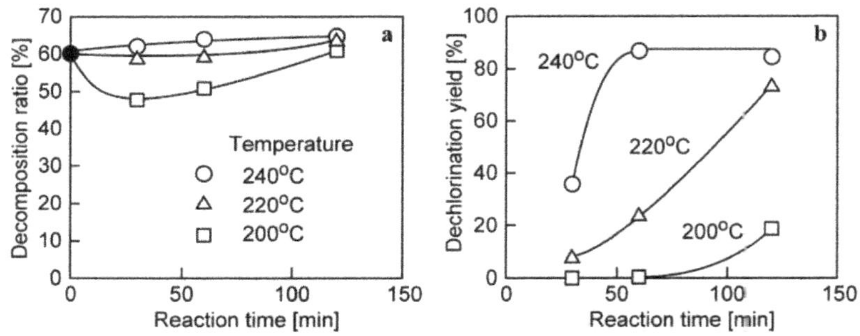

Figure 5. (**a**) Decomposition and (**b**) dechlorination ratios of PVC-P; PVC/glycerol mass ratio = 2/3.

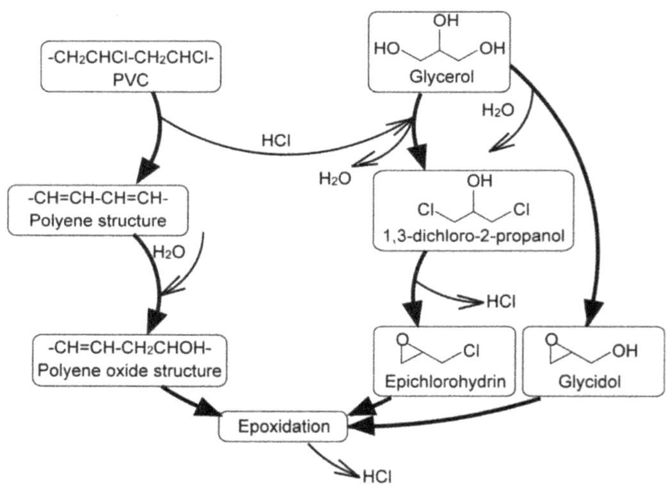

Figure 6. Schematic of dechlorination reaction of PVC in glycerol.

3.2. *Effect of PVC/Glycerol Mass Ratio on Dechlorination*

The dechlorination of the PVC-P was conducted with PVC/glycerol mass ratios of 1:3, 2:3, and 4:3. To focus on the dechlorination of the PVC, the effect of the PVC/glycerol mass ratio on the reaction characteristics was investigated at the fixed reaction time of 120 min. As shown in Figure 7a, at a PVC/glycerol mass ratio of 1:3, the decomposition ratio of the PVC-P decreased with increasing reaction temperature, despite the progress of the dechlorination. This indicated the progress of the conversion from liquid to solid via resinification with epichlorohydrin. The decomposition ratios of the PVC-P at 240 °C were 67.8, 64.9, and 62.1% for the PVC/glycerol mass ratios of 1:3, 2:3, and 4:3, respectively. The decomposition ratio of the PVC-P at 240 °C was calculated assuming that complete dechlorination proceeded and a portion of the glycerol solidified. As a result, when an amount equivalent to 30% of the starting glycerol was solidified, the calculated decomposition ratios for the PVC/glycerol mass ratios of 1:3, 2:3, and 4:3 were 66.7, 64.7, and 62.4%, respectively, which were almost the same as the individually measured values. This indicated that the PVC was converted into solid products in high yields via chlorine release, and that 30% of the glycerol was solidified by resinification. As shown in Figure 7b, the dechlorination yield tended to increase with increasing reaction temperature and was not significantly affected by the amount of PVC.

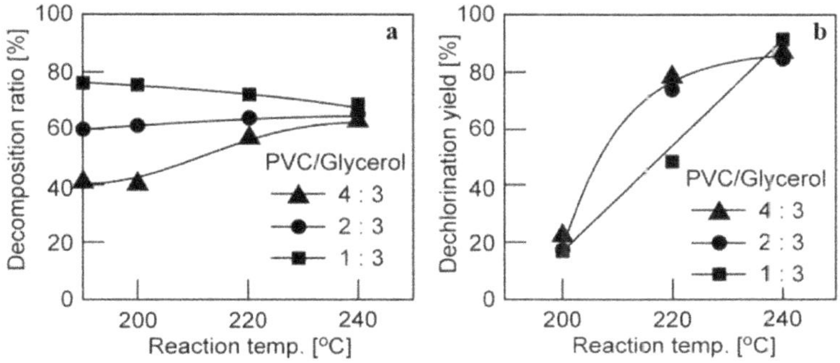

Figure 7. Effect of PVC/glycerol mass ratio on (**a**) decomposition ratio and (**b**) dechlorination yield.

3.3. Comparison with Commercially Available PVC-Sheets

The results of the elemental analysis of the unreacted PVC-P, PVC-S, and PVC-TS are listed in Table 2. The amounts of additives in the PVC-S and PVC-TS were calculated by subtracting the amount of intrinsic PVC, which was estimated by considering the Cl fraction from the total amount of PVC. The fraction of additives in the PVC-S and PVC-TS was 37% and 23%, respectively. Table 3 summarizes the decomposition ratios and dechlorination yields for the PVC-P, PVC-S, and PVC-TS at a reaction time of 120 min. The complex behavior of the decomposition ratio was influenced by the shape factor, the relaxation of the polymer network by the THF treatment, the suppression of chlorine elimination by the additives, and the decomposition of the additives. The dechlorination yield increased rapidly from 16.7 to 71.8% in the temperature range of 200–220 °C for the additive-free PVC-P, and a rapid increase in the dechlorination yield was observed in the range of 220–240 °C for the PVC-TS and PVC-S. The shift in the temperature range where dechlorination progressed was due to the suppression of zip-elimination by the additives.

Table 2. Elementary analysis of various PVC samples.

Sample	PVC-P	PVC-S	PVC-TS
C (%)	38.4	48.4	44.6
H (%)	4.8	7.1	6.4
O (%)	0	8.9	5.2
Cl (%)	56.8	35.6	43.8

Table 3. Decomposition ratio and dechlorination yield of various PVC samples.

Sample	Fraction of Additive (%)	Decomposition Ratio (%)			Dechlorination Yield (%)		
		200 °C	220 °C	240 °C	200 °C	220 °C	240 °C
PVC-P	0	61.1	63.3	64.9	16.7	71.8	82.4
PVC-TS	22.9	65.2	64.2	66.9	25.0	33.4	82.9
PVC-S	37.3	57.6	61.8	75.0	33.0	38.2	84.6

PVC/glycerol = 2:3; reaction time = 120 min.

3.4. Calorific Value of the Solid Product

Table 4 shows the effects of the PVC/glycerol ratio and the reaction temperature on the elemental composition and calorific value of the PVC-P solid product. Due to the simultaneous occurrence of carbonization and dechlorination, the carbon content increased from 41.2 to 67.1%, and the chlorine content decreased from 47.6 to 15.9% at a PVC/glycerol ratio of 2:3 in the temperature range of 200–220 °C. At the PVC/glycerol mass ratios of 4:3 and 1:3, almost the same trends were observed. In particular, at the ratio of 1:3, the carbon content reached 79.0%, and the chlorine content decreased to 3.60% at 240 °C. The oxygen content decreased with the temperature, owing to the decomposition of the glycerol. As a result, when the carbon content of the solid products exceeded 67%, the calorific values exceeded 33 MJ/kg at 240 °C. The calorific values of the solid products were higher than that of standard bituminous coal (25–32 MJ/kg).

Table 4. Elemental analysis and calorific value for PVC-P solid products.

Sample (P/G [(a)])	Temp. (%)	Weight Fraction (%)				H/C (mol/mol)	O/C (mol/mol)	Calorific Value (MJ/kg)
		C	H	O	Cl			
4:3 [(b)]	-	36.78	6.23	22.39	29.59	2.03	0.46	17.9
4:3	200	41.4	5.48	5.51	47.6	1.59	0.045	20.3
4:3	220	73.0	7.058	4.0	15.9	1.16	0.019	33.4
4:3	240	79.3	7.34	2.66	10.7	1.11	0.011	36.1
2:3 [(b)]	-	38.8	7.14	31.32	22.72	2.2	0.61	18.7
2:3	200	41.2	5.58	7.0	47.2	1.63	0.057	20.2
2:3	220	67.1	7.06	9.9	16.0	1.26	0.05	30.7
2:3	240	73.6	7.63	8.71	10.0	1.24	0.04	33.8
1:3 [(b)]	-	38.93	7.73	39.15	14.2	2.38	0.75	18.7
1:3	200	40.9	5.51	7.72	46.4	1.62	0.064	20.0
1:3	220	54.5	6.45	13.0	26.1	1.42	0.0581	25.3
1:3	240	79.0	7.12	10.3	3.6	1.08	0.044	34.9

[(a)] *P/G* indicates PVC/glycerol mass ratio. [(b)] Calculated value for mixture of PVC and glycerol. Reaction time = 120 min.

Table 5 shows the elemental composition and calorific value of the solid products of the PVC-P, PVC-S, and PVC-TS. The carbon and oxygen contents of the PVC-S and PVC-TS were higher than those of the PVC-P, owing to the presence of additives. The progress of the carbonization and dechlorination reactions of the PVC-S and PVC-TS shifted in the temperature range of 220–240 °C due to the suppression of the reactions by the additives. Compared to the PVC-P, the PVC-S had a carbon content of up to 79.1% and a chlorine content of up to 8.7% at 240 °C.

Table 5. Comparison of elemental analysis and calorific value. [(a)] Calculated value for mixture of PVC and glycerol. PVC/glycerol mass ratio = 2:3; reaction time = 120 min.

Sample	Temp. (%)	Weight Fraction (%)				H/C (mol/mol)	O/C (mol/mol)	Calorific Value (MJ/kg)
		C	H	O	Cl			
PVC-P [(a)]	-	38.8	7.14	31.32	22.72	2.2	0.61	18.7
	200	41.2	5.58	7.0	47.2	1.63	0.057	20.2
	220	67.1	7.06	9.9	16.0	1.26	0.05	30.7
	240	73.6	7.63	8.71	10.0	1.24	0.04	33.8
PVC-TS [(a)]	-	41.3	7.76	33.4	17.52	2.26	0.6	20.1
	200	45.4	6.42	6.0	42.6	1.69	0.099	22.8
	220	48.4	6.41	6.39	37.8	1.59	0.099	23.8
	240	78.3	7.49	4.89	9.75	1.15	0.047	35.7
PVC-S [(a)]	-	42.82	8.04	34.88	14.24	2.25	0.61	20.8
	200	44.4	6.35	10.8	38.0	1.72	0.182	21.9
	220	47.6	6.72	10.7	35.1	1.69	0.169	23.4
	240	79.1	7.7	6.52	8.7	1.17	0.062	36.0

4. Conclusions

PVC was dechlorinated by treatment with glycerol in an autoclave at 200–240 °C. For the PVC-P dechlorination, liquid glycerol remained in the product pyrolyzed at 200 °C for 120 min. When the reaction temperature was raised to 220 °C and 240 °C, carbonization proceeded with the disappearance of glycerol, forming a black powder.

Characterization of the solid products showed that the PVC was partially converted into a polyene structure by the zip-elimination of HCl and further reacted with water to form a polyethylene oxide structure. During the chemical conversion of the PVC, glycerol was thought to produce glycidol and epichlorohydrin via the reaction represented by Equations (4) and (5), respectively. These compounds were solidified by an epoxidation reaction with the converted PVC. Thus, the glycerol acted as a useful reaction medium in this reaction system. Analysis of the decomposition ratio and dechlorination yield showed that the PVC was converted into solid products in high yields via chlorine release, and approximately 30% of the glycerol solidified by resinification at 240 °C.

The effects of additives on dechlorination were examined by comparing commercially available PVC sheets (PVC-S and PVC-TS) and PVC-P. Although the decomposition and dechlorination of the PVC-P, PVC-TS, and PVC-S were significantly influenced by the shape factor and the additives at 200–220 °C, the dechlorination yields for all the PVCs converged to 81–85% at 240 °C. The calorific value of the solid product pyrolyzed at 240 °C exceeded 30 MJ/kg, owing to the progress of carbonization and dechlorination.

Author Contributions: Conceptualization, K.K. and Y.U.; validation, W.H.L.; formal analysis, K.I.; investigation, K.Y. and T.N.; writing—original draft preparation, K.K. and A.N.; writing—review and editing, W.H.L.; supervision, Y.U.; funding acquisition, K.K. All authors have read and agreed to the published version of the manuscript.

Funding: This research was funded by the Japan Soap and Detergent Association and the 2020 SOJO Research-Assisting Program.

Institutional Review Board Statement: Not applicable.

Informed Consent Statement: Not applicable.

Data Availability Statement: Not applicable.

Acknowledgments: The authors acknowledge financial support from the Japan Soap and Detergent Association and the 2020 SOJO Research-Assisting Program.

Conflicts of Interest: The authors declare no conflict of interest.

References

1. Poerschmann, J.; Weiner, B.; Woszidlo, S.; Koehler, R.; Kopinke, F.D. Hydrothermal carbonization of poly(vinyl chloride). *Chemosphere* **2015**, *199*, 682–689. [CrossRef]
2. Yu, J.; Sun, L.; Ma, C.; Qiao, Y.; Yao, H. Thermal degradation of PVC: A review. *Waste Manag.* **2016**, *48*, 300–314. [CrossRef]
3. López, A.; de Marco, I.; Caballero, B.M.; Laresgoiti, M.F.; Adrados, A. Dechlorination of fuels in pyrolysis of PVC containing plastic wastes. *Fuel Process. Technol.* **2011**, *92*, 253–260. [CrossRef]
4. Yoshioka, T.; Akama, T.; Uchida, M.; Okuwaki, A. Analysis of two stages dehydrochlorination of poly(vinyl chloride) using TG-MS. *Chem. Lett.* **2000**, *29*, 322–323. [CrossRef]
5. Yoshioka, T.; Saitoh, N.; Okuwaki, A. Temperature dependence on the activation energy of dechlorination in thermal degradation of polyvinylchloride. *Chem. Lett.* **2005**, *34*, 70–71. [CrossRef]
6. Yuan, G.; Chen, D.; Yin, L.; Wang, Z.; Zhao, L.; Wang, J.Y. High efficiency chlorine removal from polyvinyl chloride (PVC) pyrolysis with a gas-liquid fluidized bed reactor. *Waste Manag.* **2014**, *34*, 1045–1050. [CrossRef] [PubMed]
7. Wu, J.; Chen, T.; Luo, X.; Han, D.; Wang, Z.; Wu, J. TG/FTIR analysis on co-pyrolysis behavior of PE, PVC and PS. *Waste Manag.* **2014**, *34*, 676–682. [CrossRef]
8. Prawisudha, P.; Namioka, T.; Yoshikawa, K. Coal alternative fuel production from municipal solid wastes employing hydrothermal treatment. *Appl. Energy* **2012**, *90*, 298–304. [CrossRef]
9. Takeshita, Y.; Kato, K.; Takahashi, K. Basic study on treatment of waste polyvinyl chloride plastics by hydrothermal decomposition in subcritical and supercritical regions. *J. Supercrit. Fluids* **2004**, *31*, 185–193. [CrossRef]
10. Zhao, P.; Shen, Y.; Ge, S.; Chen, Z.; Yoshikawa, K. Clean solid biofuel production from high moisture content waste biomass employing hydrothermal treatment. *Appl. Energy* **2014**, *131*, 345–367. [CrossRef]

11. Kuhlmann, B.; Arnett, E.M.; Siskin, M. Classical organic reactions in pure superheated water. *J. Org. Chem.* **1994**, *59*, 3098–3101. [CrossRef]
12. Endo, K.; Emori, N. Dechlorination of poly(vinyl chloride) without anomalous units under high pressure and at high temperature in water. *Polym. Degrad. Stabil.* **2001**, *74*, 113–117. [CrossRef]
13. Shin, S.-M.; Yoshioka, T.; Okuwaki, A. Dehydrochlorination behavior of flexible PVC pellets in NaOH solutions at elevated temperature. *J. Appl. Polym. Sci.* **1998**, *67*, 2171–2177. [CrossRef]
14. Yoshinaga, T.; Yamaye, M.; Kito, T.; Ichiki, T.; Ogata, M.; Chen, J.; Fujino, H.; Tanimura, T.; Yamanobe, T. Alkaline dechlorination of poly(vinyl chloride) in organic solvents under mild conditions. *Polym. Degrad. Stab.* **2004**, *86*, 541–547. [CrossRef]
15. Yoshioka, T.; Kameda, T.; Imai, S. Dechlorination of poly(vinylidene chloride) in NaOH/ethylene glycol as a function of NaOH concentration, temperature, and solvent. *Polym. Degrad. Stab.* **2008**, *93*, 1979–1984. [CrossRef]
16. Huang, N.; Zhao, P.; Ghosh, S. Co-hydrothermal carbonization of polyvinyl chloride and moist biomass to remove chlorine and inorganics for clean fuel production. *Appl. Energy* **2019**, *240*, 882–892. [CrossRef]
17. Shen, Y.; Yu, S.; Ge, S.; Chen, X.; Ge, X.; Chen, M. Hydrothermal carbonization of medical wastes and lignocellulosic biomass for solid fuel production from lab-scale to pilot-scale. *Energy* **2017**, *118*, 312–323. [CrossRef]
18. Shen, Y. Dechlorination of poly(vinyl chloride) waste via hydrothermal carbonization with lignin for clean solid fuel. *Ind. Eng. Chem. Res.* **2016**, *55*, 11638–11644. [CrossRef]
19. Yao, Z.; Ma, X. Characteristics of co-hydrothermal carbonization on polyvinyl chloride wastes with bamboo. *Bioresour. Technol.* **2018**, *247*, 302–309. [CrossRef]
20. Kusakabe, K.; Steven, T.; Ong, J.Y.; Uemura, Y.; Ikenaga, K. Cohydrothermal carbonization of waste polyvinyl chloride and wood chip for dichlorination. *W3S Web Conf.* **2021**, *287*, 04008. [CrossRef]
21. Kusakabe, K.; Steven, T.; Nagai, A.; Uemura, Y.; Ikenaga, K. Solvothermal carbonization of wood chips via the dechlorination of PVC in glycerol. *J. Chem. Eng. Jpn.* **2021**, *54*, 110–115. [CrossRef]
22. Channiwala, S.A.; Parikh, P.P. A unified correlation for estimating HHV of solid, liquid and gaseous fuels. *Fuel* **2002**, *81*, 1051–1063. [CrossRef]
23. Santacesaria, E.; Tesser, R.; Di Serio, M.; Casale, L.; Verde, D. New process for producing epichlorohydrin via glycerol chlorination. *Ind. Eng. Chem. Res.* **2010**, *49*, 964–970. [CrossRef]
24. Iván, B.; Kennedy, J.P.; TKelen, T.; Tüdős, F.; Nagy, T.T.; Turcsányi, B. Degradation of PVC's obtained by controlled chemical dehydrochlorination. *J. Polym. Sci.* **1983**, *21*, 2177–2188.

processes MDPI

Article

Numerical Study of Electrostatic Desalting: A Detailed Parametric Study

Marco A. Ramirez-Argaez [1], Diego Abreú-López [1], Jesús Gracia-Fadrique [1] and Abhishek Dutta [2,*]

1 School of Chemistry, National Autonomous University of Mexico (UNAM), Av. Universidad #3000, Cd. Universitaria, Mexico City 04510, Mexico
2 Department of Chemical Engineering, Izmir Institute of Technology, Gülbahçe Campus, Urla, Izmir 35430, Turkey
* Correspondence: abhishekdutta@iyte.edu.tr

Abstract: A systematic process analysis was conducted to study the effect of the main variables in an industrial electrostatic desalter, such as electric field intensity, wash water content, droplet size, and oil viscosity, on the efficiency of the separation of water from oil. The analysis was assessed through an already published and validated CFD multiphase numerical model that considers the expression of the frequency of collisions as a function of the mentioned process variables. Additionally, the study allowed the formal optimization exercise of the operation to maximize the separation efficiency. The most significant variables were the initial water content and the electric field intensity, while the temperature (oil viscosity) had an effect to a lower extent. An increase in the electric field and temperature and a decrease in the water content improved the water separation from oil. Optimum values suggested from the factorial experimental design and the optimization implemented in this work indicated the use of an electric field of 3 kV/cm, water content of 3%, and an oil viscosity of 0.017 kg/ms. At the same time, the droplet size showed no significant effect under the conditions explored in this work.

Keywords: electrostatic desalting; computational fluid dynamics; multiphase fluid flow; variance analysis

Citation: Ramirez-Argaez, M.A.; Abreú-López, D.; Gracia-Fadrique, J.; Dutta, A. Numerical Study of Electrostatic Desalting: A Detailed Parametric Study. *Processes* **2022**, *10*, 2118. https://doi.org/10.3390/pr10102118

Academic Editors: Jie Zhang and Jun Wei Lim

Received: 28 August 2022
Accepted: 17 October 2022
Published: 18 October 2022

1. Introduction

Although the green new deal [1,2] involves the elimination of fossil fuels as soon as possible, by looking at the sources of energy currently used in the world, fossil fuels are by far the major contributors to energy production and outnumber the production of clean energies, so the world production of oil has shown an increasing trend recently [2]. Moreover, crude oil is often associated with saline water in the range of 0.8–2%, which must be eliminated early in the oil refinement if the level of salt is over 9 kg per 100 barrels [3] as this saline water may cause corrosion, scale formation, and diminish the catalyst efficiency in the subsequent stages of crude refinement [4,5]. Salty water comes along with the crude oil in a water-in-oil (W/O) emulsion due to the shear associated with the oil transportation [4]. Electrostatic desalting units are commonly employed to remove the water and salt from the crude by adding wash water and demulsifying agents; controlling temperature and the electrostatic features, such as the type of electric field, its intensity, and the geometry and arrangement of the electrodes; controlling the fluid flow in that unit to form a water-in-oil (W/O) emulsion; promoting forces acting on the water droplets (gravity, buoyancy, electrophoresis, dipolar, drag, etc.) causing high-frequency collisions of droplets and eventually flocculation and coalescence, resulting in the separation of water from oil. The factors that break the emulsion are: (a) demulsifier additions, (b) temperature increment, (c) presence of an electric field, (d) residence time of the droplets, (e) water addition, and (f) mixing [3–6]. The emulsion may be stable depending on the oil's density and electric conductivity, water content, droplet size distribution, presence of asphaltenes, pH of water, surface tension, and age of the emulsion, among other parameters, which give us an idea of the considerable complexity of this separation. The electric field could be AC, DC, or AC/DC pulsed, which exerts different forces on the droplets [5,7]. Unfortunately, the process is highly complex to monitor as it involves high-pressure and high-temperature operation conditions with high electric fields, so it is almost impossible to perform industrial trials to understand and optimize the operation of such units. Therefore, the remaining analysis tools are experimental trials at the laboratory and pilot plant scales under controlled conditions and mathematical modeling. Several process analyses have been performed in the past, where the main design and operational variables effects were determined on the separation efficiency. In all these studies, the main variables explored were the content of freshwater, pressure drop at the inlet valve, amount of demulsifier, temperature (viscosity of the oil), density of the oil, and electric field [4–6,8–12] (see Table 1).

Table 1. Summary of research works on process analysis of the electrostatic desalting.

Researchers	Nature	Variables	Conclusion
Abdul-Wahab et al. [5]	Laboratory scale model (bottle test)	Temperature, mixing time, residence time, chemical dosage, water content	Most important variables obtained are temperature, water content, and residence time
Ali Khairan Alshehri [4]	Numerical model	Oil temperature, water content, voltage, initial water content, oil flow rate, demulsifier flow rate	Minimization of wash water and final salt content.
Otaibi et al. [6]	Numerical model	Demulsifier concentration, temperature, wash water %, salt content, rate of mixing water addition	Maximizing the efficiency of water and salt removal
Bresciani et al. [10]	Analytical model	Water content, temperature, voltage, droplet size (upper and lower droplets)	Model predicts the displacement of two droplets and the time for collision under an electric field
Bresciani et al. [13]	Same model as [10] but extended with cellular automata	Water content, temperature, voltage, droplet size distribution	Predicted coalescence and validated with industrial data.
Fetter-Pruneda et al. [11]	Experimental model (industrial scale)	Water content, temperature, crude density	The optimum temperature for Mayan crude oil and recommended practices
Wilkinson et al. [12]	Numerical model	Design of baffle separator in a gravity separator of water from oil	Best design of baffle enhancing the water separation from oil
Aryafard et al. [14,15]	Numerical model, population balance model	Pressure drop, electric field, especially wash water content.	Prediction of coalescence and break up. Analysis of one and two stages of desalting processes. Improvement of separation efficiency from 96.5 to 98.5% when wash water is changed from 3 to 6% and validated in an industrial unit.
Kakhki et al. [16]	Numerical model, population balance	Pressure drop, electric field, but especially wash water content (similar to [6])	Similar results than in [13]. Increasing the rate of collision between water droplets promotes coalescence.
Mahdi et al. [8]	Experimental model (laboratory scale)	Demulsifying agent concentration, temperature, wash water dilution ratio, settling time, and mixing time with wash water	Optimum values of demulsifying agent concentration 15 ppm, temperature 77 °C, 10% wash water dilution ratio, settling time 3 min, and mixing time 9 min.
Vafajoo et al. [17]	Experimental model (laboratory scale), fuzzy logic	Temperature, injected chemicals, and the pH of the crude oil associated water	Temperature between 115 to 120 °C, best demulsifiers were C and F at levels of 50 to 100 ppm, separating 88% for water and 99% for salt. The pH has to be between 9 and 12
Bansal and Ameensayal [18]	Numerical model	Design parameters	Improve fluid flow features and increase performance
Shariff and Oshinowo [19]	Numerical model	Fluid flow analysis	Vortices formed at the inlet reduce the separation efficiency.

Table 1. *Cont.*

Researchers	Nature	Variables	Conclusion
Ilkhaani, S. [5]	Thermodynamic model	Adding a second stage of desalting to one-stage desalting to meet levels of water and salt	Improve the heat integration of the desalting process, and optimization of desalting temperature
Alhajri et al. [20]	Numerical model and plant trials	Device proposed is a static mixer at the inlet.	Turbulence is key to enhance separation of water and salt.
Favero et al. [21]	Numerical model	Population balance (CFD + DPM) to predict droplet size distribution in a duct that mimics pass over a globe valve	Good agreement between the experimental and predicted droplet sizes.
Wang et al. [22]	Numerical model	CFD + DPM in an aeroengine bearing chamber (not desalter)	Coalescence and breakup of oil droplet increases with the initial diameter of oil droplet.
Sofos [23]	Simulations with molecular dynamics	Novel electrostatic device consisting of separation cells	The proposed application could be exploited for the design of a desalination device.
Shi et al. [24]	Plant trials	Novel desalter with helical electrodes. Effect of electric field strength, frequency, water content, and fluid velocity on the performance of coalescence.	Increasing the electric field strength could contribute to the growth of small water droplets and coalescence. The study may be used for optimization

As seen in Table 1, there have been only a few attempts to simulate the process based on CFD modeling. For example, Aryafard et al. [14,15] and Kakhki et al. [16] developed numerical models to simulate the desalting in electrostatic units in one stage [15], two stages [14], and a simplified inlet and drum of a desalting process, using a population balance model to model the W/O emulsion. They found that the efficiency could be improved by 2% if wash water was increased from 3 to 6% at the inlet, a result validated by plant data, while Kakhki et al. [16] stated that the increment in the collision rate promotes the coalescence of droplets, improving the separation.

Recently, a new design of a high-efficiency static mixer located at the inlet of the desalting unit was modeled by CFD techniques and field trials by Alhajri et al. [20] in a Saudi Arabia refinery. They concluded that enhancing turbulence is the key to saving wash water and improving water and salt removal efficiency. Favero et al. [21] used a CFD model using the Eulerian multiphase approach that includes a mass population balance to account for the coalescence and breakout of water-in-oil emulsion that passes in a duct with an element that mimics the pass of such an emulsion through a globe valve as in a desalter unit. The results indicated good agreement in the droplet diameter with data from the literature. Sellman et al. [25] conducted a review in which they explained difficulties associated with separating salt from a blend of oils, particularly difficulties when the oil blend contains suspended solids that stabilizes the emulsion. They revised how the desalting units have been improved in the past, the role CFD plays in understanding the process, and the industry's benefits from such studies. Wang et al. [22] simulated the droplet size population of oil droplets using CFD under the Euler–Euler approach and using the population balance model (CFD-DPM). However, the system was an aeroengine bearing chamber, not a desalter, giving insight into a very complex system. Another attempt to simulate the separation of droplets was presented by Sofos [23], but again, not in a desalter but a novel electrostatic device consisting of separation cells. Molecular dynamics performed the simulation. Shi et al. [24] also proposed a novel electrostatic desalter device with helical electrodes, and the performance was tested with experimental trials.

Recently, our research group published an article [26] and presented [27] CFD modeling based on an industrial desalting unit in México. The study employed a correlation for the frequency of collisions between droplets, obtained from Bresciani's modeling approach [10,13]. The correlation, which includes the effect of the primary process variables, was used in Ansys Fluent with the multiphase mixture model to evaluate the separation of the salty water from the oil in the presence or the absence of an electric field. The merit of that research was to provide a tool for process analysis based on first principles. In this work, the earlier study already published was extended to perform a systematic process analysis to assess the effect of the primary process variables (electric field intensity, temperature through the oil viscosity, water content, and inlet valve pressure through the initial droplet size) on the separation efficiency. In addition, we proposed optimum conditions for maximizing the separation in an industrial desalting unit by using standard optimization subroutines.

Mathematical Modeling

A Eulerian frame of reference was used using the mixture multiphase approach to simulate the numerical isothermal fluid flow separation of the W/O emulsion through the 3D industrial unit desalting from Pemex (Mexico) shown in Figure 1, including inlets, outlets, and internal electrode plates.

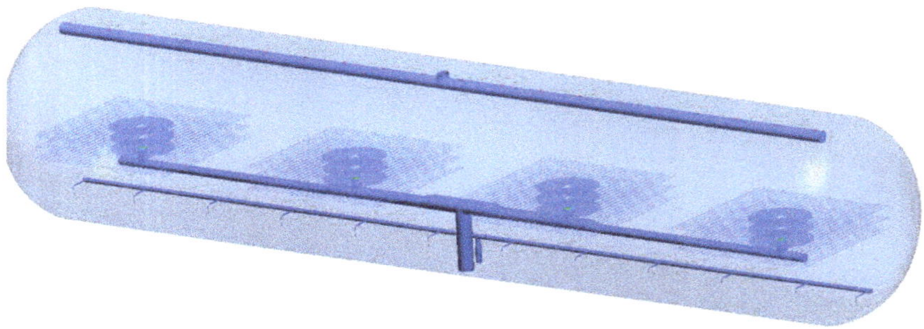

Figure 1. Schematic of the desalting unit used in this study.

The mixture algorithm involves the calculation of the physical properties of the mixture with the volume fraction of water and oil phases. It also accounts for the continuity and momentum conservation equations, while the breakup (S_{TI}) and coalescence (S_{RC}) of droplets are determined through the interfacial area concentration conservation equation, a sort of population balance expression. Continuity and momentum equations use the drift velocity, $v_{d,i}$. To account for the relative displacement of one phase to the mixture of phases and in the case of the water drift velocity, it results from a force balance between drag, turbulent dispersion, and buoyancy forces. Moreover, this slip velocity brings out additional terms in the continuity and momentum equations.

The critical feature of the interfacial area concentration is the source term related to the successful collisions producing coalescence of droplets. This term uses the correlation of the frequency of collisions, f_c developed in our previous study [26], which in turn is a function of the primary process variables, such as the electric field E_0, water content a_w, oil viscosity μ_o (temperature), and droplet size d_w. Finally, to consider the turbulence in the system, the k-ε realizable turbulence model was used.

Table 2 shows a list of major simplifying assumptions, while Tables 3 and 4 show a list of governing equations and boundary conditions.

Table 2. List of simplifying assumptions used in the development of the mathematical model.

Assumption	Consequence
Isothermal system	There are considered thermal gradients in the desalter
Constant physical properties	Both water and oil are Newtonian and incompressible fluids
Steady state	The time derivatives are zero
Non-slip and impermeable walls	All components of the velocity vector are zero at the boundary and internal static walls
Oil is the continuous phase	Water is the disperse phase in W/O emulsions. Therefore, the reverse emulsion O/W is not considered to appear in the unit.
Mixture algorithm	A single set of equations: continuity, momentum and one turbulence model to simulate the multiphase system.
k–epsilon realizable turbulence model	To represent the turbulence in the continuous phase. The disperse phase has no turbulence.
Interfacial area concentration	To account for the events of breakup and coalescence of droplets
Collision frequency	Coalescence depends directly on the frequency of the collisions between droplets, as was stated by Ramirez-Argaez et al. [26]

Table 3. Governing equations.

Name	Equation
Mixture density	$\rho_m = \sum_{i=1}^{N} \alpha_i \rho_i$
Mixture viscosity	$\mu_m = \sum_{i=1}^{N} \alpha_i \mu_i$
Mixture velocity	$v_m = \frac{\sum_{i=1}^{N} \alpha_i \rho_i v_i}{\rho_m}$
Continuity	$\nabla \cdot (\rho_m v_m) = 0$
Momentum	$\nabla \cdot (\rho_m v_m v_m) = -\nabla p_m + \nabla \cdot (\tau_m + \tau_{Tm}) + \nabla \cdot \tau_{Dm} + \rho_m g$ where the three tensors are the average viscous stress τ_m, the turbulence stress τ_{Tm} and the diffusion stress τ_{Dm} due to the phase slip: $\tau_m = \sum_{i=1}^{N} \alpha_i \tau_i$ $\tau_{Tm} = \sum_{i=1}^{N} \alpha_i \rho_i \acute{v}_i \acute{v}_i$ $\tau_{Dm} = \sum_{i=1}^{N} \alpha_i \rho_i v_{d,i} v_{d,i}$ where $v_{d,i}$ is the drift velocity, i.e., the velocity of the $i-$th phase, v_i relative to the mixture velocity, v_m: $v_{d,i} = v_i - v_m$ In the case of water droplets, according to the algebraic slip formulation by [28] where the buoyant, turbulent dispersion and drag forces acting on the water droplets are balanced: $v_w = \frac{(\rho_w - \rho_m) d_w^2}{18 \mu_o f_{drag}} g - \frac{\eta_T}{Pr_T} \left(\frac{\nabla \alpha_w}{\alpha_w} - \frac{\nabla \alpha_o}{\alpha_o} \right)$
Drag coefficient (Shiller–Nauman)	$f_{drag} = \begin{cases} 1 + 0.15 Re^{0.687} & Re \leq 1000 \\ 0.018 Re & Re > 1000 \end{cases}$
Water phase continuity	$\nabla \cdot (\rho_w v_m \alpha_w) = -\nabla \cdot (\rho_w v_{d,w} \alpha_w)$ As there are only two phases and the $\sum_{i=1}^{N=2} \alpha_i = 1$, $\alpha_o = 1 - \alpha_w$

Table 3. *Cont.*

Name	Equation
Interfacial area concentration	$\nabla\cdot(\rho_w v_w \chi_w) = \rho_w(S_{RC} + S_{TI})$ Source due to coalescence S_{RC}: $S_{RC} = -\frac{1}{108\pi}\left(\frac{\alpha_w}{\chi_w}\right)^2 n_c f_c \lambda_c$ Source due to breakup, S_{TI}: $S_{TI} = \frac{1}{108\pi}\left(\frac{\alpha_w}{\chi_w}\right)^2 n_b f_b \lambda_b$
Number of droplets per unit volume of the mixture	$n_c = 108\pi\frac{\alpha_w}{d_w^3}$
Frequency of collisions [26]	$f_c = \frac{1}{22469.1829\alpha_w^{-3.294}\mu_o^{0.968}E_0^{-2.007}d_w^{-0.013}}$ with a coalescence probability λ_c=1.
Number of eddies per unit volume [26]	$n_b = 108\pi\frac{(1-\alpha_w)}{d_w^3}$
Frequency of collision due to turbulence	$f_b = \frac{0.264\alpha_w\varepsilon^{1/3}}{d_w^{2/3}(\alpha_{w,max}-\alpha_w)}$
Breakup efficiency	$\lambda_b = exp\left(-1.37\frac{\sigma}{\rho_o d_w^{5/6}\varepsilon^{2/3}}\right)$
Water droplet diameter	$d_w = 6\frac{\alpha_w}{\chi_w}$
Turbulent kinetic energy	$\nabla\cdot(\rho_m v_m k) = \nabla\cdot\left(\rho_m\left\{\frac{\eta+\eta_T}{\sigma_{T,k}}\right\}\nabla k\right) + \rho_m(P_k - \varepsilon)$
Energy dissipation rate	$\nabla\cdot(\rho_m v_m \varepsilon) = \nabla\cdot\left(\rho_m\left\{\frac{\eta+\eta_T}{\sigma_{T,\varepsilon}}\right\}\nabla\varepsilon\right) + \rho_m\left(C_{1\varepsilon}S\varepsilon - \frac{C_{2\varepsilon}\varepsilon^2}{\{k+\sqrt{v_o\varepsilon}\}}\right)$ where $\rho_m\eta_T = C_\mu k^2/\varepsilon$, P_k is the production of turbulence, and S is the strain tensor.

Table 4. Boundary conditions.

Boundary	Condition
Non-slip conditions at the internal and external walls	Zero velocity of all components, no turbulence (standard wall functions)
Inlets	Inlet velocity of the emulsion with a volume fraction of water
Outlets	Pressure outlet (gauge pressure equal zero)

The model was implemented in the CFD software Ansys Fluent v2020R2 where the desalting unit was designed and discretized using a nonuniform mesh, as shown in Figure 2, with 3,393,218 cells. The numerical solution used the SIMPLEC velocity pressure coupling. The convection schemes for the continuity momentum and turbulent equations were second-order, while in the case of the volume fraction and the interfacial area concentration, the first-order scheme was used. Standard initialization and residual convergence level were used. A complete factorial design 2^4 was used to perform the process analysis, where the four variables: electric field, oil viscosity, water content, and initial droplet size, were tested. Table 5 shows the low and high levels of variables used, while Table 6 shows the conditions of 2^4, i.e., 16, simulations. Factors and their levels were chosen based on real plant parameters in a Mexican oil refinery. Some process variables were not considered, such as the amount of demulsifier, as the model cannot predict its effect. In contrast, other variables such as the pressure drop at the valve and the temperature were indirectly adopted through the droplet size and oil viscosity, respectively. Finally, the results were post-processed to obtain the contours of water volume fraction, mixture velocity, and turbulent kinetic energy to qualitatively analyze the variables' effect. The separation efficiency was reported for each simulation, so a multiple linear regression was applied to the results to correlate the separation efficiency as a function of the main variables and the significant double interactions between variables. This correlation was employed to perform an optimization effort using a formal optimization method called genetic algorithm NSGA-II [29], programmed in MATLAB® to obtain the optimum conditions for the desalting that minimizes the water and salt content from the oil. The model is only valid under the range of selected variables. In addition, we did not include the effect of the chemical agent (demulsifier), as the collision law does not consider this effect, only applies for DC electric strength, and is valid for the geometry of study (Pemex desalter). Despite these constraints, the methodology is valid and easily applied to any desalter unit.

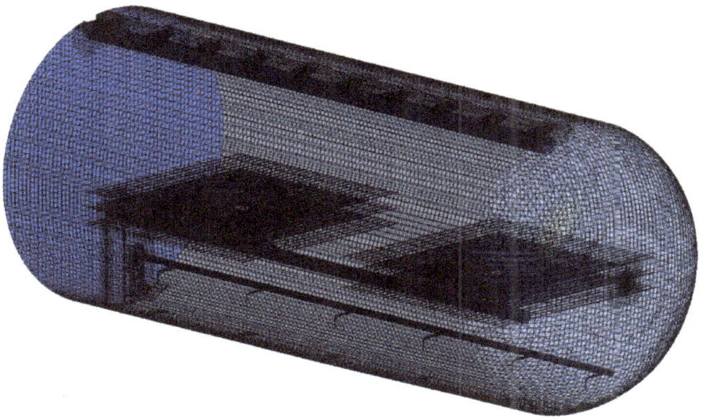

Figure 2. Optimal mesh design, after performing a grid sensitivity study, indicating a semi-structured grid with 3.4 million cells. Note that the mesh represents only half of the schematic shown in Figure 1 due to the symmetry of the system.

Table 5. Levels of the parameters used in this study.

Level/Variable	E (kV/cm)	X	μ (kg/ms)	D (μm)
(+)	3	0.12	0.071	20
(−)	0.1	0.03	0.017	1

Table 6. 2^4 full factorial experimental design.

Case Number	E	X	μ	D
1	−	−	−	−
2	−	−	−	+
3	−	−	+	−
4	−	−	+	+
5	−	+	−	−
6	−	+	−	+
7	−	+	+	−
8	−	+	+	+
9	+	−	−	−
10	+	−	−	+
11	+	−	+	−
12	+	−	+	+
13	+	+	−	−
14	+	+	−	+
15	+	+	+	−
16	+	+	+	+

2. Results

2.1. Effect of Electric Field

In this section, the effect of every studied variable on the separation of water from oil is presented through the comparison of the oil volume fraction contour plots and velocity vector plots of cases with high and low levels of each variable analyzed, keeping the other variables fixed. Then, the statistical analysis is shown and, after this quantitative analysis, the process optimization is presented.

Figure 3 shows the oil volume fraction contours in a longitudinal plane of the desalting unit, comparing the high electric field of 3 kV/cm (Figure 3a) and the low electric field of 0.1 kV/cm (Figure 3b). This variable significantly impacts the electrostatic separation even with a high amount of water (12%). Crude oil with more than 97% of oil is achieved at 3 kV/cm, while only a separation of 80% at 0.1 kV/cm is achieved. The electric field promotes the random motion of charged droplets, producing the necessary forces to approach water droplets and promoting collisions for coalescence and separation.

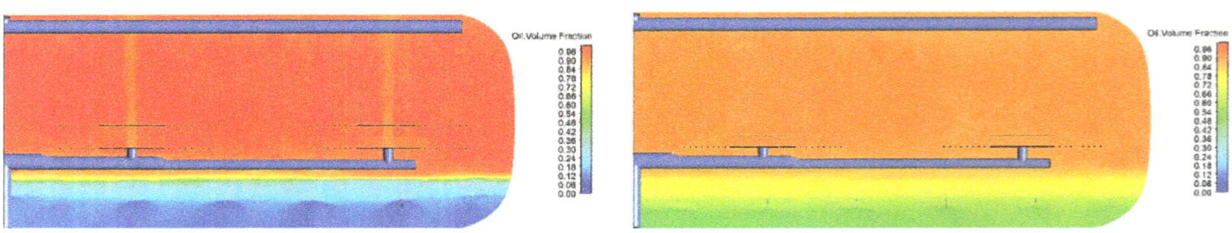

(a) (b)

Figure 3. Effect of the electric field on the oil volume fraction in a longitudinal plane for (**a**) high electric field, experiment 11 (D = 1 μm, μ = 0.071 kg/ms, x = 0.12, and E = 3 kV/cm); (**b**) low electric field, experiment 15 (D = 1 μm, μ = 0.071 kg/ms, x = 0.12, and E = 0.1 kV/cm).

2.2. Effect of Temperature (Oil Viscosity)

Figure 4 shows the oil volume fraction contours in a longitudinal plane of the desalting unit, comparing the case of high oil viscosity (low temperature) of 0.071 kg/ms (Figure 4a) and low oil viscosity (high temperature) of 0.017 kg/ms (Figure 4b). This variable has a lower effect on the electrostatic separation than the electrostatic field, but the influence is still evident. Crude oil desalting and dehydration at high temperatures (high viscosity of oil) are achieved better than at low temperatures (low viscosity of oil). One of the forces droplets feel that prevents the separation is the drag forces from the oil to the water droplet, so the higher the viscosity, the higher the drag forces and the worse the separation.

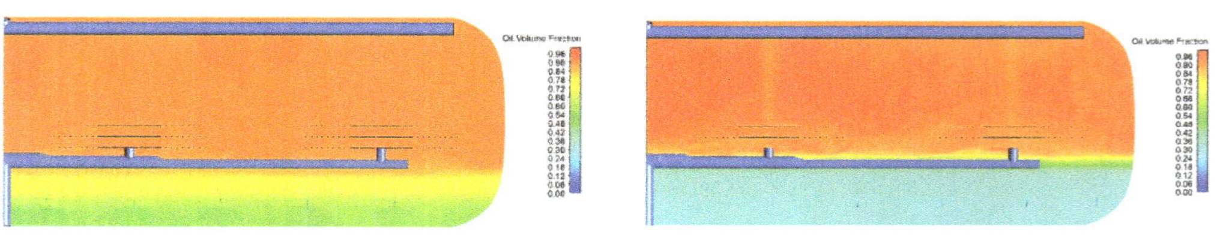

(a) (b)

Figure 4. Effect of the temperature (oil viscosity) on the oil volume fraction in a longitudinal plane for (**a**) high viscosity, experiment 12 (D = 20 μm, μ = 0.071 kg/ms, x = 0.12, and E = 0.1 kV/cm); (**b**) low viscosity, experiment 10 (D = 20 μm, μ = 0.017 kg/ms, x = 0.12, and E = 0.1 kV/cm).

2.3. Effect of Water Content

Figure 5 shows the oil volume fraction contours in a longitudinal plane of the desalting unit, comparing the case of the high water content of 12% (Figure 5a) and low water content of 3% (Figure 5b). This variable has a lower effect on the electrostatic separation than the electrostatic field but a more significant impact than the oil viscosity (temperature). Water separation is, in practice, achieved by adding extra wash water to the water already present in the emulsion. Nevertheless, adding too much water may be counterproductive and prevent water separation. Therefore, water content is one of the main variables to be optimized in practice. Increasing the water content intends to form a w/o emulsion with water droplets closer to one another and requires a lower electric field. However, adding more water than necessary may need too many collisions to achieve the coalescence, and this high number of collisions may not be achieved in the residence time of the droplets.

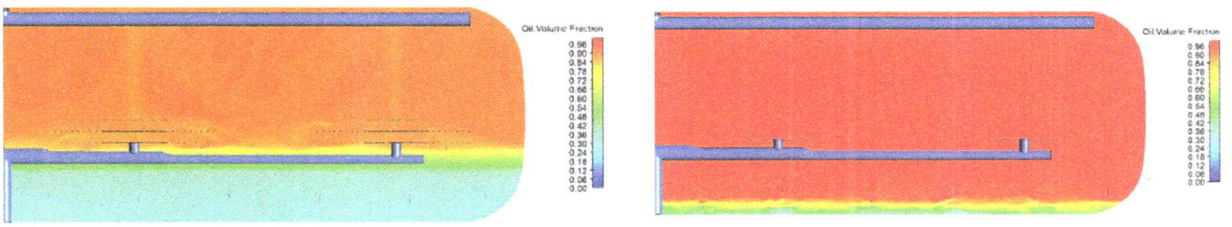

(a) (b)

Figure 5. Effect of the water content on the oil volume fraction in a longitudinal plane for (**a**) high water content, experiment 9 (D = 1 μm, μ = 0.017 kg/ms, x = 0.12, and E = 0.1 kV/cm); (**b**) low water content, experiment 1 (D = 1 μm, μ = 0.017 kg/ms, x = 0.03, and E = 0.1 kV/cm).

2.4. Effect of Droplet Size

Figure 6 shows the oil volume fraction contours in a longitudinal plane of the desalting unit, comparing the case of a large droplet size of 20 microns (Figure 6a) and a small droplet size of 1 micron (Figure 6b). This variable has the lowest effect on the electrostatic separation of all the variables explored in this study. Therefore, the droplet sizes used for the survey probably show no sensitivity to the separation. The effect of the droplet size may play a role if the size is big enough to promote buoyancy forces in the case of light crude oil, but this effect has a lower impact with heavy crude oil. The result suggests that drag and electrostatic forces dominate the motion of the droplets.

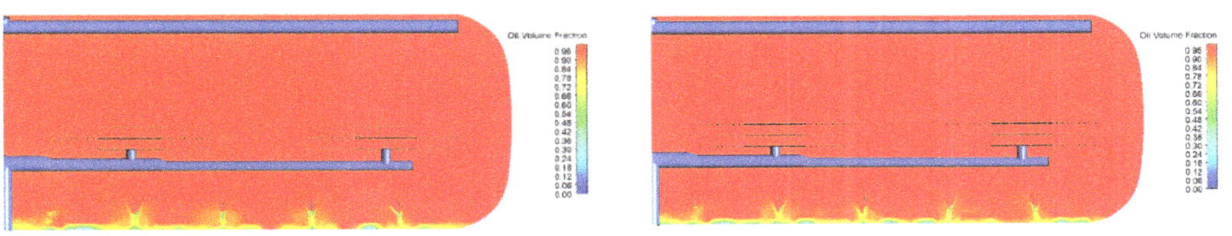

(a) (b)

Figure 6. Effect of the droplet size on the oil volume fraction in a longitudinal plane for (**a**) high droplet size, experiment 6 (D = 20 μm, μ = 0.017 kg/ms, x = 0.03, and E = 3 kV/cm); (**b**) low droplet size, experiment 5 (D = 1 μm, μ = 0.017 kg/ms, x = 0.03, and E = 3 kV/cm).

3. Discussion

3.1. Hydrodynamic Analysis

From Figure 7, it is possible to explain the behavior presented in Figures 3 and 5 on the oil volume fraction or separation of oil from water, where the effects of the electric field and water content are analyzed, which are the most significant variables in the process.

It is concluded that the separation is significantly improved by raising the electric field from 0.1 kV/cm to 3 kV/cm. By examining Figure 7c,d with electric fields of 3 kV/cm and 0.1 kV/cm, respectively, it can be seen that at 3 kV/cm, the turbulence is much lower than in the case of 0.1 kV/cm. A similar trend is found in Figure 7a,b with a water content of 0.12 and 0.03, respectively, where the turbulence is much lower in the case of better separation (x = 0.12) than in the case of worst separation (x = 0.03), showing high turbulence. These results suggest that turbulence promotes collision and prevents the coalescence of oil droplets. Then, moderate turbulence is required to encourage collisions but not too much to allow the coalescence of droplets.

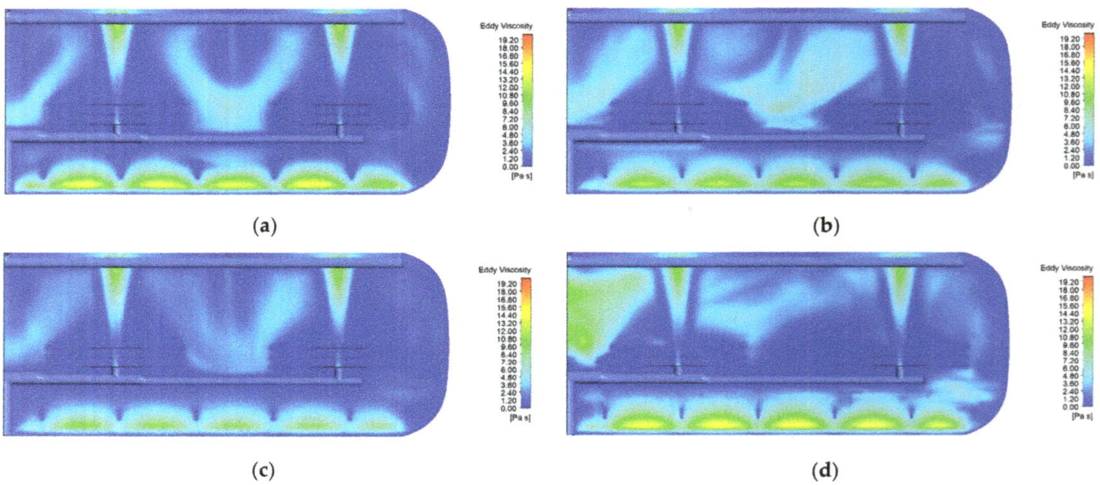

Figure 7. Effect of the water content on the Eddy viscosity in a longitudinal plane for (**a**) high water content, experiment 9 (D = 1 μm, μ = 0.017 kg/ms, x = 0.12, and E = 0.1 kV/cm); (**b**) low water content, experiment 1 (D = 1 μm, μ = 0.017 kg/ms, x = 0.03, and E = 0.1 kV/cm). Effect of the electric field on the Eddy viscosity in a longitudinal plane for (**c**) high electric field, experiment 11 (D = 1 μm, μ = 0.071 Kg/ms, x = 0.12, and E = 3 kV/cm); (**d**) low electric field, experiment 15 (D = 1 μm, μ = 0.071 Kg/ms, x = 0.12, and E = 0.1 kV/cm).

Further explanation about the differences in the separation obtained by changing the water content and the electric field comes from Figure 8, where the effect of the electric field and water content on the liquid velocity field is analyzed. In all cases, basic features of the flow field are the vertical high-velocity flow coming from the inlets, a clockwise circulation at the right lateral wall, and a low-velocity zone below the water outlets. In a high electric field of 3 kV/cm (Figure 8c), two well-defined circulation loops appear in the center of the desalter that are not defined at low electric fields of 0.1 kV/cm (Figure 8d). In the case of the high water content of x = 0.012 (Figure 8a), these circulations in the center are present but not at low water content (Figure 8b). These circulation loops seem to increase the residence time of oil droplets that help in the separation process, although the residence time curves were not considered in this work.

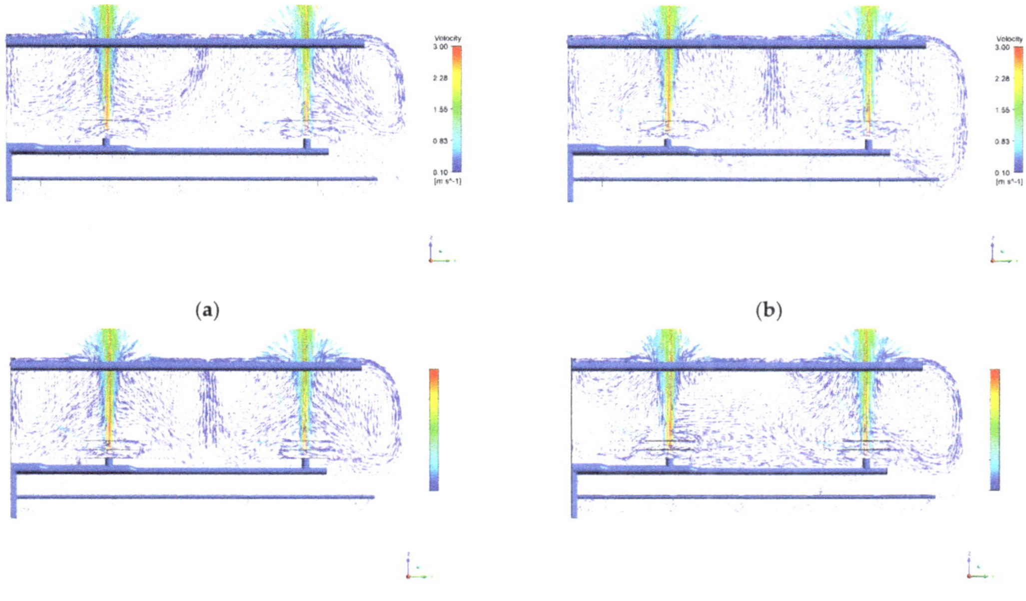

(a) (b) (c) (d)

Figure 8. Effect of the water content on the velocity in a longitudinal plane for (**a**) high water content, experiment 9 (D = 1 μm, μ = 0.017 kg/ms, x = 0.12, and E = 0.1 kV/cm); (**b**) low water content, experiment 1 (D = 1 μm, μ = 0.017 kg/ms, x = 0.03, and E = 0.1 kV/cm). Effect of the electric field on the velocity in a longitudinal plane for (**c**) high electric field, experiment 11 (D = 1 μm, μ = 0.071 kg/ms, x = 0.12, and E = 3 kV/cm); (**d**) low electric field, experiment 15 (D = 1 μm, μ = 0.071 kg/ms, x = 0.12, and E = 0.1 kV/cm).

3.2. Statistical Analysis and Optimization

Table 7 shows the statistical analysis of each main effect of the four variables, along with double effects. Figure 9 shows a Pareto graph where the single results are reported along with double and triple interactions and the statistical significance through a dashed line. All effects lower than this line may be considered statistically insignificant and are, consequently, not reported in Table 7.

Table 7. Analysis of Variance.

Term	Effect	Coefficient	Std Dev.	*p*-Value
μ	0.010134	0.005067	0.000656	0.001
E	−0.03191	−0.01596	0.000656	0.000
X	0.04239	0.02119	0.000656	0.000
μ × x	0.007805	0.003903	0.000656	0.002
E × x	−0.02950	−0.01475	0.000656	0.000

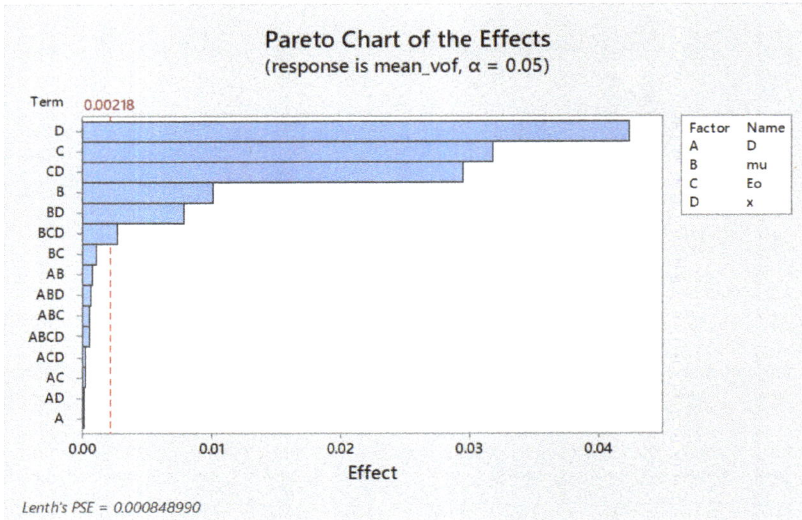

Figure 9. Pareto chart of the single effects, namely pressure drop at the inlet valve (droplet size, D), content of freshwater x, temperature (viscosity of the oil μ), and electric field E. In addition, effects of double, triple, and quadruple interactions are shown.

Table 7 and Figure 9 clearly state that only three variables and two double interactions are significant in oil separation. These are from high to low effects: water content, electric field, the combined effect of water content–electric field, and oil viscosity (temperature). If the initial water content increases from 0.03 to 0.12, water separation decreases by 4.24%. Aryafard et al. [14,15] found a similar improvement of oil content at the outlet from 96 to 98.5% when wash water is changed from 3 to 6% and validated in an industrial unit. Supposing the electric field increases from 0.1 kV/cm to 3 kV/cm, the separation of salty water is enhanced by 3.2%. Not too many studies are clear on the effect of the electric strength. Aryafard [14] et al. showed that increasing the electric field from 1.5 to 2 kV/cm increases the water removal efficiency from 96.4% to 97.8% in the crude oil. The effect of the combination of water content and the electric field is interesting, which indicates that a simultaneous increment in water and the electric field is beneficial by 3% separation. Finally, the increment in viscosity from 0.017 kg/ms to 0.071 kg/ms diminishes water separation from the oil by almost 1%. The droplet size resulted as insignificant. However, in the real process, this variable is linked somehow to the amount of fresh water in the desalting unit. Thus, the true nature of the simulation does not distinguish a case with high water content and small droplet size from a case with high water content and small droplet size. One of these two conditions may be challenging to meet in a real desalter. In the literature, many authors point out that the droplet size is essential as its size dominates the forces acting on the droplet that causes the collision and coalescence process needed for the separation [10,13,14].

The statistical analysis allowed us to obtain a response surface of the final water content at the oil outlet as a function of the main variables explored and expressed through the following equation not including the insignificant terms as:

$$\text{Mean Volume Water Fraction} = 0.01041 - 0.1221\mu + 0.003830E + 0.6304x + 4.373\mu \times x - 0.1865E \times x \quad (1)$$

The above equation was used to obtain a formal optimization analysis through the genetic algorithm built in the optimization module of MATLAB. The optimization consisted in minimizing the water content at the oil outlets, i.e., maximizing the separation of water from the crude oil. The outcome of this optimization exercise is the recognition of 25 optimum conditions (Pareto front) described in Table 8.

Table 8. Pareto front with optima.

Optimum Point	D	μ	E	x	Mean Volume Fraction of Water
1	3.06095597	0.01707254	2.99339092	0.03003833	0.02526776
2	3.52471021	0.01701073	2.99998767	0.03001266	0.02524942
3	3.55716842	0.01703218	2.98611306	0.03003497	0.02527249
4	6.1115551	0.01701526	2.99889999	0.03004764	0.02524273
5	7.33734703	0.01701039	2.9964797	0.03003354	0.02523891
6	7.7335968	0.01704495	2.99988084	0.03008657	0.0252396
7	7.83882932	0.01704015	2.94999207	0.03003251	0.0253042
8	8.06589326	0.01707272	2.91442089	0.03008012	0.02536068
9	8.52232293	0.0170021	2.99859394	0.03000871	0.02522771
10	10.0636939	0.01710667	2.99625853	0.03005619	0.0252359
11	10.2583472	0.01702999	2.96275447	0.03005567	0.02527663
12	10.3801102	0.01704703	2.99860591	0.03009518	0.02523013
13	11.7521461	0.01704481	2.99105772	0.03001022	0.02522692
14	12.2213506	0.01702969	2.98366032	0.03006916	0.02523927
15	12.4154352	0.01703886	2.99939996	0.03004532	0.02521476
16	12.9440779	0.01701494	2.99693118	0.03003288	0.02521299
17	13.0555521	0.01702826	2.90428816	0.03000283	0.02534075
18	13.6925743	0.01700749	2.97734302	0.03003171	0.0252365
19	13.70686	0.0170153	2.99584673	0.0300342	0.02521118
20	15.5920049	0.01707421	2.97423509	0.03002545	0.0252366
21	16.5123639	0.01700549	2.99958726	0.03006071	0.02519454
22	16.6602285	0.01708459	2.99414794	0.03002029	0.02520413
23	17.0696906	0.01700989	2.98707433	0.03003351	0.02520774
24	18.7182166	0.01700966	2.98791986	0.03000627	0.02519679
25	18.8514308	0.01704943	2.9998606	0.03006069	0.02518672

From the optimum variables presented in Table 8, the previous analysis can be confirmed through the comparison of the oil volume fraction contours, where the effect of every variable was discussed. Then, the optimum conditions under the range of values explored in this study are (a) electric field of 3 kV/cm, (b) high temperature with an oil viscosity of 0.017 kg/ms, and (c) initial water content of 3%, and the size of the droplet that has no effect may vary from 19 to 3 μm, ratifying that this variable showed no effect on the separation. All these conditions indicate a final water content of 2.5%, as can be seen in the last column in Table 8, where the optimum water content at the oil outlets is presented as the optimum value. Comparison against other studies is difficult because the variables are not the same and because there are only a few optimization studies. For instance, Mahdi et al. [8] reported optimum values of a demulsifying agent concentration of 15 ppm, temperature of 77 °C, 10% wash water dilution ratio, settling time of 3 min, and mixing time of 9 min. Aryafard [14] et al. proposed the following increments in the electric field from 1.5 to 2 kV/cm, wash water from 3% to 6%, and pressure drop from 20 to 30 psi.

4. Conclusions

The statistical analysis indicated that the most significant variables affecting the water separation from oil were from higher to lower impact: the water content, the electric field intensity, and the oil viscosity. At the same time, the droplet size was insignificant in the separation under the conditions explored in this work. Increasing the electric field and decreasing the water content and oil viscosity (increasing temperature), better separation is achieved. It was found that the excess turbulence promotes collisions but may prevent actual coalescence and, therefore, it does not help in the separation. Additionally, it was found that the many circulation loops found in the cases of high electric fields and low water content may increase the residence time of the droplets, giving enough time for better separation than in the cases of low electric fields and high water content. Optimum values suggested from the optimization are E0 = 3 kV/cm, μ = 0.017 kg/ms, and water content of 3%.

Author Contributions: Conceptualization, M.A.R.-A. and A.D.; methodology, D.A.-L. and M.A.R.-A.; software, D.A.-L. and A.D.; validation, M.A.R.-A., A.D. and D.A.-L.; formal analysis, A.D.; resources, M.A.R.-A.; writing—original draft preparation, M.A.R.-A.; writing—review and editing, A.D.; project administration, J.G.-F.; funding acquisition, J.G.-F. All authors have read and agreed to the published version of the manuscript.

Funding: This research was funded by SENER—CONACYT Grant number 144156, project entitled: "Alternativas tecnológicas para mejorar el sistema de desalado de crudo pesado en las refinerías".

Institutional Review Board Statement: This study did not require ethical approval.

Informed Consent Statement: Not applicable.

Conflicts of Interest: The authors declare no conflict of interest.

Abbreviation

Symbol	Meaning	Units	Symbol	Meaning
ρ	density	[kg m^{-3}]	Sub index	
α	Volume fraction	[-]	M	mixture
N	Number of phases	[-]	I	i-th phase
μ	Viscosity	[kg m^{-1} s^{-1}]	T	Turbulence
v	Average velocity	[m s^{-1}]	D	Diffusion
p	Pressure	[N m^{-2}]	d,i	Drift of i-th phase
τ	Stress tensor	[N m^{-2}]	w	Water
$\acute{v}$	Fluctuating velocity	[m s^{-1}]	o	Oil
g	Gravitational constant	9.81 [m s^{-2}]	c	Coalescence
η	Kinematic viscosity	[m^2 s^{-1}]	b	Breakup
Pr	Prandtl number	[-]		
f_{drag}	Drag coefficient	[-]		
Re	Reynolds number	[-]		
d_w	Droplet diameter	[m]		
χ_w	Interfacial area concentration	[m^2 m^{-3}]		
S_{TI}	Droplet breakup	[m^{-1} s^{-1}]		
S_{RC}	Droplet Coalescence	[m^{-1} s^{-1}]		
n_c	Number of droplets per volume of the mixture	[m^{-3}]		
f_c	Frequency of collisions	[s^{-1}]		
λ_c	Coalescence probability	[-]		
n_b	Number of eddies per volume of the mixture	[m^{-3}]		
f_b	Frequency of collisions due to turbulence	[s^{-1}]		
λ_b	Breakup efficiency	[-]		
$\alpha_{w,max}$	The maximum water volume fraction	[-]		

ε	Energy dissipation rate	$[m^2\ s^{-3}]$
k	Turbulent kinetic energy	$[m^2\ s^{-2}]$
P_k	Production of turbulence	$[m^2\ s^{-3}]$
S	Strain tensor	$[s^{-1}]$
$\sigma_{T,k}$, $\sigma_{T,\varepsilon}$, $C_{1\varepsilon}$, $C_{2\varepsilon}$, C_μ	Constants of the k-ε realizable turbulence model	[-]

References

1. Lovell, J. Climate report calls for green 'New Deal'. *Reuters* **2008**.
2. Chohan, U.W. A Green New Deal: Discursive Review and Appraisal. Notes on the 21st Century (CBRI). *Notes 21st Century CBRI* **2019**. [CrossRef]
3. Varadaraj, R.; Savage, D.W.; Brons, C.H. Chemical Demulsifier for Desalting Heavy Crude. U.S. Patent 6,168,702, 28 January 2001.
4. Alshehri, A. Modeling and Optimization of Desalting Process in Oil Industry. Master's Thesis, University of Waterloo, Waterloo, ON, Canada, 2009; p. 74.
5. Abdel-Aal, H.K.; Aggour, M.A.; Fahim, M.A. *Petroleum and Gas Field Processing*; CRC Press: Boca Raton, FL, USA, 2003.
6. Al-Otaibi, M.B.; Elkamel, A.; Nassehi, A.V.; Abdul-Wahab, S.A. A Computational Intelligence Based Approach for the Analysis and Optimization of a Crude Oil Desalting and Dehydration Process. *Energy Fuels* **2005**, *19*, 2526–2534. [CrossRef]
7. Sams, G.W.; Zaouk, M. Emulsion Resolution in Electrostatic Processes. *Energy Fuels* **1999**, *14*, 31–37. [CrossRef]
8. Mahdi, K.; Gheshlaghi, R.; Zahedi, G.; Lohi, A. Characterization and modeling of a crude oil desalting plant by a statistically designed approach. *J. Pet. Sci. Eng.* **2008**, *61*, 116–123. [CrossRef]
9. Herrmann, H.; Bucksch, H. Mud Logging. In *Dictionary Geotechnical Engineering/Wörterbuch GeoTechnik*; Springer: Berlin/Heidelberg, Germany, 2014. [CrossRef]
10. Bresciani, A.E.; Alves, R.M.B.; Nascimento, C.A.O. Coalescence of Water Droplets in Crude Oil Emulsions: Analytical Solution. *Chem. Eng. Technol.* **2010**, *33*, 237–243. [CrossRef]
11. Pruneda, E.F.; Rivero, E.; Escobedo, B.; Javier, F.; Vázquez, G. Optimum Temperature in the Electrostatic Desalting of Maya Crude Oil. *J. Mex. Chem. Soc. Vol.* **2005**, *49*, 14–19.
12. Wilkinson, D.; Waldie, B.; Nor, M.I.M.; Lee, H.Y. Baffle plate configurations to enhance separation in horizontal primary separators. *Chem. Eng. J.* **2000**, *77*, 221–226. [CrossRef]
13. Bresciani, A.E.; Mendonça, C.F.; Alves, R.M.; Nascimento, C.A. Modeling the kinetics of the coalescence of water droplets in crude oil emulsions subject to an electric field, with the cellular automata technique. *Comput. Chem. Eng.* **2010**, *34*, 1962–1968. [CrossRef]
14. Aryafard, E.; Farsi, M.; Rahimpour, M.; Raeissi, S. Modeling electrostatic separation for dehydration and desalination of crude oil in an industrial two-stage desalting plant. *J. Taiwan Inst. Chem. Eng.* **2016**, *58*, 141–147. [CrossRef]
15. Aryafard, E.; Farsi, M.; Rahimpour, M.R. Modeling and simulation of crude oil desalting in an industrial plant considering mixing valve and electrostatic drum. *Chem. Eng. Process. Process Intensif.* **2015**, *95*, 383–389. [CrossRef]
16. Kakhki, N.A.; Farsi, M.; Rahimpour, M.R. Effect of current frequency on crude oil dehydration in an industrial electrostatic coalescer. *J. Taiwan Inst. Chem. Eng.* **2016**, *67*, 1–10. [CrossRef]
17. Vafajoo, L.; Ganjian, K.; Fattahi, M. Influence of key parameters on crude oil desalting: An experimental and theoretical study. *J. Pet. Sci. Eng.* **2012**, *90*, 107–111. [CrossRef]
18. Bansal, H. Ameensayal. CFD Analysis of Horizontal Electrostatic Desalter-Influence of Header Obstruction Plate Design on Crue-Water Separation. 2015. Available online: https://www.digitalxplore.org/up_proc/pdf/51-13951413061-6.pdf (accessed on 27 August 2022).
19. Shariff, M.M.; Oshinowo, L.M. Debottlenecking Water-Oil Separation with Increasing Water Flow Rates in Mature Oil Fields. In Proceedings of the 5th Water Arabia 2017 Conference and Exhibition, Al-Khobar, Saudi, 17 October 2017; pp. 17–19.
20. Alhajri, N.A.; White, R.J.; Oshinowo, L.M. High Efficiency Static Mixer Technology for Crude Desalting. In Proceedings of the SPE Kingdom of Saudi Arabia Annual Technical Symposium and Exhibition, Dammam, Saudi Arabia, 23–26 April 2018; SPE-192388-MS. OnePetro: Richardson, TX, USA, 2018. [CrossRef]
21. Favero, J.L.; Silva, L.F.L.; Lage, P.L. Modeling and simulation of mixing in water-in-oil emulsion flow through a valve-like element using a population balance model. *Comput. Chem. Eng.* **2015**, *75*, 155–170. [CrossRef]
22. Wang, F.; Wang, L.; Chen, G.; Zhu, D. Numerical Simulation of the Oil Droplet Size Distribution Considering Coalescence and Breakup in Aero-Engine Bearing Chamber. *Appl. Sci.* **2020**, *10*, 5648. [CrossRef]
23. Sofos, F. A Water/Ion Separation Device: Theoretical and Numerical Investigation. *Appl. Sci.* **2021**, *11*, 8548. [CrossRef]
24. Shi, Y.; Chen, J.; Pan, Z. Experimental Study on the Performance of a Novel Compact Electrostatic Coalescer with Helical Electrodes. *Energies* **2021**, *14*, 1733. [CrossRef]
25. Sellman, E.; Sams, G.W.; Mandewalkar, S.P.K. Improved Dehydration and Desalting of Mature Crude Oil Fields. In Proceedings of the SPE Middle East Oil and Gas Show and Conference, Manama, Bahrain, 10 March 2013. [CrossRef]
26. Ramirez-Argaez, M.; Abreú-López, D.; Gracia-Fadrique, J.; Dutta, A. Numerical Study of Electrostatic Desalting Process Based on Droplet Collision Time. *Processes* **2021**, *9*, 1226. [CrossRef]

27. Dutta, A.; Ramírez-Argáez, M.A. Modeling of an Industrial Electrostatic Desalting Unit Using Computational Fluid Dynamics (CFD). Scientific Conference "Science, Technology and Development of Innovative Technologies". In Proceedings of the 30th Anniversary of Independence of Turkmenistan, Ashgabat, Turkmenistan, 12–13 June 2021.
28. Pericleous, K.A.; Drake, S.N. An Algebraic Slip Model of Phoenics for Multi-phase Applications. In *Numerical Simulation of Fluid Flow and Heat/Mass Transfer Processes*; Springer: Berlin/Heidelberg, Germany, 1986; pp. 375–385. [CrossRef]
29. Kar, S.; Majumder, S.; Constales, D.; Pal, T.; Dutta, A. A comparative study of multi objective optimization algorithms for a cellular automata model. *Rev. Mex. Ing. Quim.* **2019**, *19*, 299–311. [CrossRef]

processes

MDPI

Article

CO_2 Adsorption Performance on Surface-Functionalized Activated Carbon Impregnated with Pyrrolidinium-Based Ionic Liquid

Syeda Saba Fatima [1], Azry Borhan [2,*], Muhammad Ayoub [1] and Noraini Abd Ghani [3]

1 Department of Chemical Engineering, Universiti Teknologi PETRONAS, Bandar Seri Iskandar 32610, Perak, Malaysia
2 HICoE, Centre of Biofuel and Biochemical Research, Institute of Self-Sustainable Building, Department of Chemical Engineering, Universiti Teknologi PETRONAS, Bandar Seri Iskandar 32610, Perak, Malaysia
3 Department of Fundamental and Applied Sciences, Universiti Teknologi PETRONAS, Bandar Seri Iskandar 32610, Perak, Malaysia
* Correspondence: azrybo@utp.edu.my

Abstract: The serious environmental issues associated with CO_2 emissions have triggered the search for energy efficient processes and CO_2 capture technologies to control the amount of gas released into the atmosphere. One of the suitable techniques is CO_2 adsorption using functionalized sorbents. In this study, a functionalized activated carbon (AC) material was developed via the wet impregnation technique. The AC was synthesized from a rubber seed shell (RSS) precursor using chemical activation and was later impregnated with different ratios of [bmpy][Tf_2N] ionic liquid (IL). The AC was successfully functionalized with IL as confirmed by FTIR and Raman spectroscopy analyses. Incorporation of IL resulted in a reduction in the surface area and total pore volume of the parent adsorbent. Bare AC showed the largest S_{BET} value of 683 m^2/g, while AC functionalized with the maximum amount of IL showed 14 m^2/g. A comparative analysis of CO_2 adsorption data revealed that CO_2 adsorption performance of AC is majorly affected by surface area and a pore-clogging effect. Temperature has a positive impact on the CO_2 adsorption capacity of functionalized AC due to better dispersion of IL at higher temperatures. The CO_2 adsorption capacity of AC (30) increased from 1.124 mmol/g at 25 °C to 1.714 mmol/g at 40 °C.

Keywords: CO_2 capture; activated carbon; functionalization; ionic liquid; biomass; wet impregnation

Citation: Fatima, S.S.; Borhan, A.; Ayoub, M.; Ghani, N.A. CO_2 Adsorption Performance on Surface-Functionalized Activated Carbon Impregnated with Pyrrolidinium-Based Ionic Liquid. *Processes* **2022**, *10*, 2372. https://doi.org/10.3390/pr10112372

Academic Editors: Jun Wei Lim and Worapon Kiatkittipong

Received: 15 October 2022
Accepted: 10 November 2022
Published: 12 November 2022

1. Introduction

Over the years, a massive increase in greenhouse gas (GHG) emissions has led to the warming of the earth's atmosphere and these emissions are likely to increase further in future. The primary cause of GHG emissions is the rapid progress of modern civilization and industrial development [1]. Anthropogenic CO_2 is the main contributor to GHG emissions and is largely discharged from fossil fuel combustion from both vehicles and industries. Over 30 billion tons of CO_2 are released in the atmosphere every year [2]. This alarming increase of CO_2 in the atmosphere has already triggered climate issues, such as acid rain, melting of the ice caps and urban smog. Therefore, the need of the hour is to stabilize the level of CO_2 in the atmosphere to avoid irreversible and perpetual destruction of the atmosphere. This requires reducing the use of fossil fuels by using alternative fuel sources such as biofuels, promoting reforestation and using separation methods for CO_2 removal to reduce the daily release of the gas from prime sources, especially power plants and refineries [3].

Many separation techniques are available for post-combustion CO_2 capture, including absorption, membrane separation, cryogenic distillation, adsorption, chemical looping combustion and hydrate-based CO_2 capture [4]. Adsorption is a suitable separation method because of its numerous advantages, including a wide range of solid adsorbents, high CO_2 adsorption capacity, easy regeneration and good selectivity at low concentrations [5,6]. Although many adsorbent materials have been explored for CO_2 capture [7–9], researchers are still trying to come up with new and advanced adsorbents and explore the adsorption kinetics and thermodynamics for better performing materials. Despite extensive research, adsorption is still far from large-scale applications due to technical and economic difficulties. Since the adsorption process itself is a costly process, development of cheaper solid adsorbents is required to make this method economically desirable. Biomass-derived adsorbents are especially important in this context. The most common precursors include agricultural waste, woody and non-woody biomass, industrial by-products and sludge. The first two categories are important because of their high carbon content and the low ash content of the resulting activated carbon (AC) [10]. Different fruit peels and fruit stones have been used to synthesize AC with high porosity, large surface area and good adsorption properties. These include banana peel [6], olive stone [11,12], coconut shell [13] and peanut shell [14]. Other waste materials used as raw materials for preparing porous AC include rice husk [15], bamboo [16], tea shell [8], tobacco [17] and eucalyptus wood [18]. Malaysia is a leading rubber-producing country after Thailand and Indonesia. According to the Department of Statistics Malaysia (DOSM), the stocks of rubber production were recorded at 296,328 tonnes in February 2020 [19]. This huge production rate also generates large quantities of solid waste laying around due to poor waste management. Rubber seed shell (RSS) is suitable as a precursor for the synthesis of AC due to its high carbon content, low cost and availability [7].

AC can be synthesized by either physical activation or chemical activation. The chemical activation method has the advantages of shorter activation time and lower temperature, and is generally preferred over the physical activation method [20]. Chemical activation involves the impregnation of the precursor with activating agents (such as KOH, NaOH, K_2CO_3, HNO_3 or $ZnCl_2$) and carbonization under inert or limited oxygen environment, followed by washing to remove impurities to obtain porous AC. Quan et al. (2019) [8] used tea seed shells to produce highly porous AC using KOH activation and reported a CO_2 adsorption capacity of 3.15 mmol/g at 25 °C and 1 bar. Plaza et al. (2012) [15] synthesized AC from spent coffee grounds using both physical activation and chemical activation methods. AC produced by physical activation showed higher selectivity for CO_2/N_2 and CO_2/He streams. Borhan et al. (2019) [7] developed microporous AC from RSS using KOH activation. A maximum surface area of 1129 m^2/g and maximum CO_2 adsorption capacity of 54.3 mg/g at 25 °C and 1.25 bar were observed.

Although biomass-derived AC shows good CO_2 adsorption performance, it cannot compete with non-carbonaceous adsorbents, such as zeolites and metal organic frameworks (MOFs) [21]. Studies show that the performance of an adsorbent can be enhanced for CO_2 adsorption capacity, selectivity and the range of favorable adsorption conditions by treatment with suitable functional groups [22–26]. Surface functionalization introduces additional active sites and improves the CO_2 adsorption capacity. Surface functionalization can be performed by either grafting or impregnation methods. Grafted adsorbents show faster kinetics and improved thermal stability because of covalent bonding [27]. Unlike grafting, impregnation results in physical attachment of the functional groups to the surface of the adsorbent. The impregnated adsorbents are easily regenerated to their original adsorption capacity compared to the grafted ones [28]. Impregnation is usually preferred due to simpler experimental procedures and higher CO_2 adsorption capacity compared to grafting [4]. Researchers have used both functionalization techniques to prepare surface-functionalized solid adsorbents [29,30]. Amines or aminosilanes are the most common chemicals used for surface functionalization of solid adsorbents for CO_2 separation applications [24,31–36]. The reason for this is the high affinity of amino groups towards acidic gases like CO_2 and SO_2. Bezerra et al. (2011) [33] prepared monoethanolamine (MEA) impregnated AC and zeolite 13X. Prepared adsorbents showed a CO_2 adsorption capacity of 45 mg/g and 136 mg/g at 1 bar and 25 °C. It was observed that the incorporation of amine groups improves the CO_2 adsorption capacity at high temperatures (70–90 °C) due to the contribution of chemisorption, whereas physisorption was found to be dominant at low temperatures (25–30 °C) [22]. Arenillas et al. (2005) [23] reported similar findings for Polyethyleneimine (PEI) impregnated fly ash with up to a 4.5% increase in CO_2 adsorption capacity at 75 °C over room temperature. On the other hand, commercial AC relying on physisorption alone showed lower CO_2 adsorption capacity at 75 °C. Sanz-Pérez et al. (2018) [30] prepared functionalized HMS mesoporous silica grafted with diethylenetriamine-trimethoxysilane (DT) and impregnated with PEI. Impregnated silica showed a higher CO_2 adsorption capacity (2.0 mmol/g) compared to grafted silica (1.8 mmol/g) at 45 °C and 1 bar. One major disadvantage with amine solvents is the leaching during the regeneration process that results in the formation of toxic byproducts and causes serious secondary pollution [37]. It also reduces the life of the functionalized adsorbent in repeated adsorption–desorption cycles. Ahmed et al. (2017) [38] reported a gradual decrease in the CO_2 adsorption capacity of PEI functionalized MCM-41 after six adsorption–desorption cycles due to the degradation of amino groups. Xu et al. (2002) [39] studied cyclic performance of MCM-41 impregnated with 50% PEI. Experimental findings showed a 5.3% decrease in the CO_2 adsorption capacity after seven cycles using pure CO_2. Similar findings were observed for pore expanded MCM-41 loaded with diethanolamine (DEA) [40]. Prepared adsorbent showed a reduction in the CO_2 adsorption capacity (3.3% lower than the initial value) after seven adsorption–desorption cycles due to slow loss of DEA, as confirmed by mass loss measurements using thermogravimetric analysis (TGA).

Ionic liquids (ILs) have emerged as a class of liquids with CO_2 solubility comparable to that of amines. However, their high prices and high viscosities are the main limitations in their practical application. Recent studies have focused on the use of ILs for surface functionalization, thus developing cost-effective adsorbent materials called the supporting ionic liquid phase (SILP). These hybrid adsorbents contain a thin layer of IL on the surface of a porous support. This arrangement provides a high gas–IL contact area and consequently higher mass transfer coefficients while reducing the amount of IL required, thus creating the critical economic benefit required for large-scale applications [41,42]. In this context, Ban et al. (2015) [43] investigated the effect of immobilized [C_4mim][Tf_2N] towards CO_2 selectivity of ZIF-8 and reported a five times higher CO_2/CH_4 and CO_2/N_2 selectivity in the low-pressure region (0–0.2 bar). The higher CO_2 uptakes result from the high solubility of CO_2 in [C_4mim][Tf_2N] and molecular sieving capacity of functionalized ZIF-8. Yusuf et al. (2017) [44] reported that coconut shell AC impregnated with 30 wt.% choline chloride adsorbed larger amounts of CO_2 compared to bare AC in a shorter time. Zhu et al. (2018) [29] studied the CO_2 adsorption performance of porous silica grafted and impregnated with three different ILs containing the same anion. The grafted silica showed faster diffusivity compared to impregnated silica, suggesting that more IL penetrated the inner pores and hindered the diffusion of CO_2 for the latter.

Despite a large number of research articles dealing with the use of different classes of ILs in functionalized adsorbents for CO_2 capture, there is still only fragmentary literature available on the use of biomass-derived AC for SILP and on the effect of confining ILs into micropores towards CO_2 capture. The selection of IL for surface functionalization is usually done based on the CO_2 solubility, viscosity, and thermal stability of the IL. Results from various studies have confirmed that the choice of an IL anion has the most direct effect on the CO_2 solubility, whereas the effect of an IL cation is less significant, but still noticeable [45]. Among various anions, $[Tf_2N]^-$ shows the highest solubility of CO_2 due to larger number of fluoro groups [46]. Various studies have shown that ILs containing the $[Tf_2N]^-$ anion paired with some imidazolium-based cations show good performance for CO_2 capture [45,47–50]. Unfortunately, the high cost of imidazolium based ILs limits their use. For this study, [bmpy][Tf_2N] IL with the $[Tf_2N]^-$ anion and the pyrrolidinium substituted 1-buthyl-1methylpyrrolidinium bis[trifluoromethylsulfonyl]imide cation was selected for functionalization of RSS-based AC. Literature shows a higher solubility of CO_2 (3.3 mol-CO_2/kg-IL at 20 °C and 3.5 MPa) compared to H_2 (0.03 mol-H_2/kg-IL) in this

IL [51]. This opens the possibility of using a relatively cheaper IL for the functionalization of AC. This study aimed to prepare IL functionalized AC derived from RSS using six different ratios of IL, and to investigate the effect of IL loading and temperature on the CO_2 adsorption performance of the prepared adsorbent.

2. Materials and Methods

Rubber seeds were collected from Kampung Tersusun Tanah Hitam, Chemor, Perak 32100, Malaysia. The shells were separated after removing the inner white mass. The rubber seed shells (RSS) were then washed, dried, ground and sieved to obtain particles between 0.1 to 0.25 mm for further use.

Other chemicals including [bmpy][Tf_2N] (99.5%), K_2CO_3 (>99%) and methanol (99.8%) were purchased from Sigma Aldrich (M) Sdn. Bhd., Petaling Jaya, Selangor, Malaysia. Pure CO_2 (99.98%) for adsorption experiments was provided by Linde Malaysia Sdn. Bhd., Petaling Jaya, Selangor, Malaysia.

2.1. Synthesis of RSS Derived AC

A total of 30 g of ground RSS was activated by mixing it with 100 mL of 0.2 M aqueous K_2CO_3 solution. The mixture was left overnight to ensure that the K_2CO_3 had penetrated the RSS. The activated RSS was dried in an oven at 105 °C for 24 h, followed by carbonization in a tube furnace under N_2 atmosphere at 800 °C for 120 min. The nitrogen flow was set at 100 mL/min. AC was thus obtained was mixed with 0.01 M HCl to remove traces of salt and ash. The mixture was filtered and washed repetitively with distilled water to remove undissolved salt until a neutral pH was obtained. The AC was dried at 105 °C for 24 h. The dried sample was labelled as AC-fresh and stored in an airtight container. Ground and sieved RSS and prepared AC are shown in Figure 1.

(a) (b)

Figure 1. (**a**) Ground and sieved RSS with particle sizes of 0.1–0.25 mm dried at 105 °C (**b**) RSS AC activated by K_2CO_3 and carbonized at 800 °C for 180 min.

2.2. Preparation of Functionalized AC

RSS-derived AC was functionalized with six different concentrations of IL. Impregnation solution was prepared by mixing the desired amount of IL (10–60 wt.%) in 20 mL methanol. A fixed amount of AC was mixed with the impregnation solutions and stirred for 30 min at 350 rpm. The mixture was left overnight to ensure successful functionalization. Afterward, the AC samples were dried in an oven at 60 °C for 8 h to remove the solvent and stored in airtight bottles for further use. All functionalized AC samples were labelled as AC (x) with x representing the wt.% of the IL in the impregnation solution.

2.3. Characterization of Prepared Adsorbent

The AC samples were analyzed by a Field Emission Scanning Electron Microscope (FESEM) (Zeiss Supra 55 VP, Germany) using different magnification scales to study the structure and morphological properties before and after IL impregnation. The FESEM was equipped with an Energy Dispersive X-ray (EDX) analyzer to study the composition and distribution of the elements of interest. Surface functionalization was confirmed using

Fourier Transform Infrared Spectroscopy (FTIR) (Frontier 01, PerkinElmer, USA). IR spectra of fresh and functionalized AC were collected by the KBr pelleting method between the wavelength of 400 cm^{-1} to 4000 cm^{-1}. Thermal stability of the prepared AC samples was studied by TGA using a PerkinElmer, STA6000 instrument. The TGA profiles of all AC samples were obtained under N_2 flow for the temperature range of 30–750 °C and a constant heating rate of 10 °C/min. The actual IL loading was quantified using individual TGA profiles. Surface properties of the AC before and after functionalization were estimated by performing surface area and porosity (SAP) analysis using a Brunauer–Emmett–Teller (BET) surface area analyzer (Micromeritics, Tristar 3020 Plus). The relevant surface properties, namely the BET specific surface area (S_{BET}), total pore volume (V_p), microporous surface area (S_{micro}), percentage of micropores and mesopores and the average pore diameter (n), were estimated from the N_2 adsorption isotherms obtained at 77 K. A Raman spectrometer (Horiba, Jobin Yvon HR) was used to examine the extent of the functionalization of the AC. The peaks appear between 200 to 2000 cm^{-1}. The spectrum of AC, like other carbon-based materials, is characterized by graphite (G) and disoriented (D) bands. The intensity ratio (I_D/I_G) was used to quantify the maturity of sp^2 hybridized carbon, which determines the extent of functionalization and increases with disorder in the structure.

2.4. CO_2 Adsorption Test

The adsorption performance of bare and functionalized AC for pure CO_2 (99.98%) was studied using the High Pressure Volumetric Analyzer (HPVA II). The experimental setup of the HPVA system is shown in Figure 2a. Each sample was weighed and placed into the sample holder. Before the adsorption experiment, the prepared sample was degassed under a vacuum for 4 h at 150 °C to ensure a complete removal of impurities. Upon completion of degassing, the sample was cooled to room temperature and transferred to the analysis port. First, helium gas was introduced for free space volume measurement at the ambient and analysis temperatures. Purified helium was injected into the manifold. The gas was allowed to expand into the sample cell once pressure was equilibrated by opening valve (1) while all other valves were kept closed. The temperature and pressure readings were monitored throughout. The amount of helium dosed was obtained using Equation (1) and the free space volume was calculated using Equation (2).

$$n_D = \frac{P_A V_m}{T_A Z_A R} - \frac{P_B V_m}{T_B Z_B R} \tag{1}$$

$$V_{AFS} = V_s + V_{xL} + V_{xU} \tag{2}$$

Figure 2b shows that the upper stem was placed inside the temperature-controlled zone, and therefore its volume (V_{xU}) was assumed to be at the manifold temperature (T_B) after helium dosing. The lower stem volume (V_{xL}) was exposed to ambient temperature (i.e., 298 K). The free space volume at the analysis temperature (V_s) consisted of the volume of the sample cylinder and the stem volumes, was set to the targeted temperature. For free-space measurement at ambient temperature, V_s and V_{xL} were assumed to be at the same temperature as represented in Equation (3).

$$V_{sxL} = V_s + V_{xL} \tag{3}$$

The overall mass balance for calculating V_{sxL} is represented by Equation (4).

$$n_D + \frac{P_{s0} V_{xU}}{T_A Z_{xU0} R} + \frac{P_{s0} V_{sxL}}{T_{s0} Z_{s0} R} = \frac{P_{s1} V_{xU}}{T_B Z_{xU1} R} + \frac{P_{s1} V_{sxL}}{T_{s1} Z_{s1} R} \tag{4}$$

P_{s0} is assumed to be zero since the system was evacuated before the start of the experiment. The net mass balance after combining Equations (3) and (4) is presented in Equation (5).

$$n_D = \frac{P_{s1} V_{xU}}{T_B Z_{xU} R} + \frac{P_{s1} V_{sxL}}{T_{s1} Z_{s1} R} \tag{5}$$

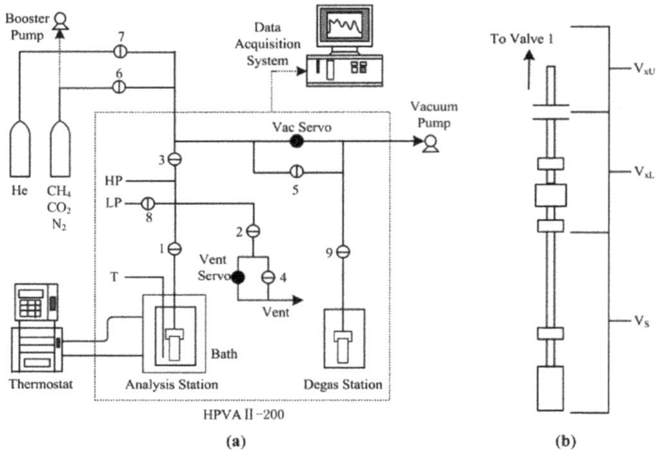

Figure 2. (**a**) Experimental set-up of HPVA equipment used for CO_2 adsorption study. HP, high pressure transducer; LP, low pressure transducer; T, temperature probe; (1), analysis port valve; (2) vent valve; (3) manifold valve; (4) external vent valve; (5) full vacuum valve; (6) inlet valve for $CH_4/CO_2/N_2$ gas; (7) inlet valve for He gas; (8) 1000 torr isolation valve; (9) degassing port valve. (**b**) Parts of HPVA system below valve 1 under different temperature zones [52].

Once V_{sxL} and n_D are known, free space volume is calculated using Equation (2). For free space measurement at the targeted temperature, V_s and V_{xL} were not at the same temperature. Therefore, V_s was estimated using the material balance in Equation (6) and later the free space volume was obtained from Equation (2).

$$n_D = \frac{P_{s1}V_{xU}}{T_B Z_{xU} R} + \frac{P_{s1}(V_{sxL} - V_s)}{T_{AM} Z_{xL1} R} + \frac{P_{s1}V_s}{T_{s1}Z_{s1}R} \tag{6}$$

After free space measurement, the sample cell was evacuated to remove the helium gas completely and the adsorbate (CO_2) was dosed for the adsorption test. The adsorbate was similarly expanded into the sample cell once the pressure was stable. After the adsorption test was complete, desorption was carried out by reducing the pressure in the sample cell at the same intervals. The adsorption study was conducted at 25 °C and 40 °C. The temperature was maintained by coupling the system with a Julabo recirculating water bath. The pressure and temperature data were recorded in a comprehensive data analysis package using Microsoft Excel macro (v.22.0.6) software. The compressibility factor was determined using NIST Reference Fluid Thermodynamic and Transport Properties incorporated in the software. The moles of the adsorbate (CO_2) dosed were calculated using Equation (7).

$$n_{Dosed} = \frac{P'_A V_m}{T'_A Z'_A R} - \frac{P'_B V_m}{T'_B Z'_B R} \tag{7}$$

The remaining amount of adsorbate (i.e., the amount not adsorbed) was calculated using Equation (8). Finally, the amount of adsorbate (CO_2) adsorbed by the sample was estimated by a simple mass balance shown in Equation (9). Each adsorption test was repeated twice. The experimental adsorption capacity is reported as mmol CO_2/g adsorbent.

$$n_{Nadsorb} = \frac{P_s}{R}\left(\frac{V_{sxL}}{Z_s T_s} + \frac{V_{xU}}{Z_{xU}T_{xU}}\right) \tag{8}$$

$$n_{Adsorb} = n_{dosed} - n_{NAdsorb} \tag{9}$$

3. Results and Discussion

Prepared AC samples were analyzed by various characterization techniques to study their structure and CO_2 adsorption performance. The effect of key variables, including IL loading and adsorption temperature on CO_2 adsorption performance, were studied. The details are presented in the following section.

3.1. FESEM and EDX Mapping

The surface chemistry of an adsorbent plays an important role in determining its adsorption properties. A porous surface indicates more active sites and yields higher CO_2 adsorption capacity. The structure and morphological properties of fresh functionalized AC were examined through FESEM under different magnification scales and the respective images are presented in Figure 3, which clearly shows the difference in the structure of AC before and after functionalization. Fresh AC has a highly porous surface with a combination of smaller and larger pores. This fine porosity of AC has resulted in high S_{BET} and V_p. On the contrary, functionalized AC has fewer surface pores due to the pore filling effect of the IL. This effect is more pronounced in AC samples impregnated with higher amounts of IL. At lower IL loading (10–20 wt.%), solvent molecules penetrate the narrow channels of AC, and surface porosity remains intact. As the IL loading is increased, inner pores start to fill up, which in turn reduces the available surface area. At very high IL loading (40–60 wt.%), a layer of bulk liquid is prominent on the surface, blocking most of the pores. This pore clogging and the presence of a thick liquid layer leads to a decrease in CO_2 adsorption capacity due to poor diffusion and low S_{BET}. Aside from examining the structure, the elemental composition of AC before and after functionalization was also analyzed using EDX mapping analysis. The elemental composition is presented in Table 1. Carbon and oxygen are the key elements with maximum composition present in all samples. Small amounts of potassium were also found due to the K_2CO_3 used for the chemical activation of the AC. The quantity of potassium is small and is attributed to the traces of salt left behind after washing. All functionalized AC samples have some quantities of fluorine and sulfur. The absence of both of these elements in the fresh AC confirms that their origin is from the IL and indicates the presence of IL functionalities in functionalized AC samples. The amount of both elements generally increases with increasing IL loading until saturated at 50 wt% loading. The maximum amounts of fluorine and sulfur are found in AC (50) as 12.97% and 3.72%, respectively.

Table 1. Elemental composition of RSS derived activated carbon (AC).

Elements		AC-Fresh	AC (10)	AC (20)	AC (30)	AC (40)	AC (50)	AC (60)
Carbon	Weight (%)	77.42	70.25	61.97	69.44	67.50	65.15	67.78
	Atomic (%)	84.33	78.49	73.60	78.61	76.37	74.58	76.90
Oxygen	Weight (%)	16.79	19.80	20.63	14.73	16.76	13.99	14.49
	Atomic (%)	13.77	16.61	18.39	12.52	14.23	12.02	12.34
Potassium	Weight (%)	5.79	4.91	11.58	4.37	3.33	2.58	2.91
	Atomic (%)	1.94	1.68	4.22	1.52	1.16	0.91	1.02
Fluorine	Weight (%)		3.84	3.89	8.51	10.20	12.97	11.82
	Atomic (%)		2.72	2.92	6.09	7.30	9.38	8.48
Sulfur	Weight (%)		1.20	1.93	2.74	2.22	3.72	3.00
	Atomic (%)		0.50	0.86	1.16	0.94	1.60	1.27

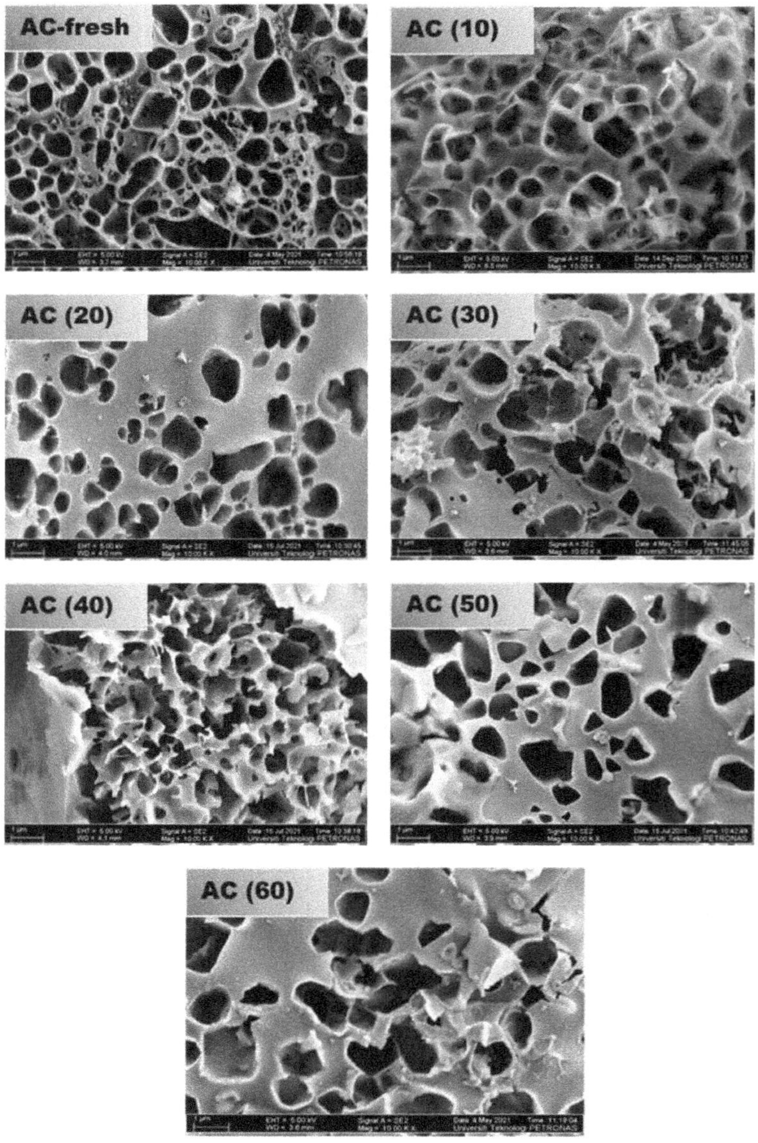

Figure 3. FESEM images of fresh activated carbon (AC) and functionalized with different concentrations of IL obtained at 1 µm (10,000× magnification scale).

Elemental mapping was also performed to study the spatial distribution of key elements. The relevant results are presented in Figure 4. Fresh AC (Figure 4a) has a homogeneous presence of carbon and oxygen as the building blocks of its structure. A small amount of potassium from the residual traces of K_2CO_3 is also uniformly scattered. Functionalized AC (Figure 4b) also shows a homogeneous presence of carbon, oxygen and potassium. In addition, significant amounts of fluorine and sulfur are present due to functionalization. Both elements are evenly distributed and confirm a homogeneous dispersion of IL in the form of a liquid layer of uniform thickness.

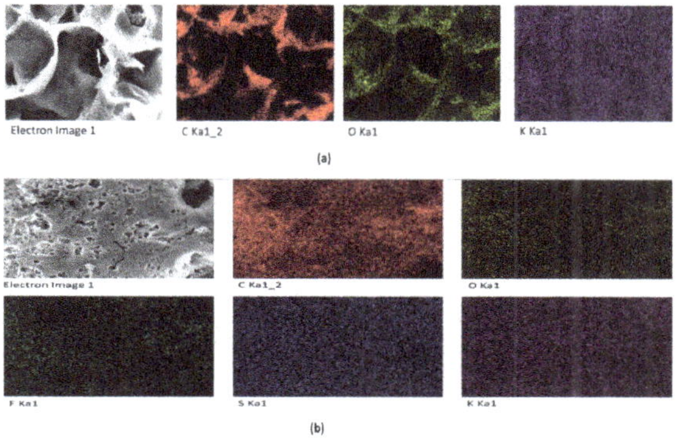

Figure 4. (**a**) EDX elemental mapping of fresh AC for C, O and K. (**b**) EDX elemental mapping of IL functionalized AC for C, O, F, S and K.

3.2. TGA

TGA measures variation in the weight of a sample with temperature and is used to estimate the thermal stability of the prepared adsorbent material. The TGA profiles of fresh and functionalized AC samples are presented in Figure 5. There are three regions of weight loss. The first stage accounts for changes in the mass of adsorbent below 200 °C and occurs due to the presence of surface bounded moisture. The second stage of weight loss, also called the rapid weight loss region, is from 200 °C to 600 °C. Major mass loss occurs during this stage due to the decomposition of tethered groups. The third stage of mass loss is above 600 °C. The weight loss in this region is small and usually occurs slowly due to thermal decomposition of the solid. The weight loss of fresh and functionalized AC samples in all three thermal stability regions is presented in Table 2. The weight loss during the first stage was less than 10% for all functionalized samples due to the hydrophobic nature of the IL and AC. The weight loss in the second stage (200–600 °C) is due to the decomposition of IL functionalities. The difference in the degradation temperature and the weight loss during the second stage indicates that each functionalized sample had a different amount of IL functional groups attached to the surface of AC. The weight changes in this stage are used to calculate the amount of IL loaded using Equation (10).

$$\gamma_{IL} = \frac{(\Delta wt\%_{imp} - \Delta wt\%_{raw})}{100 MW_{IL}} \tag{10}$$

where γ_{IL} is the specific molar amount of IL impregnated on the support (mmol/g). $\Delta wt\%_{imp}$ and $\Delta wt\%_{raw}$ are the percentage mass loss during the rapid weight loss region for the functionalized and fresh AC, respectively. MW_{IL} is the molecular weight of the IL (422.41 g/mol).

Table 2. Thermal stability regions and respective weight loss for fresh and functionalized AC.

Thermal Stability Regions	AC-fresh	AC (10)	AC (20)	AC (30)	AC (40)	AC (50)	AC (60)
Moisture Evaporation Region (Below 200 °C)	9.08	9.69	8.46	6.06	1.88	4.94	1.78
Rapid Weight Loss Region (200–600 °C)	52.00	25.54	32.81	32.30	38.37	48.17	51.64
Decomposition Region (Above 700 °C)	36.31	6.76	11.62	9.36	14	14.76	28.87

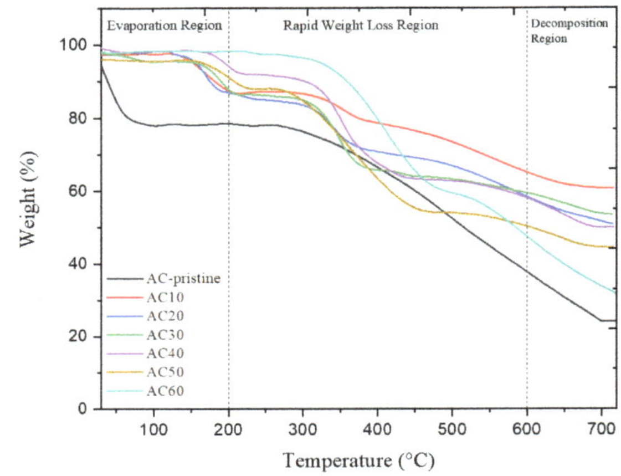

Figure 5. TGA profiles for fresh and functionalized AC under N_2 flow.

Weight loss in the third stage is associated with the decomposition of the graphite structure of the AC. Weight loss in the final stage for fresh AC was 36.3%, which is higher than functionalized AC. The lower relative stability of fresh AC at high temperatures is due to numerous surface defects in its structure, which accelerate combustion of the graphite network. Overall, the incorporation of IL improved the thermal stability of the AC. Similar findings were presented by Lemus et al. (2011) for imidazolium-based IL impregnated on three different supports [42].

3.3. FTIR Analysis

Surface functionalization improves the surface capturing properties of an adsorbent and results in higher uptake of the adsorbate due to additional active sites containing affinity groups. The study of surface groups was performed using FTIR analysis. IR spectra of fresh AC and AC functionalized with different amounts of IL are presented in Figure 6. A comparison between these spectra shows the presence of additional peaks in the functionalized AC samples. The additional peaks in functionalized samples resulted from the interactions between the IL functional groups and the porous support. The characteristic peaks are identified using the available literature on FTIR spectroscopy analysis and are similar to the findings reported previously [53].

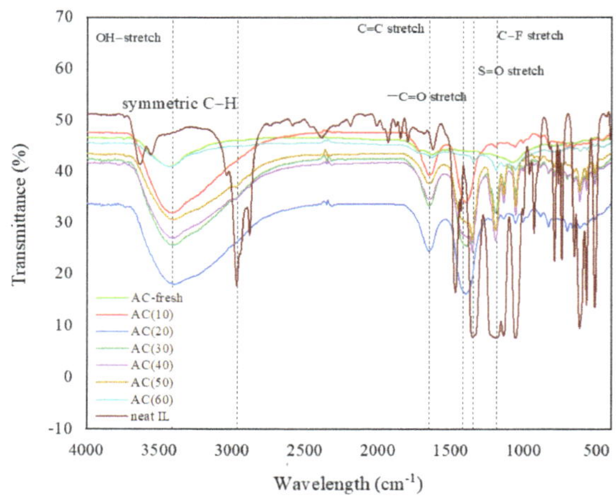

Figure 6. FTIR spectra of fresh and functionalized AC.

A broad peak observed for fresh AC at 3434 cm^{-1} is assigned to OH-stretching vibrations caused by the presence of surface hydroxyl groups. Similar peaks are observed in functionalized AC samples at 3432 cm^{-1}, 3399 cm^{-1}, 3400 cm^{-1}, 3401 cm^{-1}, 3400 cm^{-1} and 3432 cm^{-1} with reduced intensity. Because of the hydrophobic nature of IL, functionalized samples retain smaller amounts of moisture, hence a lower intensity of the hydroxyl group peak. The neat IL spectrum has distinguished peaks between 2900 cm^{-1} and 2500 cm^{-1} because of the C-H stretching of alkanes. Alkenes form the backbone of carbon structure and are indicated by the presence of the C=C stretching at 1600 cm^{-1} in the fresh AC spectrum. The same peak appears in the impregnated samples at 1650 cm^{-1}, 1650 cm^{-1}, 1648 cm^{-1}, 1653 cm^{-1}, 1648 cm^{-1} and 1631 cm^{-1}, which confirms that the structure of the AC remains intact after the incorporation of IL. For functionalized AC samples, sharp peaks are observed near 1468 cm^{-1} and 1352 cm^{-1}, indicating the presence of the S=O stretching of the sulfonyl group, while the smaller peaks between 1100–1000 cm^{-1} show the presence of the fluoro group. The presence of both functional groups is confirmed from the neat IL spectrum and are originated from the $[Tf_2N]^-$ anion of the IL. The intensity of these peaks is higher in AC samples impregnated with a higher dosage of IL, which indicates a higher level of surface functionalization in these samples.

3.4. SAP Analysis

The shape of the adsorption isotherm provides useful information regarding the adsorption process. Nitrogen adsorption isotherms of fresh and functionalized AC obtained at 77 K are presented in Figure 7. According to the classification of the International Union of Pure and Applied Chemistry (IUPAC), all isotherms are identified as type I, representing microporous solids [54]. The N_2 uptake increases gradually in the region of low P/P_0, followed by a plateau until $P/P_0 = 0.5$. Higher uptake is observed in the high-pressure region. These findings agree with those reported previously since physisortion is favorable under high pressure and low temperatures. The N_2 adsorption capacity is reduced for functionalized AC with increasing IL loading. After incorporating IL, surface active sites are covered with IL functionalities. N_2 does not interact with IL functionalities; therefore, less surface active sites are available for N_2 to be adsorbed and the adsorption capacity of AC is reduced. This effect becomes pronounced with increasing IL concentration until surface saturation is achieved. For the RSS AC prepared in this study, surface saturation was observed at an IL concentration of 60 wt.%, where negligible pore volume resulted in no N_2 adsorbed on the surface.Since N_2 is non-reactive, the lower N_2 uptake for functionalized samples is consistent with consistently low S_{BET}.

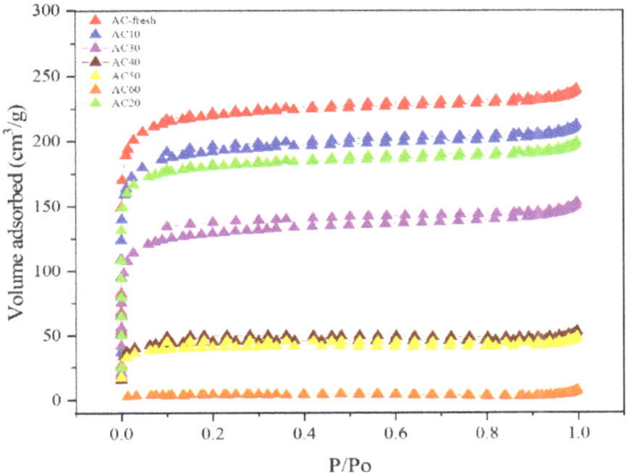

Figure 7. N_2 adsorption–desorption isotherm of fresh AC and functionalized with different amounts of IL at 77 K.

Surface properties play a significant role in estimating the adsorption capacity of an adsorbent. The relevant surface properties of fresh and functionalized AC were estimated by BET analysis and are presented in Table 3. CO_2 is adsorbed on fresh AC by only a physical attachment to the surface. Therefore, a porous surface is desirable because it provides more surface-active sites. The high porosity of fresh AC has resulted in high S_{BET} and V_p of 683 m^2/g and 0.37 cm^3/g, respectively. Functionalized AC shows a reduction in both S_{BET} and V_p with higher IL loading. At lower IL loading (10–20 wt.%), the reduction in S_{BET} is small. This shows that IL has penetrated the inner pores rather than sitting on the surface as a bulk liquid. However, as the concentration of the IL in the impregnation solution is increased above 30 wt.%, the inner pores of the AC are filled and a liquid layer covers most of the surface porosity. As a result, a very rapid decrease in the surface area is observed at higher IL loading. These surface area values are derived from micropores, which is evident from the high values of S_{micro} for all samples. AC (60) with maximum IL loading only has a surface area of 14 m^2/g and negligible pore volume. A massive reduction of V_p from 0.07 cm^3/g to merely 0.01 cm^3/g for an increase of IL loading from 50 to 60 wt.% indicates surface saturation. Values of n are consistent and did not change significantly after functionalization.

Table 3. Surface properties of fresh and functionalized activated carbon.

Sample	Specific Surface Area (S_{BET})	T-Plot Micropore Area (S_{micro})	External Surface Area (S_{ext})	Total Pore Volume (V_p)	Average Pore Diameter (D)	Percent Micropores
	(m^2/g)	(m^2/g)	(m^2/g)	(cm^3/g)	nm	(%)
AC-fresh	683	579	104	0.37	2.16	79.6
AC (10)	599	496	103	0.33	2.18	77.2
AC (20)	559	478	81	0.31	2.18	79.7
AC (30)	404	319	86	0.23	2.31	69.5
AC (40)	140	116	23	0.08	2.27	74.7
AC (50)	126	103	23	0.07	2.31	72.2
AC (60)	14	7	7	0.01	2.73	37.6

3.5. Raman Spectroscopy Analysis

Raman spectroscopy is widely used for the characterization of carbonaceous adsorbents to confirm the heterogeneous structure of microporous carbon. Raman spectra of fresh AC and functionalized AC samples with the highest CO_2 adsorption capacity (AC 30) are shown in Figure 8. For investigation of the structural properties of carbon, disoriented (D) band at 1380 cm^{-1} and graphite (G) band at 1590 cm^{-1} are important. The D-band is attributed to the defects of highly ordered sp^3 carbon and aromatics with less than six rings, whereas the G-band represents the vibrations in the graphite structure due to ordered sp^2 carbon.

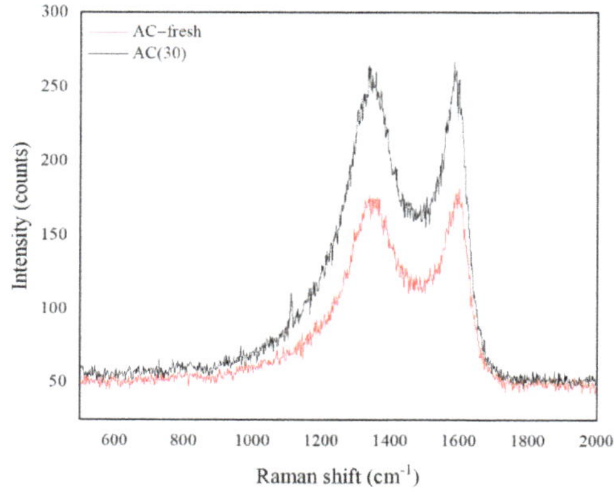

Figure 8. Raman spectra of fresh AC and functionalized AC (30).

The degree of functionalization is estimated using an I_D/I_G ratio. The value for fresh and functionalized AC is 0.918 and 0.967, respectively. A higher I_D/I_G ratio for functionalized AC indicates a change in symmetry. The interactions between the IL functional groups and carbon structure create a strain that creates a disorder in the structure and a change in symmetry from sp^2 to sp^3 carbon.

3.6. CO_2 Adsorption Study

The screening of adsorption performance was performed at 25 °C and 1 bar. Figure 9 depicts the CO_2 adsorption capacity of fresh and functionalized AC samples. Esteves et al. (2008) demonstrated that 25 °C is the ideal temperature for CO_2 adsorption because of the exothermic nature of physisorption that takes place for most of the physical adsorbents like AC [55]. Moreover, the selected conditions often reported in the literature are 0–1 bar pressure and 25 °C temperature. Therefore, the selected temperature allows for an easy comparison with earlier findings. Overall, the CO_2 adsorption capacity for functionalized AC increases with IL loading until reaching a maximum value at 30 wt.% IL loading. Afterwards, the CO_2 adsorption capacity remains nearly the same for the last three samples, indicating surface saturation. The CO_2 adsorption capacity measured at 25 °C is in the following order: AC-fresh (2.165 mmol/g) > AC (30) (1.124 mmol/g) > AC (20) (1.055 mmol/g) > AC (10) (0.758 mmol/g) > AC (40) (0.546 mmol/g) > AC (50) (0.533 mmol/g) > AC (60) (0.421 mmol/g). A similar trend is seen for CO_2 adsorption obtained at 40 °C. It is observed that S_{BET} has the most significant effect on the CO_2 adsorption capacity. The S_{BET} of the samples followed the order: AC-fresh (683.4 m^2/g) > AC (10) (599 m^2/g) > AC (20) (559 m^2/g) > AC (30) (404 m^2/g) > AC (40) (140 m^2/g) > AC (50) (126 m^2/g) > AC (60) (14 m^2/g). As S_{BET} is reduced with increasing IL loading, fewer active sites are available, leading to lower adsorption of CO_2. The influence of the surface area is crucial in functionalized samples with higher IL loading. For IL loading (10–30 wt.%), the CO_2 adsorption capacity increases despite a reduction in the surface area, as is evident from a higher CO_2 adsorption capacity for AC (30) compared to AC (20) and AC (10). This indicates that other factors are affecting CO_2 adsorption. Besides surface area, IL loading also affects the CO_2 adsorption capacity of AC. For functionalized AC, CO_2 is adsorbed by a combination of chemisorption by IL functionalities and physisorption by the support.

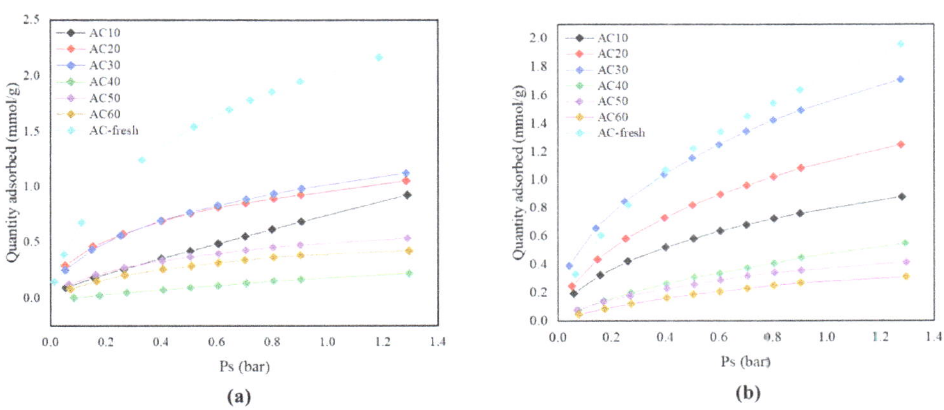

Figure 9. CO_2 adsorption capacity of fresh and functionalized AC obtained at (**a**) 25 °C and (**b**) 40 °C.

The effect of pressure on CO_2 adsorption is studied where higher adsorption takes place at higher pressure. Higher pressure pushes the CO_2 molecules into the pores. The percentage of CO_2 uptake of AC samples at both adsorption temperatures is presented in Figure 10. Fresh AC showed the highest uptake of 9.52 wt.% at 25 °C and 8.63 wt.% at

40 °C and 1.3 bar, whereas functionalized AC (30) showed the highest uptake of 4.748 wt.% at 25 °C and 7.54 wt.% at 40 °C and 1.3 bar.

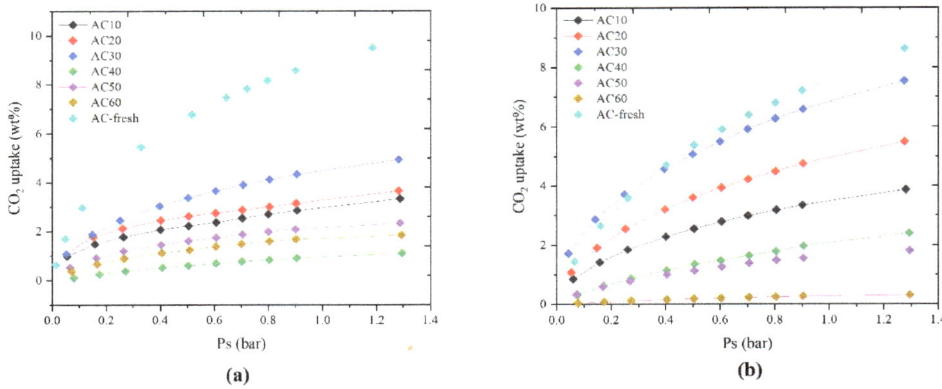

Figure 10. Percentage CO_2 uptake for fresh and functionalized AC obtained at (**a**) 25 °C and (**b**) 40 °C.

3.7. Effect of IL Loading

A reduction in the surface area and pore volume is observed with increasing IL concentration in the impregnation solution. The distribution of V_P with IL loading for each AC sample is presented in Figure 11. At IL loading (10–30 wt.%), IL penetrates the inner pores and there is a small decrease in V_p and S_{BET} values, as seen in Table 3. As the IL concentration is increased, the pore-clogging effect becomes prominent. For IL concentrations above 30 wt.%, inner pores are filled, and surface-active sites are reduced. There is a sharp decrease in pore volume at higher IL loadings (40–60 wt.%). AC (60) has a negligible pore volume, indicating the limitation of the amount of IL that can be impregnated.

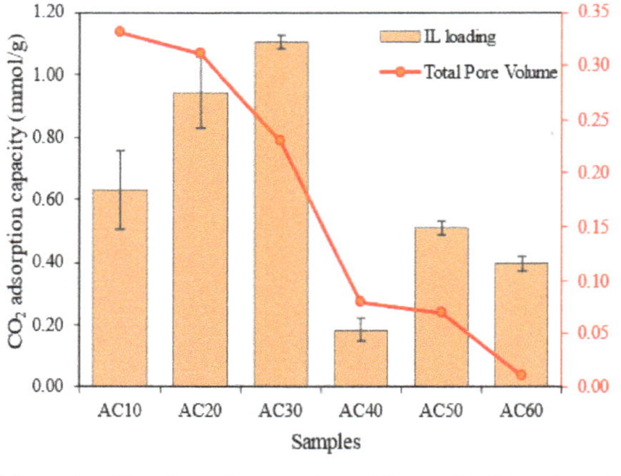

Figure 11. CO_2 adsorption capacity at 1 bar and 25 °C and total pore volume of AC functionalized with different amounts of IL.

The influence of IL loading on the CO_2 adsorption capacity was observed to identify the optimum IL loading with the best adsorption performance. The CO_2 adsorption capacity first increased for IL loading (10–30 wt.%) despite a reduction in the pore volume. AC (30) shows the highest CO_2 adsorption capacity of 1.124 mmol/g. For higher IL loadings (40–60 wt.%), the CO_2 adsorption capacity was reduced due to pore-clogging effect and poor diffusion due to the active phase layer of the IL. The interaction between the IL functional groups and CO_2 is much stronger at lower IL loading and there is significant contribution of physisorption by the support. CO_2 molecules have stronger interactions with an anion of IL. Due to the larger size of the $[Tf_2N]^-$ anion, maximum surface coverage was achieved at much lower loading. Afterward, IL molecules started depositing in a multilayer formation and increased the thickness of the active phase layer of the IL. At a higher IL loading (>30 wt.%), the surface area was reduced and the contribution of physisorption by the support was minimal. Furthermore, a bulkier liquid layer makes the diffusion of CO_2 molecules difficult. Consequently, lower CO_2 adsorption capacity was observed at higher IL loadings. Overall, AC impregnated with 30 wt% IL showed the best performance among other functionalized samples.

3.8. Effect of Adsorption Temperature

The CO_2 adsorption capacity of all samples was tested at 25 °C and 40 °C, and the results are presented in Figure 12. The temperature had a pronounced effect on CO_2 adsorption capacity. For fresh AC, the CO_2 adsorption capacity was reduced from 2.16 mmol/g at 25 °C to 1.96 mmol/g at 40 °C. CO_2 was adsorbed on fresh AC by physisorption involving weak van der Waals forces. This interaction is favorable at low temperatures. At high temperatures, the CO_2 molecules move faster and the surface adsorption energy is increased, leading to desorption. Ammendola et al. (2017) reported that at high temperatures, the adsorbate molecules adsorbed on the surface gain sufficient energy and can easily escape the surface and diffuse back to the gaseous phase [56]. Tamilarasan and Ramaprabhu (2012) also reported similar findings and ascribed them to the kinetic energy of adsorbate molecules [57]. At low adsorption temperatures, CO_2 molecules interact with the surface and are adsorbed by a multilayer formation. However, at elevated adsorption temperatures, CO_2 molecules have a much higher velocity to overcome the weak forces of interaction and eventually escape from the pores of the AC. Therefore, lower CO_2 adsorption capacity is observed for fresh AC at 40 °C than at 25 °C.

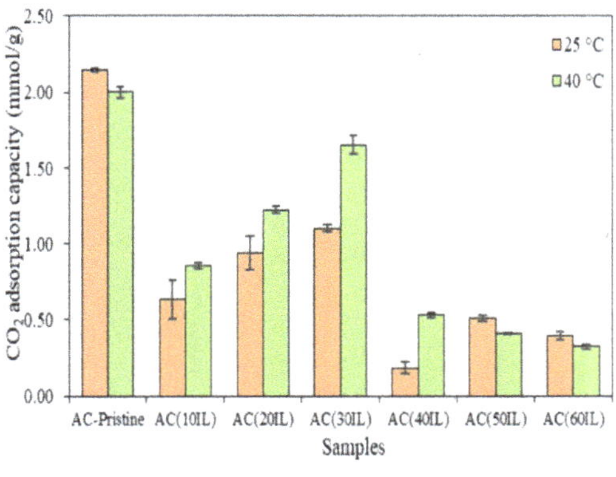

Figure 12. Effect of adsorption temeprature on the CO_2 adsorption capacity.

For IL functionalized AC under dilute impregnation concentrations, the difference in the CO_2 adsorption capacity with respect to the fresh AC was reduced as the temperature was increased. High temperature has a favorable effect on the dispersion of IL in the pores of AC. This implies that as the temperature rises, a better distribution of IL on the surface of AC (due to reduced surface tension of IL) makes it easier for the CO_2 molecules to reach the inner active sites. Hence, it can be said that CO_2 adsorption for IL functionalized AC is more favorable at higher temperatures. For AC (30), a higher amount of CO_2 was adosrbed at 40 °C compared to 25 °C. Similar findings were reported for amine impregnated fly ash, where the CO_2 adsorption capacity increased at high temperatures [24,52]. Zhang et al. (2019) [58] prepared amine-functionalized SBA-15 and observed a high CO_2 adsorption capacity at high temperatures due to higher reactivity of amine functionalities. Erto et al. (2015) [53] used two ILs for surface functionalization of AC and reported that the IL functionalized AC showed comparable CO_2 adsorption capacity as the raw AC at 80 °C despite showing a poor performance at room temperature. This phenomenon is attributed to a complex effect of temperature on the dispersion of IL. Although functionalized AC did not adsorb a high amount of CO_2 at 25 °C compared to fresh AC, it performed effectively at 40 °C. Based on this observation, we can expect a much higher CO_2 adsoprtion capacity for AC (30) at higher temperatures (70–120 °C).

3.9. Comparison with Previous Studies

The CO_2 adsorption performance of the prepared AC samples was compared with other biomass materials reported in literature. A summary of this comparison is presented in Table 4. RSS-derived AC has noticeably higher CO_2 adsorption capacity compared to some other biomass precursors including rice husk, palm kernels and bagasse pith. K_2CO_3 performs better than KOH for the same precursor (RSS) as evidenced by a higher CO_2 adsorption capacity for the former. The applicability of RSS for the development of AC is practical to replace traditional non-carbonaceous adsorbents like zeolites and silica. It is also a potential mechanism for solid waste management by utilizing agricultural waste.

Table 4. A comparison of CO_2 adsorption performance of various adsorbents.

Adsorbent	Functionalization Method	Chemicals	CO_2 Adsorption Capacity at 25 °C (mmol/g)	Ref.
RSS AC	wet impregnation	K_2CO_3	2.160	This study
RSS AC	wet impregnation	KOH	1.231	[7]
Rice husk	wet impregnation	$ZnCl_2$	1.330	[15]
Bagasse pith	wet impregnation	$ZnCl_2$	1.747	[59]
Palm kernel shell	physical activation	CO_2	1.660	[60]
RSS AC	wet impregnation	[bmpy][Tf_2N]	1.124	This study
Palm shell AC	wet impregnation	PEI	0.759	[34]
SBA-15	wet impregnation	[bmim][BF_4]	0.816	[61]
AC Norit CGP Super	wet impregnation	DETA	1.760	[22]
SBA-15	wet impregnation	[bmim][Ac]	0.926	[62]
SiO_2	wet impregnation	[Emim][HSO_4]	1.095	[63]
SiO_2	grafting	[P8883][TFSI]	0.99	[29]
MCM-41	water-aided grafting	[$(MeO)_3$Sipmim][Cl]	1.48	[64]

The AC (30) prepared in this study performed better than many functionalized adsorbents including silica, zeolites and AC. The obtained CO_2 adsorption capacity was also higher than KOH-activated RSS and some other biomass precursors. The CO_2 adsorption capacity obtained for AC (30) in this study was better compared to other functionalized adsorbents. The [Tf_2N] anion performed better compared to $[BF_4]^-$, $[Cl]^-$ and $[HSO_4]^-$ anions, as evidenced by a higher CO_2 adsorption capacity.

4. Conclusions

In this study, synthesis of surface-functionalized RSS-derived AC impregnated with [bmpy][Tf_2N], its detailed characterization and the CO_2 adsorption study were carried out. The synthesis of this adsorbent material not only aids in reducing the rising amount of CO_2, but also resolves the issue of land space covered by the selected biomass waste.

- FTIR and BET analyses of fresh and functionalized AC confirm successful functionalization with the selected IL. BET analysis shows that both S_{BET} and V_p are reduced with increasing IL concentration. The reduction of both surface properties is prominent at higher IL loading (40–60 wt%).
- The N_2 adsorption isotherm analysis shows type I isotherms for fresh and functionalized AC and suggests the properties of a microporous material.

- Raman spectroscopy confirms that the structure of AC is unharmed by the incorporation of IL. The improved thermal stability of the functionalized adsorbents compared to the fresh AC is confirmed through TGA.
- The CO_2 adsorption occurs by physisorption for fresh AC and is mainly affected by available surface area. The CO_2 adsorption capacity of functionalized AC is affected by IL loading and adsorption temperature. The CO_2 adsorption capacity initially increases with IL loading up to 30 wt.% and reduces afterward due to low surface area and resultant pore obstruction.
- The performance of functionalized AC is improved at higher adsorption temperature due to a better dispersion of IL.

Author Contributions: Conceptualization, A.B.; methodology, A.B., S.S.F. and M.A.; validation, A.B., M.A. and N.A.G.; formal analysis, A.B. and S.S.F.; investigation, S.S.F.; resources, A.B. and M.A.; data curation, S.S.F.; writing—original draft preparation, S.S.F.; writing—review and editing, A.B., M.A. and N.A.G.; supervision, A.B., M.A. and N.A.G.; funding acquisition, A.B. All authors have read and agreed to the published version of the manuscript.

Funding: This research work is funded by the Yayasan Universiti Teknologi PETRONAS, Malaysia, grant number (YUTP-FRG 015LC0-068).

Data Availability Statement: Original high-resolution images of FESEM and EDX analysis presented in Figures 3 and 4 are provided as supplementary information.

Acknowledgments: The authors graciously thank the HICoE, Centre of Biofuel and Biochemical Research (CBBR) and Institute of Self-Sustainable Building Universiti Teknologi PETRONAS (UTP) Malaysia for providing technical and financial assistance.

Conflicts of Interest: The authors declare no conflict of interest.

Nomenclature

n_D	moles of helium dosed into the sample cell (mol)
n_{Dosed}	moles of adsorbate dosed into the sample cell (mol)
$n_{Nadsorb}$	moles of adsorbate remaining in the sample cell (mol)
n_{adsorb}	moles of adsorbate adsorbed by the sample (mol)
P_A, P_B	pressure of the manifold before and after dosing (bar)
P_{s0}, P_{s1}	pressure of sample cell before and after dosing Helium (bar)
P'_A, P'_B	pressure of sample cell before and after dosing adsorbate (bar)
P_s	pressure of sample cell for adsorption (bar)
T_{AM}	ambient temperature (298 K)
T_A, T_B	temperature of the manifold before and after dosing (K)
T_{s0}, T_{s1}	temperature of sample before and after dosing Helium (K)
T'_A, T'_B	temperature of sample cell before and after dosing adsorbate (K)
T_s	temerature of sample cell for adsorption (K)
T_{xU}	manifold temperature (K)
Z_A	gas compressibility of helium at T_A and P_A
Z_B	gas compressibility of helium at T_B and P_B
Z_{s0}, Z_{s1}	gas compressibility at (T_{s0}, P_{s0}) and (T_{s1}, P_{s1}) respectively
Z_{xU0}, Z_{xU1}	gas compressibility at (T_A, P_{s0}) and (T_B, P_{s1}) respectively
Z_{xL0}, Z_{xL1}	gas compressibility at (T_{AM}, P_{s0}) and (T_{AM}, P_{s1}) respectively
Z'_A, Z'_B	gas compressibility at (T'_A, P'_A) and (T'_B, P'_B) respectively
Z_{xU}	gas compressibility at (T_{xU}, P_s)
R	universal gas constant $\left(83.14\ cm^3 \cdot bar \cdot mol^{-1} \cdot K^{-1}\right)$
V_m	LP manifold volume $\left(43.9311\ cm^3\right)$
V_{AFS}	free space volume at ambient temperature (cm^3)
V_{sxL}	manifold volume outside the temperature-controlled zone (cm^3)
V_s	free space volume at analysis temperature (cm^3)
V_{xL}	lower stem volume (cm^3)
V_{xU}	upper stem volume (3.5 cm^3)

References

1. Sher, F.; Iqbal, S.Z.; Albazzaz, S.; Ali, U.; Mortari, D.A.; Rashid, T. Development of biomass derived highly porous fast adsorbents for post-combustion CO_2 capture. *Fuel* **2020**, *282*, 118506. [CrossRef]
2. Li, B.; Duan, Y.; Luebke, D.; Morreale, B. Advances in CO_2 capture technology: A patent review. *Appl. Energy* **2013**, *102*, 1439–1447. [CrossRef]
3. Leung, D.Y.C.; Caramanna, G.; Maroto-Valer, M.M. An overview of current status of carbon dioxide capture and storage technologies. *Renew. Sustain. Energy Rev.* **2014**, *39*, 426–443. [CrossRef]
4. Fatima, S.S.; Borhan, A.; Ayoub, M.; Ghani, N.A. Development and progress of functionalzied silica-based adsorbents for CO_2 capture. *J. Mol. Liq.* **2021**, *338*, 116913. [CrossRef]
5. Borhan, A.; Hoong, P.K.; Taha, M.F. Biosorption of heavy metal ions, oil and grease from industrial waste water by banana peel. *Appl. Mech. Mater.* **2014**, *625*, 749–752. [CrossRef]
6. Borhan, A.; Thangamuthu, S.; Taha, M.F.; Ramdan, A.N. Development of activated carbon derived from banana peel for CO_2 removal. In Proceedings of the AIP Conference Proceedings 2015, Negeri Sembilan, Malaysia, 24–26 February 2015.
7. Borhan, A.; Yusup, S.; Lim, J.W.; Show, P.L. Characterization and modelling studies of activated carbon produced from rubber-seed shell using KOH for CO_2 adsorption. *Processes* **2019**, *7*, 855. [CrossRef]
8. Quan, C.; Jia, X.; Gao, N. Nitrogen-doping activated biomass carbon from tea seed shell for CO_2 capture and supercapacitor. *Int. J. Energy Res.* **2019**, *44*, 1218–1232. [CrossRef]
9. Kumar, K.; Kumar, A. Enhanced CO_2 adsorption and separation in ionic-liquid-impregnated mesoporous silica MCM-41: A molecular simulation study. *J. Phys. Chem. C* **2018**, *122*, 8216–8227. [CrossRef]
10. Ao, W.; Fu, J.; Mao, X.; Kang, Q.; Ran, C.; Liu, Y.; Zhang, H.; Gao, Z.; Li, J.; Liu, G.; et al. Microwave assisted preparation of activated carbon from biomass: A review. *Renew. Sustain. Energy Rev.* **2018**, *92*, 958–979. [CrossRef]
11. Hazzaa, R.; Hussein, M. Adsorption of cationic dye from aqueous solution onto activated carbon prepared from olive stones. *Environ. Technol. Innov.* **2015**, *4*, 36–51. [CrossRef]
12. Yakout, S.M.; Sharaf El-Deen, G. Characterization of activated carbon prepared by phosphoric acid activation of olive stones. *Arab. J. Chem.* **2016**, *9*, 1155–1162. [CrossRef]
13. Chen, J.; Yang, J.; Hu, G.; Hu, X.; Li, Z.; Shen, S.; Radosz, M.; Fan, M. Enhanced CO_2 capture capacity of nitrogen-doped biomass-derived porous carbons. *ACS Sustain. Chem. Eng.* **2016**, *4*, 1439–1445. [CrossRef]
14. Deng, S.; Hu, B.; Chen, T.; Wang, B.; Huang, J.; Wang, Y.; Yu, G. Activated carbons prepared from peanut shell and sunflower seed shell for high CO_2 adsorption. *Adsorption* **2015**, *21*, 125–133. [CrossRef]
15. Plaza, M.G.; Gonzalez, A.S.; Pevida, C.; Pis, J.J.; Rubiera, F. Volarisation of spent coffee grounds as CO_2 adsorbents for postcombustion capture applications. *Appl. Energy* **2012**, *99*, 272–279. [CrossRef]
16. Wei, H.; Deng, S.; Hu, B.; Chen, Z.; Wang, B.; Huang, J.; Yu, G. Granular bamboo-derived activated carbon for high CO_2 adsorption: The dominant role of narrow micropores. *ChemSusChem* **2012**, *5*, 2354–2360. [CrossRef]
17. Sha, Y.; Lou, J.; Bai, S.; Wu, D.; Liu, B.; Ling, Y. Facile preparation of nitrogen-doped porous carbon from waste tobacco by a simple pre-treatment process and their application in electrochemical capacitor and CO_2 capture. *Mater. Res. Bull.* **2015**, *64*, 327–332. [CrossRef]
18. Heidari, A.; Younesi, H.; Rashidi, A.; Ghoreyshi, A.A. Evaluation of CO_2 adsorption with eucalyptus wood based activated carbon modified by ammonia solution through heat treatment. *Chem. Eng. J.* **2014**, *254*, 503–513. [CrossRef]
19. Department of Statistics Malaysia. Press Release Monthly rubber statistics Malaysia, January 2020. Available online: https://www.dosm.gov (accessed on 13 March 2020).
20. Tran, H.N.; You, S.-J.; Chao, H.-P. Fast and efficient adsorption of methylene green 5 on activated carbon prepared from new chemical activation method. *J. Environ. Manag.* **2017**, *188*, 322–336. [CrossRef]
21. Houshmand, A.; Wan Daud, W.M.A.; Shafeeyan, M.S. Exploring potential methods for anchoring amine groups on the surface of activated carbon for CO_2 adsorption. *Sep. Sci. Technol.* **2011**, *46*, 1098–1112. [CrossRef]
22. Plaza, M.G.; Pevida, C.; Arenillas, A.; Rubiera, F.; Pis, J.J. CO_2 capture by adsorption with nitrogen enriched carbons. *Fuel* **2007**, *86*, 2204–2212. [CrossRef]
23. Arenillas, A.; Smith, K.M.; Drage, T.C.; Snape, C.E. CO_2 capture using some fly ash-derived carbon materials. *Fuel* **2005**, *84*, 2204–2210. [CrossRef]
24. Maroto-Valer, M.M.; Lu, Z.; Zhang, Y.; Tang, Z. Sorbents for CO_2 capture from high carbon fly ashes. *Waste Manag.* **2008**, *28*, 2320–2328. [CrossRef]
25. Wei, J.; Shi, J.; Pan, H.; Zhao, W.; Ye, Q.; Shi, Y. Adsorption of carbon dioxide on organically functionalized SBA-16. *Microporous Mesoporous Mater.* **2008**, *116*, 394–399. [CrossRef]
26. He, X.; Zhu, J.; Wang, H.; Zhou, M.; Zhang, S. Surface functionalization of activated carbon with phosphonium ionic liquid for CO_2 adsorption. *Coatings* **2019**, *9*, 590. [CrossRef]
27. Girish, C.R. Various impregnation methods used for the surface modification of the adsorbent: A review. *Int. J. Eng. Technol.* **2018**, *7*, 330–334. [CrossRef]
28. Jain, S.; Bansiwal, A.; Biniwale, R.B.; Milmile, S.; Das, S.; Tiwari, S.; Antony, P.S. Enhancing adsorption of nitrate using metal impregnated alumina. *J. Environ. Chem. Eng.* **2015**, *3*, 232–2349. [CrossRef]

29. Zhu, J.; He, B.; Huang, J.; Li, C.; Ren, T. Effect of immobilization methods and the pore structure on CO_2 separation performance in silica-supported ionic liquids. *Microporous Mesoporous Mater.* **2018**, *260*, 190–200. [CrossRef]
30. Sanz-Perez, E.S.; Arencibia, A.; Calleja, G.; Sanz, R. Tuning the textural properties of HMS mesoporous silica. Functionalization towards CO_2 adsorption. *Microporous Mesoporous Mater.* **2018**, *260*, 235–244. [CrossRef]
31. Houshmand, A.; Wan Daud, W.M.A.; Lee, M.-G.; Shafeeyan, M.S. Carbon dioxide capture with amine-grafted activated carbon. *Water Air Soil Pollut.* **2012**, *223*, 827–835. [CrossRef]
32. Maroto-Valer, M.M.; Tang, Z.; Zhang, Y. CO_2 capture by activated and impregnated anthracites. *Fuel Process. Technol.* **2005**, *86*, 1487–1502. [CrossRef]
33. Bezerra, D.P.; Oliveira, R.S.; Vieira, R.S.; Cavalcante, C.L., Jr. Adsorption of CO_2 on nitrogen-enriched activated carbon and zeolite 13X. *Adsorption* **2011**, *17*, 235–246. [CrossRef]
34. Aroua, M.K.; Wan Daud, W.M.A.; Yin, C.Y.; Adinata, D. Adsorption capacities of carbon dioxide, oxygen, nitrogen and methane on carbon molecular basket derived from polyethyleneimine impregnation on microporous palm shell activated carbon. *Sep. Purif. Technol.* **2008**, *62*, 609–613. [CrossRef]
35. Zhang, C.; Song, W.; Sun, G.; Xie, L.; Wang, J.; Li, K.; Sun, C.; Liu, H.; Snape, C.E.; Drage, T. CO_2 capture with activated carbon grafted by nitrogenous funtional groups. *Energy Fuels* **2013**, *27*, 4818–4823. [CrossRef]
36. Grondein, A.; Bélanger, D. Chemical modification of carbon powders with aminophenyl and aryl-aliphatic amine groups by reduction of in situ generated diazonium cations: Applicability of the grafted powder towards CO_2 capture. *Fuel* **2011**, *90*, 2684–2693. [CrossRef]
37. McDonald, J.D.; Kracko, D.; Doyle-Eisele, M.; Garner, C.E.; Wegerski, C.; Senft, A.; Knipping, E.; Shaw, S.; Rohr, A. Carbon capture and sequestration: An exploratory inhalation toxicity assessment of amine-trapping solvents and their degradation products. *Environ. Sci. Technol.* **2014**, *48*, 10821–10828. [CrossRef]
38. Ahmed, S.; Ramli, A.; Yusup, S. Development of polyethylenimine-functionalized mesoporous Si-MCM-41 for CO_2 adsorption. *Fuel Process. Technol.* **2017**, *167*, 622–630. [CrossRef]
39. Xu, X.; Song, C.; Andresen, J.M.; Miller, B.G.; Scaroni, A.W. Novel polyethylenimine-modified mesoporus molecular seive of MCM-41 type as high-capacity adsorbent for CO_2 capture. *Energy Fuels* **2002**, *16*, 1463–1469. [CrossRef]
40. Franchi, R.S.; Harlick, P.J.E.; Sayari, A. Applications of pore-expanded mesoporous silica. 2. Development of a high-capacity, water-tolerant adsorbent for CO_2. *Ind. Eng. Chem. Res.* **2005**, *44*, 8007–8013. [CrossRef]
41. Hasib-ur-Rahman, M.; Siaj, M.; Larachi, F. Ionic liquids for CO_2 capture-Development and progress. *Chem. Eng. Process.* **2010**, *49*, 313–322. [CrossRef]
42. Lemus, J.; Palomar, J.; Gilarraanz, M.A.; Rodriguez, J.J. Characterization of supported ionic liquid phase (SILP) materials prepared from different supports. *Adsorption* **2011**, *17*, 561–571. [CrossRef]
43. Ban, Y.; Li, Z.; Li, P.D.Y.; Peng, Y.; Jin, H.; Wenmei, J.; Guo, A.; Wang, P.; Yang, P.D.Q.; Zhong, P.D.C.; et al. Confinement of ionic liquids in nanocages: Tailoring the molecular sieving properties of ZIF-8 for membrane based CO_2 capture. *Angew. Chem.* **2015**, *54*, 15483–15487. [CrossRef] [PubMed]
44. Yusuf, N.Y.; Masdar, M.S.; Isahak, W.N.R.W.; Nordin, D.; Husaini, T.; Majlan, E.H.; Rejab, S.A.M.; Chew, C.L. Ionic-liquid impregnated activated carbon for biohydrogen purification in an adsorption unit. In Proceedings of the IOP Conference Series: Materials Science and Engineering, 29th Symposium of Malaysian Chemical Engineers (SOMChE) 2016, Miri, Sarawak, Malaysia, 1–3 December 2016; IOP Publishing Ltd.: Bristol, UK, 2017.
45. Aki, S.N.V.K.; Mellein, B.R.; Saurer, E.M.; Brennecke, J.F. High pressure phase behavior of carbon dioxide with imidazolium-based ionic liquids. *J. Phys. Chem. B* **2004**, *108*, 20355–20365. [CrossRef]
46. Ramdin, M.; de Loos, T.W.; Vlugt, T.J.H. State-of-the-art of CO_2 capture with ionic liquids. *Ind. Eng. Chem. Res.* **2012**, *51*, 8149–8177. [CrossRef]
47. Blanchard, L.A.; Hancu, D.; Beckman, E.J.; Brennecke, J.F. Green processing using ionic liquids and CO_2. *Nature* **1999**, *399*, 28–29. [CrossRef]
48. Revelli, A.-L.; Mutelet, F.; Jaubert, J.N. High carbon dioxide solubilities in imidazolium-based ionic liquids and in poly(ethylene glycol) Dimethyl ether. *J. Phys. Chem. B* **2010**, *114*, 12908–12913. [CrossRef]
49. Cadena, C.; Anthony, J.L.; Shah, J.K.; Morrow, T.I.; Brennecke, J.F.; Maginn, E.J. Why is CO_2 so soluble in imidazolium-based ionic liquids? *J. Am. Chem. Soc.* **2004**, *126*, 5300–5308. [CrossRef]
50. Jung, Y.-H.; Jung, J.-Y.; Jin, Y.-R.; Lee, B.-C.; Baek, I.-H.; Kim, S.-H. Solubility of carbon dioxide in imidazolium-based ionic liquids with a methanesulfonate anion. *J. Chem. Eng. Data* **2012**, *57*, 3321–3329. [CrossRef]
51. Kumelan, J.; Tuma, D.; Kamps, A.P.-S.; Maurer, G. Solubility of the single gases carbon dioxide and hydrogen in the ionic liquid [bmpy][Tf_2N]. *J. Chem. Eng. Data* **2010**, *55*, 165–172. [CrossRef]
52. Zhang, Y.; Xing, W.; Liu, S.; Liu, Y.; Yang, M.; Zhao, J.; Song, Y. Pure methane, carbon dioxide, and nitrogen adsorption on anthracite from china over a wide range of pressures and temperatures: Experiments and modeling. *RSC Adv.* **2015**, *5*, 52612–52623. [CrossRef]
53. Erto, A.; Silvestre-Albero, A.; Silvestre-Albero, J.; Rodriguez-Reinoso, F.; Balsamo, M.; Lancia, A.; Fabio, M. Carbon-supported ionic liquids as innovative adsorbents for CO_2 separation from synthetic flue-gas. *J. Colloid Interface Sci.* **2015**, *448*, 41–50. [CrossRef]

54. Singh, K.S.W.; Everett, D.H.; Haul, R.A.W.; Moscou, L.; Pierotti, R.A.; Rouquerol, J.; Siemeiniewska, T. Reporting physisorption data for gas/solid systems with special reference to the determination of surface area and porosity (recommendations 1984). *Pure Appl. Chem.* **1985**, *57*, 603–619. [CrossRef]
55. Esteves, I.A.A.C.; Lopes, M.S.S.; Nunes, P.M.C.; Mota, J.P.B. Adsorption of natural gas and biogas components on activated carbon. *Sep. Purif. Technol.* **2008**, *62*, 281–296. [CrossRef]
56. Ammendola, P.; Raganati, F.; Chirone, R. CO_2 adsorption on a fine activated carbon in a sound assisted fluidized bed: Thermodynamics and kinetics. *Chem. Eng. J.* **2017**, *322*, 302–313. [CrossRef]
57. Tamilarasan, P.; Ramaprabhu, S. Polyaniline-magnetite nanocapsules based nanocomposite for carbon dioxide adsorption. *Int. J. Greenh. Gas Control.* **2012**, *10*, 486–493. [CrossRef]
58. Zhang, W.; Gao, E.; Li, Y.; Bernards, M.T.; He, Y.; Shi, Y. CO_2 capture with polyamine-based protic ionic liquid functionalized mesoporous silica. *J. CO_2 Util.* **2019**, *34*, 606–615. [CrossRef]
59. Boonpoke, A.; Chiarakorn, S.; Laosiripojana, N.; Towprayoon, S. Synthesis of activated carbon and MCM-41 from bagasse and rice husk and their carbon dioxide adsorption. *J. Sustain. Energy Environ.* **2011**, *2*, 77–81.
60. Nasri, N.S.; Hamza, U.D.; Ismail, S.N.; Ahmed, M.M.; Mohsin, R. Assessment of porous carbons derived from sustainable palm solid waste for carbon dioxide capture. *J. Clean. Prod.* **2014**, *71*, 148–157. [CrossRef]
61. Xu, Q.-Q.; Yin, J.-Z.; Zhou, X.-L.; Yin, G.-Z.; Liu, Y.-F.; Cai, P.; Wang, A.-Q. Impregnation of ionic liquids in mesoporous silica using supercritical carbon dioxide and co-solvent. *RSC Adv.* **2016**, *6*, 101079–101086. [CrossRef]
62. Yin, J.-Z.; Zhen, M.-Y.; Cai, P.; Zhou, D.; Li, Z.-J.; Zhu, H.-Y.; Xu, Q.-Q. Supercritical CO_2 preparation of SBA-15 supported ionic liquid and its adsorption for CO_2. *Mater. Res. Express* **2018**, *5*, 065060. [CrossRef]
63. Marliza, T.S.; Yarmo, M.A.; Hakim, A.; Abu Tahari, M.N.; Hisham, M.W.M.; Taufiq-Yap, Y.H. CO_2 capture on NiO supported imidazolium-based ionic liquid. In Proceedings of the AIP Conference Proceedings, 2017, Selangor, Malaysia, 28–30 November 2016.
64. Aquino, A.S.; Bernard, F.L.; Borges, J.V.; Mafra, L.; Vecchia, F.D.; Vieira, M.O.; Ligabue, R.; Seferin, M.; Chaban, V.V.; Cabrita, E.J.; et al. Rationalizing the role of the anion in CO_2 capture and conversion using imidazolium based ionic liquid modified mesoporous silica. *RSC Adv.* **2015**, *5*, 64220–64227. [CrossRef]

processes

MDPI

Article

Model Forecasting Development for Dengue Fever Incidence in Surabaya City Using Time Series Analysis

Mahmod Othman [1], Rachmah Indawati [2,*], Ahmad Abubakar Suleiman [1,3], Mochammad Bagus Qomaruddin [2] and Rajalingam Sokkalingam [1]

[1] Department of Fundamental and Applied Sciences, Universiti Teknologi Petronas, Seri Iskandar 32610, Perak Darul Ridzuan, Malaysia
[2] Faculty of Public Health, Universitas Airlangga, Surabaya 60115, Indonesia
[3] Department of Statistics, Kano University of Science and Technology, Wudil 713281, Nigeria
* Correspondence: rachmah.indawati@fkm.unair.ac.id

Abstract: Dengue hemorrhagic fever (DHF) is one of the most widespread and deadly diseases in several parts of Indonesia. An accurate forecast-based model is required to reduce the incidence rate of this disease. Time-series methods such as autoregressive integrated moving average (ARIMA) models are used in epidemiology as statistical tools to study and forecast DHF and other infectious diseases. The present study attempted to forecast the monthly confirmed DHF cases via a time-series approach. The ARIMA, seasonal ARIMA (SARIMA), and long short-term memory (LSTM) models were compared to select the most accurate forecasting method for the deadly disease. The data were obtained from the Surabaya Health Office covering January 2014 to December 2016. The data were partitioned into the training and testing sets. The best forecasting model was selected based on the lowest values of accuracy metrics such as the root mean square error (RMSE), mean absolute error (MAE), and mean absolute percentage error (MAPE). The findings demonstrated that the SARIMA (2,1,1) (1,0,0) model was able to forecast the DHF outbreaks in Surabaya City compared to the ARIMA (2,1,1) and LSTM models. We further forecasted the DHF cases for 12 month horizons starting from January 2017 to December 2017 using the SARIMA (2,1,1) (1,0,0), ARIMA (2,1,1), and LSTM models. The results revealed that the SARIMA (2,1,1) (1,0,0) model outperformed the ARIMA (2,1,1) and LSTM models based on the goodness-of-fit measure. The results showed significant seasonal outbreaks of DHF, particularly from March to September. The highest cases observed in May suggested a significant seasonal correlation between DHF and air temperature. This research is the first attempt to analyze the time-series model for DHF cases in Surabaya City and forecast future outbreaks. The findings could help policymakers and public health specialists develop efficient public health strategies to detect and control the disease, especially in the early phases of outbreaks.

Keywords: ARIMA model; SARIMA model; forecasting; dengue fever; time-series analysis

Citation: Othman, M.; Indawati, R.; Suleiman, A.A.; Qomaruddin, M.B.; Sokkalingam, R. Model Forecasting Development for Dengue Fever Incidence in Surabaya City Using Time Series Analysis. *Processes* **2022**, *10*, 2454. https://doi.org/10.3390/pr10112454

Academic Editor: Philippe Bogaerts

Received: 11 October 2022
Accepted: 17 November 2022
Published: 19 November 2022

1. Introduction

Dengue is a mosquito-borne disease caused by the dengue virus, affecting most tropical regions worldwide [1]. According to [2], dengue infection has been defined as dengue without warning signs, dengue with warning signs, and severe dengue. Therefore, dengue fever is more likely to be dengue without warning signs, however, a few cases of dengue fever may present warning signs. DHF will include dengue with warning signs, especially plasma leakage and severe dengue. It is estimated that between 50 and 500 million people worldwide are infected with dengue each year [3,4]. Between 10,000 and 20,000 people die each year, and about two and a half billion people are in danger of infection [5–7]. Recent figures have shown that 60% of the world's population would be susceptible to dengue disease by 2080 [8]. According to this estimation, 10,000 people have died from dengue in more than 125 countries worldwide. Even though dengue deaths are 99% preventable, case fatality rates significantly greater than 1% have been recorded globally [9].

Dengue fever is one of the most severe and common health problems in Indonesia. Since 1968, the number of cases and the transmission of dengue fever have been rising [10]. The growing population, rapid urbanization, and modern transportation have significantly contributed to the spread of the disease. Indonesia is a tropical country with a high population density, especially in urban areas, which could serve as a habitat for dengue viruses. Dengue viruses are transmitted through the bite of the Aedes aegypti and Ae. Albopictus causes a high fever, red spots on the skin, and pain in the muscles. There is no vaccine available for preventing DHF. The current disease prevention plan is not effective as the only available treatments for DHF patients are supportive symptomatic care such as antipyretics, antiemetics, and IV fluids [11]. According to information from the Surabaya Health Office, Surabaya City is a dengue-endemic area. In 2010, there were 3379 cases of dengue, with an incidence rate (IR) of 116.03 per 100,000 people and a case fatality rate (CFR) of 0.8%. In 2011, there were 1008 dengue cases with 36.22 IR and 0.3% CFR. In 2012, there were 1091 dengue cases with 38.60 IR and 0.55% CFR. In 2013, there were 2207 dengue cases with 78.35 IR and 0.86% CFR. The government has implemented a variety of initiatives through a variety of programs to prevent the rise in incidence including preventive and promotive efforts. Preventive efforts can be described by instilling clean living habits (such as not littering, hoarding junks, and not allowing any containers to be the breeding ground for larvae). This activity is known as mosquito nest eradication (MNE). This approach, however, cannot appropriately recognize changes in prevalence [12].

A forecast-based early warning system is required to reduce the incidence rate of this disease. There are several other forecasting models in use globally and even in Indonesia. However, these models are not efficient enough to accommodate all the characteristics of the DHF data. Hence, there is a dire need to develop more flexible forecasting models that can provide better results than the existing ones. This will assist policymakers to make better decisions. Time-series forecasting is a statistical method that has been used extensively in numerous fields, particularly in the study of infectious disease epidemiology [13,14]. Several studies have used statistical models developed with the aim of forecasting dengue in various settings [15–23]. Due to the time-varying behavior, seasonal pattern, secular trend, and rapid fluctuations in time-series exhibited by DHF data, it is feasible to forecast the incidence of DHF with time-series methods to enable an early response to the disease [24]. Autoregressive integrated moving average (ARIMA) is a popular time-series forecasting technique in health science research [25]. It enables us to identify hidden behaviors in the data. However, the ARIMA model is inappropriate for time-series data containing seasonality [10]. The ARIMA model is also a tedious method requiring computational skill, despite producing efficient results [7]. Consequently, the ARIMA method is modified for seasonal data and is known as SARIMA. The SARIMA model combines seasonal and non-seasonal autoregressive and moving average models [3,5]. The SARIMA is a method for identifying the patterns from seasonal time-series data for forecasting future values of the time-series and has received the most attention in recent years [26,27]. DHF data exhibit both linear and nonlinear behavior [28]. However, the SARIMA model is only used for modeling linear time-series data and cannot handle nonlinear behavior. A modern method of deep learning algorithms has been developed for prediction applications. This method can handle nonlinearity and complexity in time-series forecasting. LSTM is one deep learning method that allows for the processing of longer temporal sequences.

Time-series analysis can be used to examine past trends of DHF outbreaks and improve the present prevention and control measures. This approach is one of the most accurate statistical models that can be developed using the data to predict future DHF epidemics. The time-series forecasting approach has not previously been considered to forecast DHF prevalence in Surabaya City. Hence, this research aimed to propose time-series models that can forecast future values of DHF outbreaks in Surabaya City.

2. Materials and Methods

In this section, we highlighted the study area where the data on DHF cases were collected. Next, the concepts and main definitions of the ARIMA, SARIMA, and LSTM models are introduced. In addition, the non-stationary models test is presented and followed by the algorithm of the proposed method.

2.1. Study Area

Surabaya is the capital city of the Indonesian province of East Java and the second-largest city in Indonesia after Jakarta, located at 7°14′45″ S, 112°44′16″ E, and covers an area of $911/\text{km}^2$. It has a total population of 2,874,314 (2020) and a density of $7134/\text{km}^2$. Surabaya is an endemic area for DHF cases with the highest incidence of dengue in the country. Figure 1 shows a map of the study area.

Figure 1. A map of the study area showing Surabaya City, East Java, Indonesia.

2.2. Data Collection

This study used monthly DHF cases from January 2014 to December 2016 obtained from the Surabaya Health Office. The monthly dataset contains a total of 36 monthly observations. We used the DHF cases covering 2014–2016 due to the availability of these datasets recorded from the Surabaya Health Office at the time of this project and their sufficient size for model testing to validate this research. The data were processed, coded, and entered into R-studio version 4.2.1 and evaluated for normality. The R software was created by Ross Ihaka and Robert Gentleman at the University of Auckland, New Zealand, and is currently developed by the R Development Core Team. The data were partitioned into training and testing sets. The training set was applied for the model development, while the testing set was applied for the validation of the developed model. The best model was selected and used for out-of-sample forecasting.

2.3. ARIMA Model

The ARIMA model is suitable for modeling stationary time-series data, although most time-series data from real-life phenomena exhibit non-stationary patterns. However, the model assumes that a non-stationary series could become stationary through a differencing approach [29–31]. The generic version of the ARIMA model is given in Equation (1):

$$Y_t = C + \phi_1 Y_{t-1} + \phi_2 Y_{t-2} + \ldots + \phi_p Y_{t-p} + \theta_1 \varepsilon_{t-1} + \theta_2 \varepsilon_{t-2} + \ldots + \theta_q \varepsilon_{t-q} + \varepsilon_t \quad (1)$$

This model is denoted by

$$\text{ARIMA } (p, d, q) \tag{2}$$

where Y_t is a dependent variable on t that might have been differenced once or more; p is the order of the autoregressive part; d is the degree of differencing; q is the order of the moving average part; ε_t is the error term; C is the constant number of the model.

2.4. SARIMA Model

SARIMA means the seasonal ARIMA. In time-series analysis, seasonality is a regular pattern of variations that repeats over s periods, where s denotes the number of periods until the pattern repeats. For instance, there is seasonality in the monthly data for which high values tend to occur in some months, while low values tend to occur in other months. The general form of the SARIMA model is given in Equation (3):

$$\Phi_P(B^s)\varphi(B)\nabla_s^D\nabla^d x_t = \Theta_Q(B^s)\theta(B)w_t \tag{3}$$

This model can be denoted by

$$\text{SARIMA } (p, d, q)(P, D, Q)_s \tag{4}$$

where x_t is the nonstationary time-series; w_t is the usual Gaussian white noise process; and s is the period of the time-series. The autoregressive and moving average components are represented by polynomials Φ_P and Θ_Q of orders p and q. The seasonal autoregressive and moving average components are $\Phi_P(B^s)$ and $\Theta_Q(B^s)$, where P and Q are their orders. ∇^d and ∇_s^D are the ordinary and seasonal difference components. B is the backshift operator. The expressions are shown as follows:

$$\begin{aligned}
&\varphi(B) = 1 - \varphi_1 B - \varphi_2 B^2 - \ldots - \varphi_p B^p \\
&\Phi_P(B^s) = 1 - \Phi_1 B^s - \Phi_2 B^{2s} - \ldots - \Phi_P B^{Ps} \\
&\theta(B) = 1 + \theta_1 B + \theta_2 B^2 + \ldots + \theta_q B^q \\
&\Theta_Q(B^s) = 1 + \Theta_1 B^s + \Theta_2 B^{2s} + \Theta_Q B^{Qs} \\
&\nabla^d = (1 - B)^d \\
&\nabla_s^D = (1 - B^s)^D \\
&B^k x_t = x_{t-1}
\end{aligned}$$

This study focused on the monthly confirmed DHF cases. If the seasonal period of the series $s = 12$, then we can rewrite Equation (3) as:

$$\Phi_P\left(B^{12}\right)\varphi(B)\nabla_{12}^D\nabla^d x_t = \Theta_Q\left(B^{12}\right)\theta(B)w_t \tag{5}$$

2.5. LSTM Model

One of the advancements in neural networks that can learn long-term dependence is known as LSTM. The architecture of an LSTM is composed of three gates: forget gates, input gates, and output gates [32].

The forget gate G_t determines the specific information that is deleted from the memory cells (cell state). Forget gates use a sigmoid activation function as their activation function, where the result is between 0 and 1. If the output is 1, all the information will be retained. If it is 0, all the information will be discarded. It is given by

$$G_t = \sigma(W_g P_{t-1} + W_g X_t) \tag{6}$$

where W_g is the forget gate weight; P_{t-1} represents the previous state or state at time $t-1$; X_t denotes the input at time t; and σ denotes the sigmoid activation function.

The input gate V_t is responsible for determining the information that is added to the cell state (s_t). This process is broken down into two distinct processes. In the first step of the process, the candidate value Ψ, which can be added to the cell states, is calculated. The activation values V_t of the input and gates are determined in the second step in the process, where W_v denotes the input gate weight and W_s denotes the cell state weight. The two processes are, respectively, given by

$$V_t = \sigma(W_v P_{t-1} + W_v X_t) \tag{7}$$

$$\psi = tanh(W_s P_{t-1} + W_s X_t) \tag{8}$$

The new cell states S_{t-1} are determined based on the outcomes of the previous processes. The formula can be broken down as follows:

$$s_t = (G_t * s_{t-1} + v_t * \psi) \tag{9}$$

After the memory cell has passed the input gate and the forget gate, the output gate m_t will generate output (n_t). Two gates will be implemented at the output gates. The first gates will use a sigmoid layer to determine which parts of the cell state will be output. Utilizing the *tanh* activation function will result in the storage of a value within the memory cell. Finally, the two gates are multiplied together to produce a value that will be distributed (n_t). Applying the following formula, we have

$$m_t = \sigma(W_m P_{t-1} + W_m X_t) \tag{10}$$

$$n_t = m_t * tanh(s_t) \tag{11}$$

2.6. Non-Stationary Test

The statistical check was performed using the Augmented Dickey–Fuller (ADF) test and the Kwiatkowski–Phillips-Perron Unit Root (KPSS) test to assess stationarity in both the original dataset and the differenced time-series. The null hypothesis of both ADF and KPSS assumed that the time-series was non-stationary. In the ADF and KPSS tests, if the p-value was less than 5% or there was a 0.05 level of significance for a time-series, then we rejected the null hypothesis and inferred that the series was assumed to be stationary.

The proposed model consists of five steps, as shown in Figure 2. The first step is data processing, the second step is model identification, the third step is model development, the fourth step is evaluating models, and the last step is forecasting.

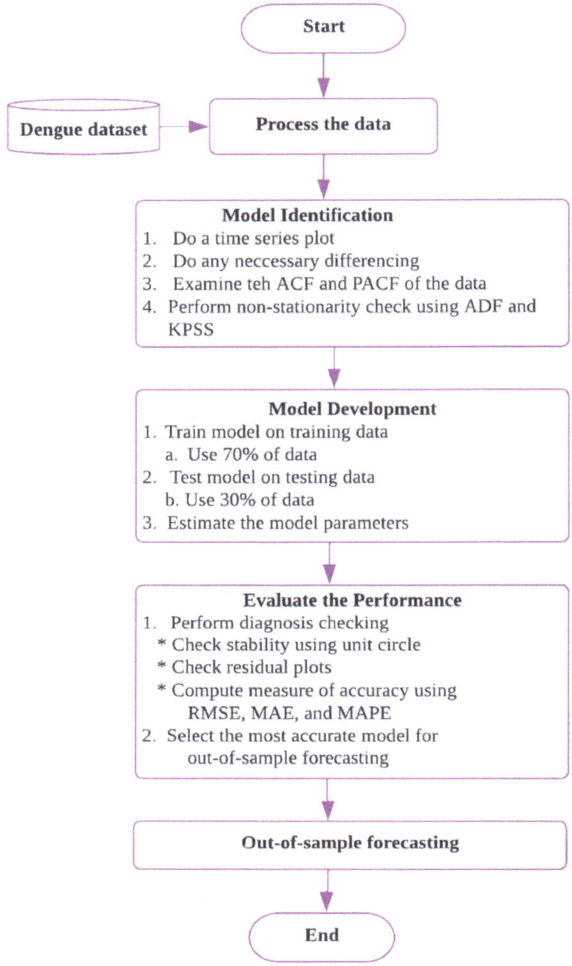

Figure 2. Framework for the model development and forecasting.

3. The Proposed Model

This section discusses the descriptive statistics of the DHF cases, model identification, stationarity testing as well as training and testing models.

3.1. Descriptive Statistics

Table 1 displays the DHF reported cases from 2014 to 2016. The maximum (164) number of cases was reported in April 2016, while the minimum (4) number of cases were obtained each in November 2015 and 2016, respectively. Table 2 provides a descriptive summary of the Surabaya City dengue time data. From Table 2, there was a rise in the incidence in 2016. Table 2 also shows that the dataset was positively and negatively skewed, indicating non-normality. Graphical inspection of the disease indicated time-varying patterns in Figure 3. The series tended to rise and fall gradually, with no noticeable outliers. Higher numbers showed a trend, especially in April, when the rate was at its maximum in 2016. Regular fluctuations and the time-varying behavior of the trend revealed that the ARIMA models were appropriate [7].

Table 1. Average of DHF cases in Surabaya City from 2014 to 2016

Month	2016	2015	2014	Total	Mean
January	60	46	36	142	12
February	114	109	46	269	23
March	134	107	71	312	26
April	164	119	94	377	33
May	141	95	127	363	31
June	119	78	110	307	26
July	89	40	82	211	18
August	66	21	71	158	14
September	24	9	71	104	9
October	18	5	42	65	6
November	4	4	37	45	4
December	5	7	29	41	4
Grand Total	640	640	816	2394	200

Table 2. Descriptive statistics of the annual confirmed DHF cases in Surabaya from 2014 to 2016.

Years	Min.	Max.	Q1	Q2	Q3	Mean	S.D.	Skewness	Kurtosis
2016	4.00	164.00	22.50	77.50	122.75	78.17	56.71914	−0.00655	−1.6732
2015	4.00	119.00	8.50	43.00	98.00	53.33	45.52189	0.213275	−1.8190
2014	29.00	127.00	40.75	71.00	85.00	68.00	31.34848	0.4051189	−1.2137

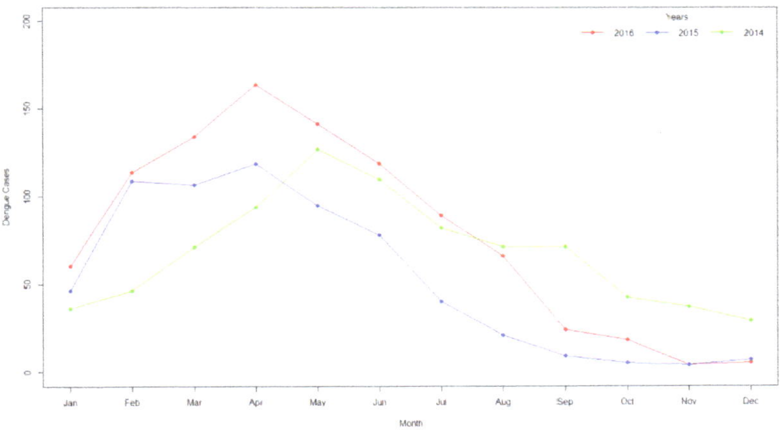

Figure 3. Plot of the monthly DHF case data from 2014–2016, showing maximum peaks from April to May each year.

3.2. Model Identification

To identify a suitable ARIMA model for time-series forecasting, it must be free of trend and seasonality. A time-series may be influenced by the trend and seasonality components at distinct periods [33]. The time-series data must be stationary to develop a model that is effective in forecasting future values. Several tests can be used to determine whether a series is stationary [34]. These tests include the Augmented Dickey–Fuller unit root test (ADF test), the Kwiatkowski–Phillips–Schmidt–Shin (KPSS) test, the partial correlation function (PACF), and the autocorrelation function (ACF) [32]. A time-series is assumed to be stationary in the ADF test if the *p*-value is less than a 5% level of significance. On the other hand, if the time-series is non-stationary, then it is necessary to look at the time-series graph and differentiate the data appropriately [35].

The time-series had a seasonality effect, as shown by the repeating cycles in Figure 4. The ACF and PACF of the original dataset are shown in Figure 5. Observing the ACF in Figure 5, it is reasonable to see that the worst outbreaks occurred every six months, which indicates a seasonal pattern. This suggests that the incidence of DHF in Surabaya City was strongly affected by the seasons. Using a differencing approach, non-stationarity in the time-series data can be corrected [36]. A similar seasonal effect was seen in the original dataset after the first differencing, necessitating the second differencing. The ACF and PACF plots of second-order differenced time-series are shown in Figure 6. It can be seen that the ACF decreased to zero exponentially, indicating stationary behavior [37]. Therefore, the SARIMA model could be used to fit the deseasonalized data [36].

Decomposing the time-series data assists in discovering various hidden behavior within the time-series. In general, a time-series consists of four components: random, seasonal, trend, and cyclic. Three of these components are indicated in Figure 7. It can be observed that the seasonal component exhibited seasonal fluctuations in the dataset. However, the trend and random components seemed to be stationary.

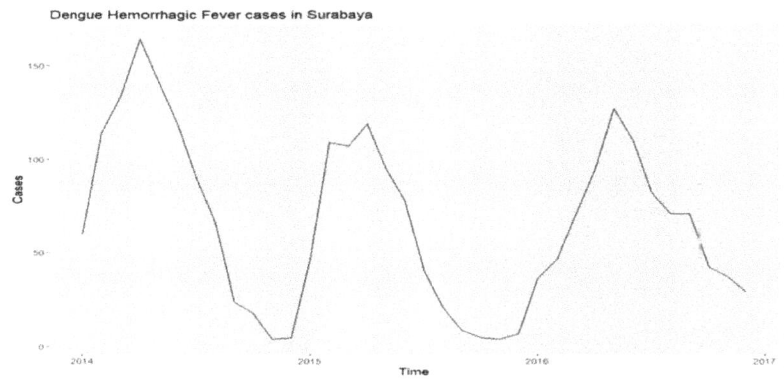

Figure 4. Monthly collected DHF cases in Surabaya City, 2014–2016, showing the time-series pertaining to the time-varying patterns and seasonality across the study period.

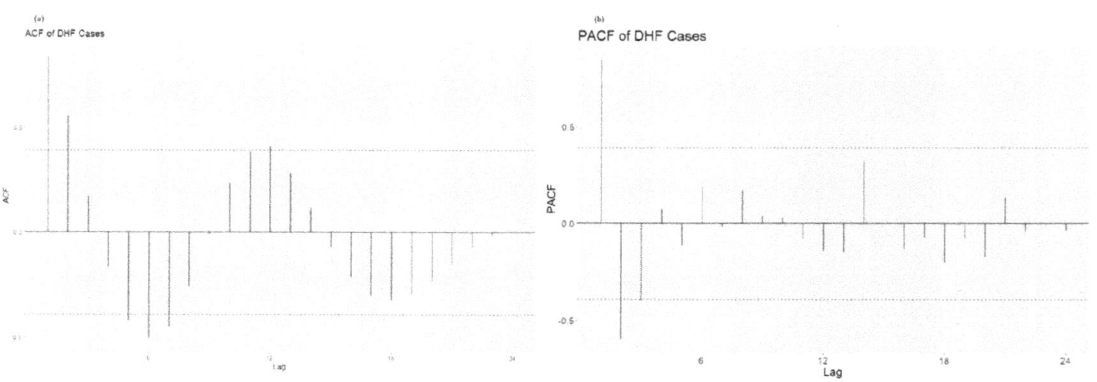

Figure 5. (**a**) Autocorrelation function. (**b**) Partial autocorrelation for the original time-series of monthly confirmed DHF cases in Surabaya City.

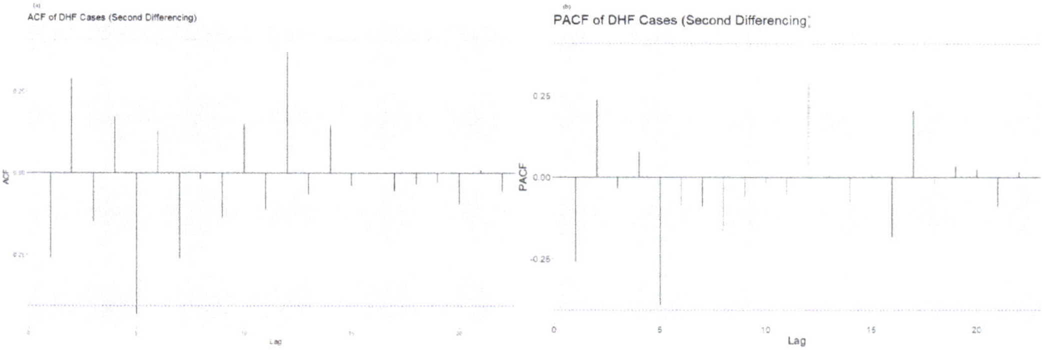

Figure 6. (**a**) Autocorrelation function. (**b**) Partial autocorrelation for second-ordered differencing time-series of monthly confirmed DHF cases in Surabaya City.

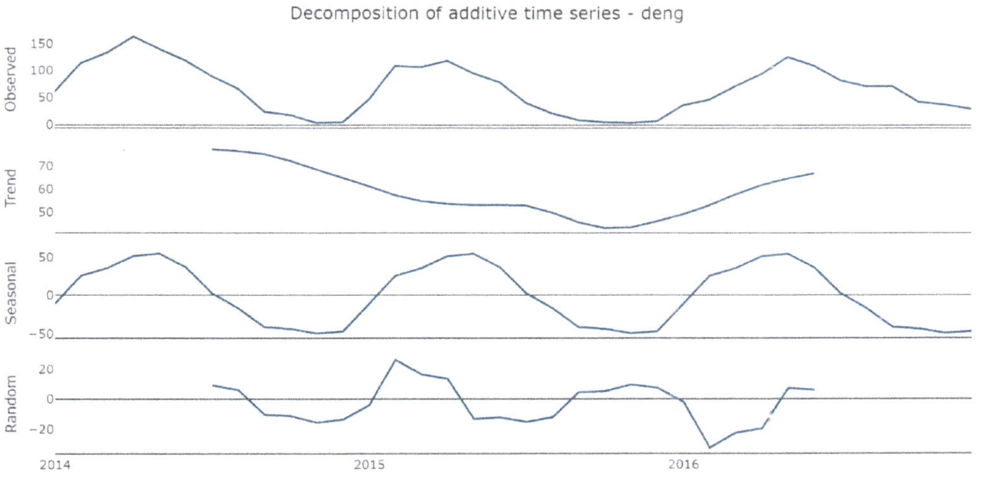

Figure 7. Decomposition of Surabaya City monthly DHF cases time-series data.

3.3. Non-Stationary Test

Table 3 presents the results of the ADF for both the original and second-order differenced datasets. It can be observed that the p-value for the original dataset was greater than 0.05, and we can say that we failed to reject the null hypothesis and concluded that the original dataset was nonstationary. In contrast, the p-value for the second differenced time-series was less than 0.05, so we can say that the time-series is stationary. Similarly, it can be observed in Table 4 that the p-value obtained from the KPSS test for the second differenced dataset was less than 0.05, which means that the time-series is stationary. Interestingly, the time-series data under consideration is non-stationary in its original form and becomes stationary when it is second differenced. The level of significance of the ADF and KPSS tests for the second differenced series were 5% and 1%, respectively. Therefore, the data can be suitable for time-series analysis.

Table 3. Augmented Dickey–Fuller test for the original and second differenced of the Surabaya City monthly DHF case time-series data.

Dataset	Critical Value	p-Value
Original	−3.2892	0.08936
Second-order differencing	−3.7968	0.03287

Table 4. Kwiatkowski–Phillips–Perron unit root test for the original and second differenced of the Surabaya City monthly DHF case time-series data.

Dataset	Critical Value	p-Value
Original	−12.042	0.3605
Second-order differencing	−43.200	0.0100

3.4. Training and Test Models

It is important to determine that the models are adequate to forecast l future values with high accuracy. Therefore, our first step is to split the sample into training and test sets. We selected the data points from January 2014 to December 2015 for model training, consisting of 24 data points. The data points from January 2016 to December 2016 were for model testing.

Fitting the ARIMA model to the training data point will enable the model to learn from the time-series dataset. We used an ARIMA diagnosis plot on the training dataset to determine what lag to use in the model. We examined the properties of the data using residuals plots of the ACF and PACF. The statistical criteria used to evaluate these models are the root mean square error (RMSE), mean absolute error (MAE), and mean absolute percentage error (MAPE), which are expressed in Equations (12)–(14), respectively. More details on these measures can be found in [38,39].

$$RMSE = \sqrt{\frac{\sum_{i=1}^{N} (Y_i - F_i)^2}{N}} \tag{12}$$

$$MAE = \frac{\sum_{i=1}^{N} |Y_i - F_i|}{N} \tag{13}$$

$$MAPE = \frac{1}{N}\sum_{i=1}^{N}\left|\frac{Y_i - F_i}{Y_i}\right| \times 100\% \tag{14}$$

where Y_i is the actual data; F_i is the forecasted values; and N is the total number of observations in the data.

It can be observed from Figure 8 that the ACF geometrically declined, indicating a signal of the error process. The PACF in the plot had two significant lags. As a result, we developed an ARIMA (2,1,1) model, which means that we included two AR lags, one difference, and one MA lag. Thus, ARIMA (2,1,1) passed this test. Next, we tested the residuals of the model. The residuals plot in Figure 9 indicates that there was one lag at the 12 periods that was significant. This suggests that there is seasonality in the model.

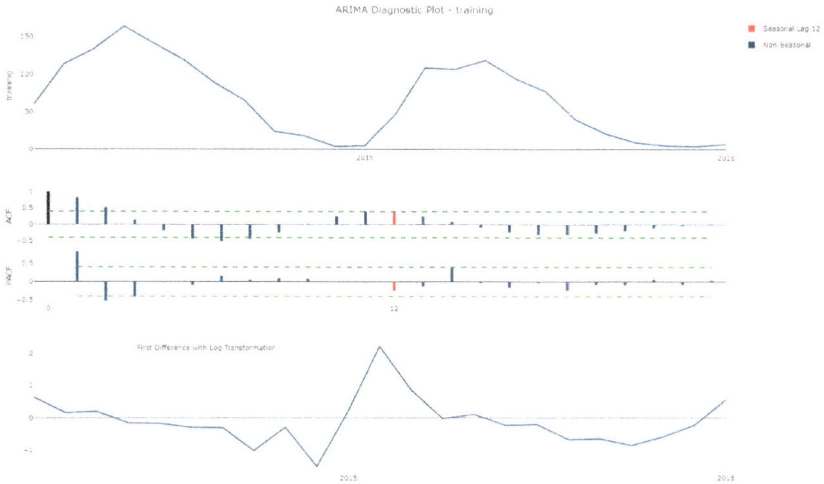

Figure 8. The ARIMA (2,1,1) diagnosis plot in the training dataset.

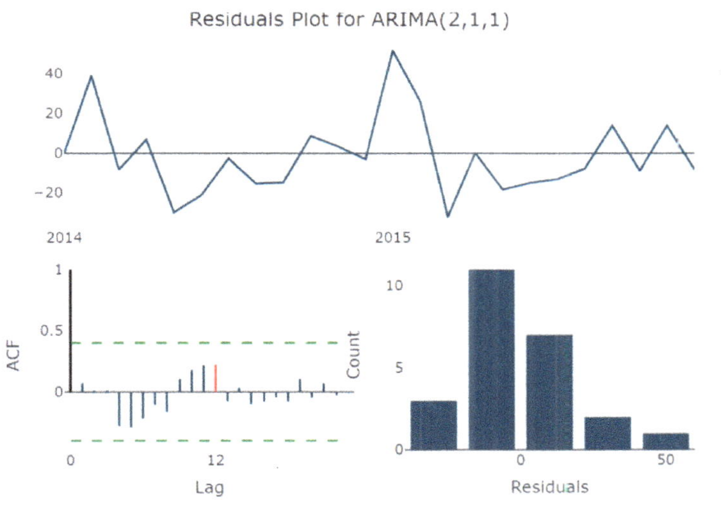

Figure 9. The residual plots of ARIMA (2,1,1) on the training dataset.

4. Results and Discussion

In this section, we discuss the results obtained from the model development. The results of the ARIMA (2,1,1), SARIMA (2,1,1) (1,0,0), and LSTM models were compared. The out-of-sample forecast for the twelve month horizons were obtained based on the three competitive models.

After identifying and evaluating each parameter, the ARIMA (2,1,1), SARIMA (2,1,1) (1,0,0), and LSTM models can be used to predict future values for both the training and test datasets. We found that the SARIMA (2,1,1) (1,0,0) model outperformed the ARIMA (2,1,1) and LSTM models in terms of the accuracy metrics reported in Tables 5 and 6, respectively. Figures 10–12 depict the predicted dengue outbreaks based on the training datasets. From these figures, we can note that the SERIMA (2,1,1) (1,0,0) model predicts the dengue outbreak values more accurately than the ARIMA (2,1,1) and LSTM models.

Table 5. Comparison of ARIMA (2,1,1), SARIMA (2,1,1) (1,0,0), and LSTM models on the training dataset using accuracy metrics.

Model.	RMSE	MAE	MAPE
ARIMA (2,1,1)	19.66198	15.10929	60.09833
SARIMA (2,1,1) (1,0,0)	13.07250	10.27440	53.47647
LSTM	15.35412	12.54872	56.67457

Table 6. Comparison of the ARIMA (2,1,1), SARIMA (2,1,1) (1,0,0), and LSTM models on the testingdataset using accuracy metrics.

Model	RMSE	MAE	MAPE
ARIMA (2,1,1)	77.42119	71.02259	104.95288
SARIMA (2,1,1) (1,0,0)	38.00313	33.54329	54.565850
LSTM	45.37822	41.45632	76.983421

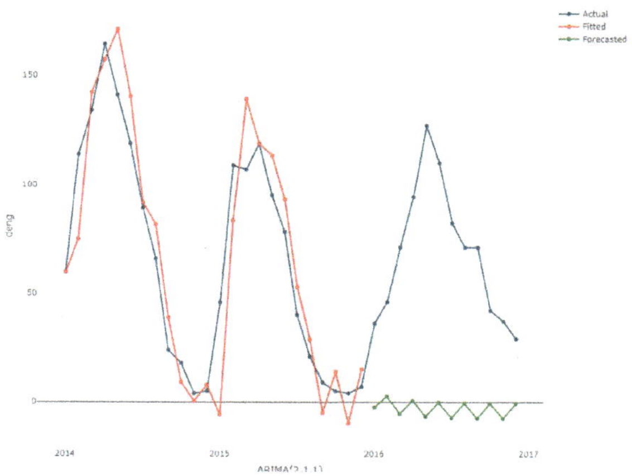

Figure 10. Plot of the actual, fitted, and forecasted values using ARIMA (2,1,1) on the training dataset.

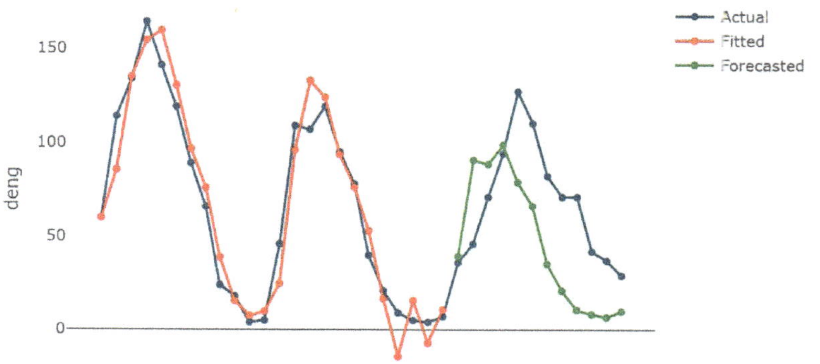

Figure 11. Plot of the actual, fitted, and forecasted values using SARIMA (2,1,1) (1,0,0) on the training dataset.

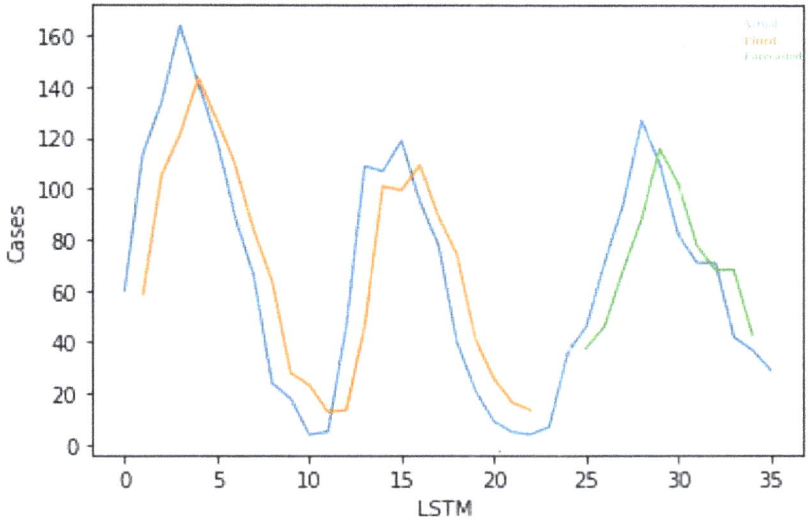

Figure 12. Plot of the actual, fitted, and forecasted values using SARIMA (2,1,1) (1,0,0) on the training dataset.

The results of the training set indicate that the SARIMA (2,1,1) (1,0,0) is more appropriate to perform the out-of-sample forecasting of dengue outbreaks in Surabaya City in the next 12 months.

Out-of-Sample Forecasting

Finally, a 12-month out-of-sample forecasting of dengue cases was conducted from January 2017 to December 2017. According to the findings, SARIMA (2,1,1) (1,0,0) performed well in forecasting future outbreaks. Figure 13 shows that this model is stable since all the data points lie within the unit root circle. The findings of the monthly future values of the outbreaks from January 2017 to December 2017 are presented in Table 7. From this table, the SARIMA (2,1,1) (1,0,0) model revealed the smallest value of the RMSE metric in comparison to the ARIMA (2,1,1) and LSTM models. Thus, the SARIMA (2,1,1) (1,0,0) model can be selected as the best forecasting model for the monthly DHF cases in Surabaya City.

From Table 7, the forecasted results showed that the number of monthly dengue outbreaks began to increase in March and continued through September, with the maximum incidence occurring in May. This time-dynamic pattern suggested that there was a significant seasonal effect in the forecast. It is noteworthy that the highest forecasted incidence in May was like the outbreaks that occurred in 2014, which had the highest incidence in May. Figure 14 confirms the findings of Table 7 to show the trend pattern in the number of DHF cases across the forecasting horizons. Therefore, this forecast using the SARIMA (2,1,1) (1,0,0) model provides the future values of dengue outbreaks and could be used to assist authorities and public health professionals in designing effective public health measures to prevent and control the disease, particularly during the early stages of the outbreaks.

Table 7. The out-of-sample forecasted value for monthly DHF outbreaks in Surabaya City from January 2017 to December 2017 using ARIMA (2,1,1), SARIMA (2,1,1) (1,0,0), and LSTM models.

Model	Month	January	February	March	April	May	June	July	August	September	October	November	December	RMSE
ARIMA	Forecast	6	9	26	18	37	33	25	16	13	8	6	4	31.21
SARIMA	Forecast	47	46	63	71	91	80	67	59	61	44	43	38	11.35
LSTM	Forecast	13	14	46	101	99	109	89	74	41	26	16	13	20.13

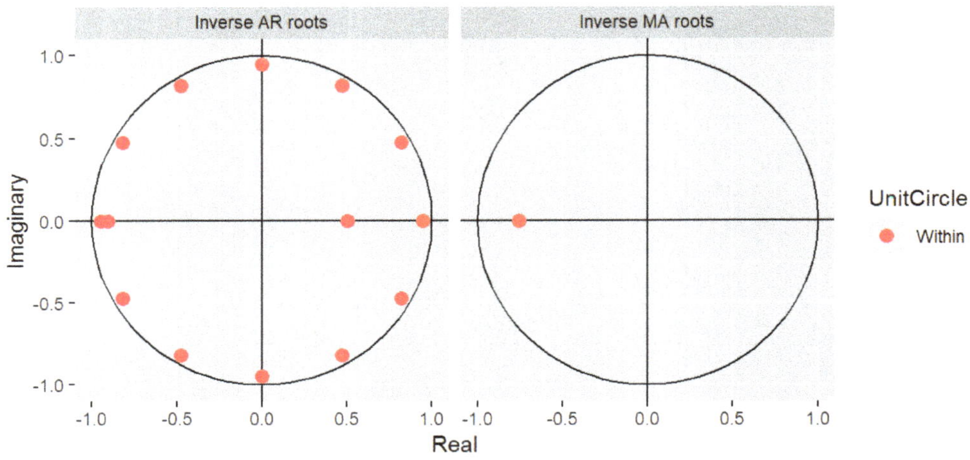

Figure 13. Unit root circle plot of the out-of-sample forecasting analysis.

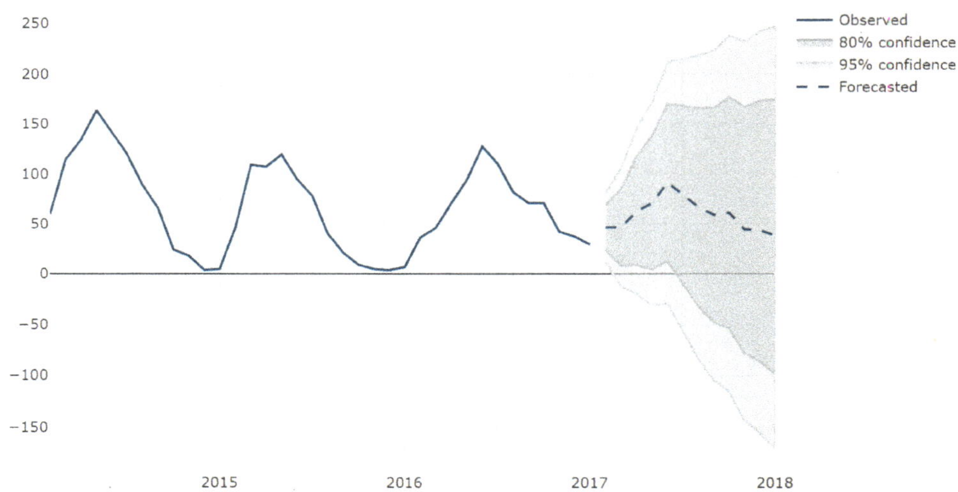

Figure 14. Out-of-sample forecasted values of the monthly dengue outbreak in Surabaya City from January 2017 to December 2017 using the SARIMA (2,1,1) (1,0,0) model.

DHF is a serious public health issue in Surabaya. This is mostly due to an unprecedented rise in the number of cases recorded each year. Consequently, only a few preventative measures have been implemented to prevent outbreaks. Instead of responding in advance to safeguard against disease, the health sector responds after it emerges. The present research attempted to forecast the DHF incidence using an effective forecasting method to assist the public and authorities in properly adjusting to outbreak and provide enough preparations in creating public awareness. The objective was to forecast the DHF outbreaks using a time-series model from 2014 to 2016 for twelve months covering January 2017 to December 2017 using time-series analysis consisting of monthly confirmed DHF cases from 2014 to 2016. The series showed that the disease had a strong seasonal effect with the maximum rates occurring between March and September. The forecast showed peaks of incidence in May with the highest incidence forecasted to be 91 cases in May 2017.

Similarly, researchers from Rajasthan designed a forecasting model for DHF using time-series data from the past decade to forecast monthly dengue fever/dengue hemorrhagic incidence for 2011. The SARIMA model was employed for statistical modeling. Dengue fever/dengue hemorrhagic cases that were reported between January 2001 and December 2010 displayed a cyclical pattern with seasonal variation. The forecast for 2011 indicated a seasonal peak in October with a predicted 546 cases [12]. Another study conducted in Ribeirao Preto, Sao Paulo State, Brazil also applied SARIMA to fit in a model of monthly reported cases of DHF from 2000 to 2008. The forecasted values for the incidence for 2009 were obtained and compared with the results of the original number of cases. The researchers found that the SARIMA model effectively forecast the number of DHF cases and is a reliable approach for disease control and prevention [40]. In a study recently conducted in Jeddah, Saudi Arabia, the SARIMA model was used to forecast DHF mortality and morbidity using time-series from 2006 to 2016 for the years 2017 to 2019. According to the study, incidence rates increased from May to September, having the highest rate in 2012, suggesting a strong seasonality [7].

The prevalence of DHF has been associated with climate variations. High temperatures promote mosquito reproduction, while an increase in rainfall contributes to the availability of vector habitat [41,42]. DHF epidemics often occur seasonally and tend to spike in the summer and spring when the weather is hot and humid. However, the association between incidence and climate remains poorly understood and typically varies among locations due to local climate heterogeneity and virus–host interactions, which are all factors in the spread of the disease. Globally, every year, 50–100 million humans are infected by a female Aedes aegypti mosquito that has fed on infected human blood. In many tropical and subtropical countries, dengue disease is seasonal. This is because rainfall provides breeding sites and stimulates egg hatching, and temperature affects the mosquito's survival, development, and reproduction. Temperature enhances the mosquitoes' capacity to spread the dengue virus; higher temperatures boost transmission rates.

5. Conclusions

Dengue is the fastest-spreading vector-borne disease in the world. Monthly confirmed DHF cases in Surabaya City were obtained from 2014 to 2016 for this study to forecast disease outbreaks in the early phases and enable quick response. To develop effective forecasting models, monthly DHF occurrence patterns were studied. The prevalence of DHF in 2017 was then forecast using the best models. The results of DHF incidence revealed a significant seasonal effect. These findings showed an increase in DHF cases from March to September. Additionally, the dengue outbreaks appeared to spike more in May every three years, in 2014 and 2017. The air temperature is suspected to be the significant factor associated with DHF cases in Surabaya City.

6. Recommendations

To help with health care planning, public health officials want a means to forecast when epidemics will occur. To develop such a system, they need to understand the factors that lead to epidemics. The presence of mosquito larvae is the most important risk factor for dengue fever in several parts of Surabaya City. Hence, it is necessary for the government to exercise early efforts to eradicate and minimize DHF cases by conducting dengue surveillance, which aims to monitor the trends. In future studies, we will incorporate correlation studies of DHF cases with meteorological data, explore more current data for DHF cases, and consider several models for dengue incidence forecasting.

Author Contributions: Conceptualization, M.O., R.I. and M.B.Q.; Methodology, R.I., M.O., R.S., M.B.Q., A.A.S. and R.S.; Software, M.O. and A.A.S.; Validation, M.O., R.I. and A.A.S.; Formal analysis, M.O., R.I., A.A.S. and R.S.; Writing—original draft preparation, M.O., A.A.S., R.I., M.B.Q. and R.S.; Discussion about behavior data, M.O., R.I. and M.B.Q.; Writing—and editing, M.O., R.I., A.A.S. and R.S. All authors have read and agreed to the published version of the manuscript.

Funding: This project is supported by Universitas Airlangga.

Data Availability Statement: Not applicable.

Acknowledgments: The authors would like to thank the Universiti Teknologi Petronas, Universiti Airlangga, and the Surabaya Health Office, who provided support to this project.

Conflicts of Interest: The authors declare no conflict of interest.

References

1. World Health Organization. WHO Guidelines Approved by the Guidelines Review Committee. In *Dengue: Guidelines for Diagnosis, Treatment, Prevention and Control: New Edition*; World Health Organization Copyright © 2009; World Health Organization: Geneva, Switzerland, 2009.
2. WHO. *Dengue: Guidelines for Diagnosis, Treatment, Prevention and Control*; WHO Library: Geneva, Switzerland, 2009; pp. 10–12.
3. Guha-Sapir, D.; Schimmer, B. Dengue fever: New paradigms for a changing epidemiology. *Emerg. Epidemiol.* **2005**, *2*, 1. [CrossRef]
4. Sutriyawan, A.; Herdianti, H.; Cakranegara, P.A.; Lolan, Y.P.; Sinaga, Y. Predictive Index Using Receiver Operating Characteristic and Trend Analysis of Dengue Hemorrhagic Fever Incidence. *Open Access Maced. J. Med. Sci.* **2022**, *10*, 681–687. [CrossRef]
5. Bhatt, S.; Gething, P.W.; Brady, O.J.; Messina, J.P.; Farlow, A.W.; Moyes, C.L.; Drake, J.M.; Brownstein, J.S.; Hoen, A.G.; Sankoh, O. The global distribution and burden of dengue. *Nature* **2013**, *496*, 504–507. [CrossRef]
6. Stanaway, J.D.; Shepard, D.S.; Undurraga, E.A.; Halasa, Y.A.; Coffeng, L.E.; Brady, O.J.; Hay, S.I.; Bedi, N.; Bensenor, I.M.; Castañeda-Orjuela, C.A. The global burden of dengue: An analysis from the Global Burden of Disease Study 2013. *Lancet Infect. Dis.* **2016**, *16*, 712–723. [CrossRef]
7. Abualamah, W.A.; Akbar, N.A.; Banni, H.S.; Bafail, M.A. Forecasting the morbidity and mortality of dengue fever in KSA: A time series analysis (2006–2016). *J. Taibah Univ. Med. Sci.* **2021**, *16*, 448–455. [CrossRef]
8. Messina, J.P.; Brady, O.J.; Golding, N.; Kraemer, M.U.; Wint, G.; Ray, S.E.; Pigott, D.M.; Shearer, F.M.; Johnson, K.; Earl, L. The current and future global distribution and population at risk of dengue. *Nat. Microbiol.* **2019**, *4*, 1508–1515. [CrossRef] [PubMed]
9. Aziz, A.T.; Al-Shami, S.A.; Mahyoub, J.A.; Hatabbi, M.; Ahmad, A.H.; Md Rawi, C.S. Promoting health education and public awareness about dengue and its mosquito vector in Saudi Arabia. *Parasites Vectors* **2014**, *7*, 487. [CrossRef] [PubMed]
10. Khaira, U.; Utomo, P.E.P.; Aryani, R.; Weni, I. A comparison of SARIMA and LSTM in forecasting dengue hemorrhagic fever incidence in Jambi, Indonesia. *J. Phys. Conf. Ser.* **2020**, *1566*, 012054. [CrossRef]
11. Riaz, M.M.; Mumtaz, K.; Khan, M.S.; Patel, J.; Tariq, M.; Hilal, H.; Siddiqui, S.A.; Shezad, F. Outbreak of dengue fever in Karachi 2006: A clinical perspective. *J. Pak. Med. Assoc.* **2009**, *59*, 339. [PubMed]
12. Bhatnagar, S.; Lal, V.; Gupta, S.D.; Gupta, O.P. Forecasting incidence of dengue in Rajasthan, using time series analyses. *Indian J. Public Health* **2012**, *56*, 281.
13. Siregar, F.A.; Makmur, T.; Saprin, S. Forecasting dengue hemorrhagic fever cases using ARIMA model: A case study in Asahan district. In Proceedings of the IOP Conference Series: Materials Science and Engineering, Medan, Indonesia, 21–23 August 2017; p. 012032.
14. Suleiman, A.A.; Suleiman, A.; Abdullahi, U.A.; Suleiman, S.A. Estimation of the case fatality rate of COVID-19 epidemiological data in Nigeria using statistical regression analysis. *Biosaf. Health* **2021**, *3*, 4–7. [CrossRef]
15. Cummings, D.A.; Iamsirithaworn, S.; Lessler, J.T.; McDermott, A.; Prasanthong, R.; Nisalak, A.; Jarman, R.G.; Burke, D.S.; Gibbons, R.V. The impact of the demographic transition on dengue in Thailand: Insights from a statistical analysis and mathematical modeling. *PLoS Med.* **2009**, *6*, e1000139. [CrossRef] [PubMed]
16. Johansson, M.A.; Reich, N.G.; Hota, A.; Brownstein, J.S.; Santillana, M. Evaluating the performance of infectious disease forecasts: A comparison of climate-driven and seasonal dengue forecasts for Mexico. *Sci. Rep.* **2016**, *6*, 33707. [CrossRef]
17. Xu, H.-Y.; Fu, X.; Lee, L.K.H.; Ma, S.; Goh, K.T.; Wong, J.; Habibullah, M.S.; Lee, G.K.K.; Lim, T.K.; Tambyah, P.A. Statistical modeling reveals the effect of absolute humidity on dengue in Singapore. *PLoS Negl. Trop. Dis.* **2014**, *8*, e2805. [CrossRef]
18. Honório, N.A.; Nogueira, R.M.R.; Codeço, C.T.; Carvalho, M.S.; Cruz, O.G.; de Avelar Figueiredo Mafra Magalhães, M.; de Araújo, J.M.G.; de Araújo, E.S.M.; Gomes, M.Q.; Pinheiro, L.S. Spatial evaluation and modeling of dengue seroprevalence and vector density in Rio de Janeiro, Brazil. *PLoS Negl. Trop. Dis.* **2009**, *3*, e545. [CrossRef] [PubMed]
19. Nishiura, H. Mathematical and statistical analyses of the spread of Dengue. *Dengue Bull.* **2006**, *30*, 51–67.
20. Reiner, R.C., Jr.; Perkins, T.A.; Barker, C.M.; Niu, T.; Chaves, L.F.; Ellis, A.M.; George, D.B.; Le Menach, A.; Pulliam, J.R.; Bisanzio, D. A systematic review of mathematical models of mosquito-borne pathogen transmission: 1970–2010. *J. R. Soc. Interface* **2013**, *10*, 20120921. [CrossRef]

21. Reich, N.G.; Shrestha, S.; King, A.A.; Rohani, P.; Lessler, J.; Kalayanarooj, S.; Yoon, I.-K.; Gibbons, R.V.; Burke, D.S.; Cummings, D.A. Interactions between serotypes of dengue highlight epidemiological impact of cross-immunity. *J. R. Soc. Interface* **2013**, *10*, 20130414. [CrossRef]
22. Lauer, S.A.; Sakrejda, K.; Ray, E.L.; Keegan, L.T.; Bi, Q.; Suangtho, P.; Hinjoy, S.; Iamsirithaworn, S.; Suthachana, S.; Laosiritaworn, Y. Prospective forecasts of annual dengue hemorrhagic fever incidence in Thailand, 2010–2014. *Proc. Natl. Acad. Sci. USA* **2018**, *115*, E2175–E2182. [CrossRef]
23. Buczak, A.L.; Baugher, B.; Moniz, L.J.; Bagley, T.; Babin, S.M.; Guven, E. Ensemble method for dengue prediction. *PLoS ONE* **2018**, *13*, e0189988. [CrossRef]
24. Banditvilai, S.; Anansatitzin, S. Comparative study of three time series methods in forecasting dengue hemorrhagic fever incidence in thailand. In Proceedings of the International Academic Conferences, Sevilla, Spain, 5–8 March 2018.
25. Nájera, J. World Health Organiziation Global Partnership to Roll Back Malaria. Malaria Control: Achievements, Problems and Strategies; WHO/CDS/RBM/99.10. 1999. Available online: www.who.int/malaria/publications/atoz (accessed on 25 August 2022).
26. Luz, P.M.; Mendes, B.V.; Codeço, C.T.; Struchiner, C.J.; Galvani, A.P. Time series analysis of dengue incidence in Rio de Janeiro, Brazil. *Am. J. Trop. Med. Hyg.* **2008**, *79*, 933–939. [CrossRef]
27. Wu, P.-C.; Guo, H.-R.; Lung, S.-C.; Lin, C.-Y.; Su, H.-J. Weather as an effective predictor for occurrence of dengue fever in Taiwan. *Acta Trop.* **2007**, *103*, 50–57. [CrossRef]
28. Chen, K.-Y.; Wang, C.-H. A hybrid SARIMA and support vector machines in forecasting the production values of the machinery industry in Taiwan. *Expert Syst. Appl.* **2007**, *32*, 254–264. [CrossRef]
29. Conejo, A.J.; Plazas, M.A.; Espinola, R.; Molina, A.B. Day-ahead electricity price forecasting using the wavelet transform and ARIMA models. *IEEE Trans. Power Syst.* **2005**, *20*, 1035–1042. [CrossRef]
30. Suresh, K.; Krishna Priya, S. Forecasting sugarcane yield of Tamilnadu using ARIMA models. *Sugar Tech.* **2011**, *13*, 23–26. [CrossRef]
31. Idrees, S.M.; Alam, M.A.; Agarwal, P. A prediction approach for stock market volatility based on time series data. *IEEE Access* **2019**, *7*, 17287–17298. [CrossRef]
32. Siami-Namini, S.; Tavakoli, N.; Namin, A.S. A Comparison of ARIMA and LSTM in Forecasting Time Series. In Proceedings of the 2018 17th IEEE International Conference on Machine Learning and Applications (ICMLA), Orlando, FL, USA, 17–20 December 2018; pp. 1394–1401.
33. Narasimha Murthy, K.; Saravana, R.; Vijaya Kumar, K. Modeling and forecasting rainfall patterns of southwest monsoons in North–East India as a SARIMA process. *Meteorol. Atmos. Phys.* **2018**, *130*, 99–106. [CrossRef]
34. Thevakumar, P.; Jayathilaka, R. Exchange rate sensitivity influencing the economy: The case of Sri Lanka. *PLoS ONE* **2022**, *17*, e0269538. [CrossRef]
35. Chang, X.; Gao, M.; Wang, Y.; Hou, X. Seasonal autoregressive integrated moving average model for precipitation time series. *J. Math. Stat.* **2012**, *8*, 500–505. [CrossRef]
36. Jain, G.; Mallick, B. A study of time series models ARIMA and ETS. *SSRN* **2017**, 2898968. [CrossRef]
37. Han, P.; Wang, P.; Wang, Y. Drought forecasting based on the Standardized Precipitation Index at different temporal scales using ARIMA models. *Agr. Res. Arid Areas* **2008**, *26*, 212–218.
38. Alyousifi, Y.; Othman, M.; Faye, I.; Sokkalingam, R.; Silva, P.C.L. Markov Weighted Fuzzy Time-Series Model Based on an Optimum Partition Method for Forecasting Air Pollution. *Int. J. Fuzzy Syst.* **2020**, *22*, 1468–1486. [CrossRef]
39. Dalah, C.M.; Singh, V.; Abdullahi, I.; Suleiman, A. The study of HIV/AIDS trend in Yobe state for the prescribed period (1999–2019). *Int. J. Stat. Appl.* **2020**, *10*, 10–16.
40. Martinez, E.Z.; Silva, E.A.S.d.; Fabbro, A.L.D. A SARIMA forecasting model to predict the number of cases of dengue in Campinas, State of São Paulo, Brazil. *Rev. Da Soc. Bras. De Med. Trop.* **2011**, *44*, 436–440. [CrossRef] [PubMed]
41. Johansson, M.A.; Cummings, D.A.; Glass, G.E. Multiyear climate variability and dengue—El Nino southern oscillation, weather, and dengue incidence in Puerto Rico, Mexico, and Thailand: A longitudinal data analysis. *PLoS Med.* **2009**, *6*, e1000168. [CrossRef]
42. Morales, I.; Salje, H.; Saha, S.; Gurley, E.S. Seasonal Distribution and Climatic Correlates of Dengue Disease in Dhaka, Bangladesh. *Am. J. Trop. Med. Hyg.* **2016**, *94*, 1359–1361. [CrossRef]

processes

MDPI

Article

Developing Design Approaches for Tile Pattern Designs Inspired by Traditional Textile Patterns

Arus Kunkhet [1,2] and Disaya Chudasri [1,2,*]

1 Arts Program in Animation and Visual Effects, College of Arts, Media and Technology, Chiang Mai University, Chiang Mai 50200, Thailand
2 Advanced Research Center for Computational Simulation, Chiang Mai University, Chiang Mai 50200, Thailand
* Correspondence: disaya.c@cmu.ac.th; Tel.: +66-(0)53-920299

Abstract: This article presents multidisciplinary research that involved design (i.e., textiles, tiles, pattern design), mathematics (i.e., symmetry and seven frieze groups) and a viewpoint on product design and development for business opportunities. This research comprised a design experiment and a survey. In the design experiment, two design approaches were created to translate the characteristics of traditional textile patterns into new pattern designs for floor tiles. These two design approaches were entitled: "partial replication", and "combination and simplification". The seven frieze groups were used as a transformation rule in both design approaches, resulting in two sets of frieze patterns. Although they were derived from the same origin they looked different. A survey was conducted with 61 respondents to gain outsiders' perspectives on these new pattern designs. The findings include: (i) positive responses to applying traditional textile patterns to other products, (ii) plausible products for pattern designs, (iii) preferences for design approaches and frieze patterns and (iv) opportunities for design research and education with other disciplines. This paper concludes with theoretical and practical implications for further research.

Keywords: cultural significance; design structure; frieze group; symmetry; textile pattern; tile pattern

Citation: Kunkhet, A.; Chudasri, D. Developing Design Approaches for Tile Pattern Designs Inspired by Traditional Textile Patterns. *Processes* **2022**, *10*, 2744. https://doi.org/10.3390/pr10122744

Academic Editors: Jun Wei Lim and Worapon Kiatkittipong

Received: 31 October 2022
Accepted: 13 December 2022
Published: 19 December 2022

1. Introduction

This article presents multidisciplinary research that involved design (i.e., textiles, tiles, pattern design), mathematics (i.e., symmetry and seven frieze groups) and a viewpoint on product design and development for business opportunities. This research was initiated by us as we have a common interest in pattern design in relation to cultural adherence, continuity and change. Chudasri is a design researcher who has been researching indigenous woven textiles in Thailand and has recently applied visual and symmetrical analysis in design research to examine traditional textile patterns [1,2]. In the long run, Chudasri intends to develop design methods and digital tools that invigorate cultural revitalization. Kunkhet is a design researcher specializing in shape grammar and character design for 3D animation. Previously, we developed a framework for Harmonized Shape Grammar (HSG) that was applied to a real case study, namely a specific context in which traditional textile patterns were regenerated [2]. It was a design process that involved symmetry analysis, i.e., seven frieze groups. Having conducted that, we realised that this mathematical theory also enabled us to design a variety of new patterns, which were not limited to textile patterns. Designers often begin their creative work by searching for information and redesigning successful elements from preceding collections, e.g., traditional textiles and tiles [3]. In this article, we conducted a design experiment in which the characteristics of traditional textile patterns were translated through a design process into pattern designs for floor tiles. We also developed two design approaches, both of which involved seven frieze groups that could generate a variety of new patterns for floor tiles. Finally, we conducted an online survey with 61 respondents in order to gain outsiders' perspectives on the design approaches to invigorating cultural significance.

2. The Core Components

2.1. Traditional Patterns and Design Structure of "Sin Tin Chok"

The word "pattern" in this research context refers to a decorative design. A pattern involves the repetition of a motif, or motifs, at regular intervals using rigid motions to fill a flat surface [3,4] (p. 1). "Motifs are any kind of figure, image or illustrative drawing, produced by artists or graphic designers" [3]. Motifs may be either symmetrical or asymmetrical. A symmetrical motif is a figure which is comprised of two or more units of identical size, shape and content [4] (p. 1). Traditional patterns are those considered by indigenous informants to have been produced for several generations by using similar techniques, and those patterns have not undergone significant change even when they have been produced in recent times [5] (p. 581). In any specific cultural group, the people will consistently use specific symmetries and design structures in their cultural design systems [4] (pp. 2, 8, 55), e.g., the patterns in "sin tin chok" from Long district, Phrae province in northern Thailand (Figure 1). Handwoven textiles and cloth-making are meaningful to the Long people, who recognise these as an intangible cultural heritage. "Sin tin chok" was originally designed to comprise three parts: a waistband, a body section and a lower section. Its lower section ("tin chok") is a woven fabric comprising frieze patterns (Figure 1: right). "Tin chok" is designed to have four parts: a supplementary part, a main part, a supplementary part and a hem (Figure 1: right). We referred to these four parts as "a pattern set". It is a set of four frieze patterns placed in "tin chok" according to this design structure.

Figure 1. "Sin tin chok" from Komol Phaboraan Museum, Long district, Phrae province. (Photo by Chudasri).

2.2. Textile Patterns and Tiles

Textile patterns and tiling with tiles share a similar practice, by which motifs/units are regularly repeated on a flat surface [3]. Tiling is the regular repetition of tiles with defined geometric forms without gaps or overlaps between them [3]. In the last two decades, a number of scholars have researched previous collections of tiles and textile patterns, with attempts to understand and describe the structural designs of patterns and tiling, and developed methods, systems and tools for use in redesigning textiles and tile patterns [3,6–8]. There is the possibility in the design process to integrate both types of patterns and interchange their characteristics. A tile is a flat piece of hardened material, such as baked clay or stone, that is used for covering a wall, floor or roof [9]. In building construction, tiles can be classified into two main types: wall tiles and floor tiles. Floor tiles are produced from stronger materials than wall tiles. Floor tiles are important materials in building construction and floor decoration [10]. Tiles can be designed in various geometric shapes such as squares, rectangles, diamonds and hexagons. However, a square is seen as a standardised tile shape [11]. A tile shape can inform the tile layout. A layout is an arrangement of diverse units [12]. For example, Figure 2 depicts a square tile that can be arranged in five different layouts: grid/stacked, checkerboard, diagonal/diamond, old English and hopscotch [13]. A tile layout is used as a guide to divide and organise a space and inform the quantity of tiles required for the space.

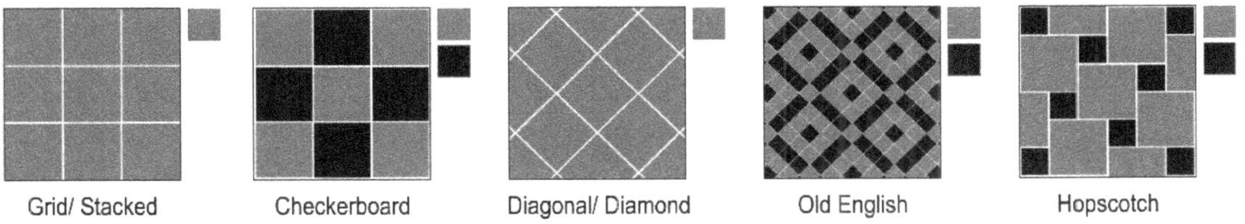

Figure 2. A square tile arranged in various layouts. (Illustration by Kunkhet).

2.3. Frieze Patterns and Seven Frieze Groups

Frieze patterns can be found in many places and products such as architecture, pottery, handwoven textiles and embroidered patterns on clothes. In mathematics, a frieze pattern is a two-dimensional image (with width and length) identified with a motif that is repeated at regular intervals in one direction (e.g., in a horizontal direction) across a flat surface. Every symmetric motif consists of at least two asymmetric units of the same size, shape and content [4] (pp. 1–2). All frieze patterns can be classified into seven groups based on their symmetries [14] (pp. 44, 58–59, 83), [15] (pp. 24–27), [16]. These are known as the seven frieze groups. A frieze pattern is generated as a result of the movement of the original motif using rigid motion(s) on a flat surface in one direction [15] (p. 24). A rigid motion is a transformation of a flat surface where the original motif and its image are congruent in size, shape and content [17] (p. 159). There are four kinds of rigid motions: translation, twofold rotation, reflection and glide reflection, which can generate any pattern [15] (p. 24), [16,17] (p. 159). All frieze patterns necessarily contain translations, but they may or may not contain twofold rotations, reflections or glide reflections. Figure 3 depicts four rigid motions and seven frieze groups. There are many naming systems for frieze groups besides a notable group from a scientific union, i.e., the International Union of Crystallography (IUCr) [18]. Here, we referred to the crystallographic symbols of Washburn and Crowe [14] to classify the seven frieze groups. However, we also compared this naming system to the IUCr, as shown in Figure 3.

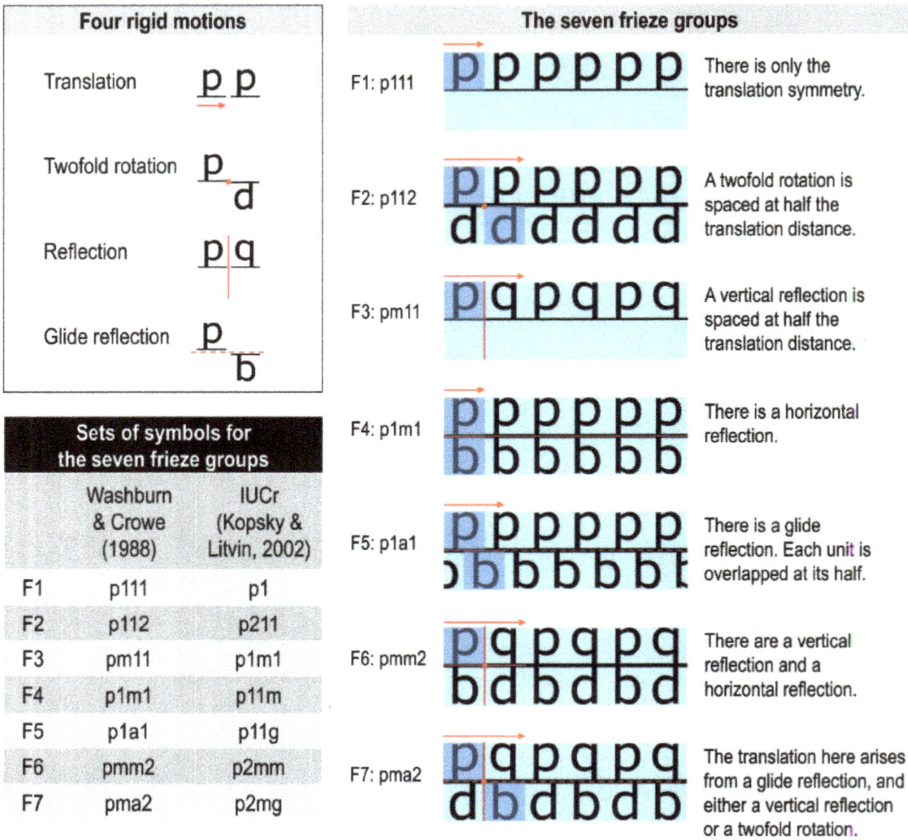

Figure 3. Seven frieze groups. Red arrows indicate translation distance. Red lines indicate axes of symmetry (horizontal and vertical). Dashed red lines indicate horizontal glide reflection axes of symmetry. Red dots indicate twofold rotation axes in each part. (Illustration by Kunkhet and Chudasri).

3. Research Methodology

The pattern sets in "tin chok" are frieze patterns, as these patterns are two-dimensional designs that are woven incrementally along the warp threads of a loom, so the fabric's length increases in one direction [1]. We aimed to design the tile patterns by taking inspiration from the traditional textile patterns in "tin chok". Here we posed a question: how can the traditional textile patterns in "tin chok" be translated into tile patterns? In this design experiment, we narrowed down the research scope to embrace three conditions. We applied the seven frieze groups as a set of transformation rules in the design process. A square was the only tile shape in this design experiment because it is a standardised tile shape, and several motifs in "tin chok" are identified as square or rectangular shapes. A grid/stacked layout is the only arrangement of tiles that we employed to place square tiles on a flat surface. The research methodology comprised two phases: the design process and the online survey of design derivations because we wanted to gain outsiders' perspectives (Figure 4). In business and marketing, the terms "inside-out" and "outside-in" refer to opposite strategies for creating value for customers and achieving success [19].

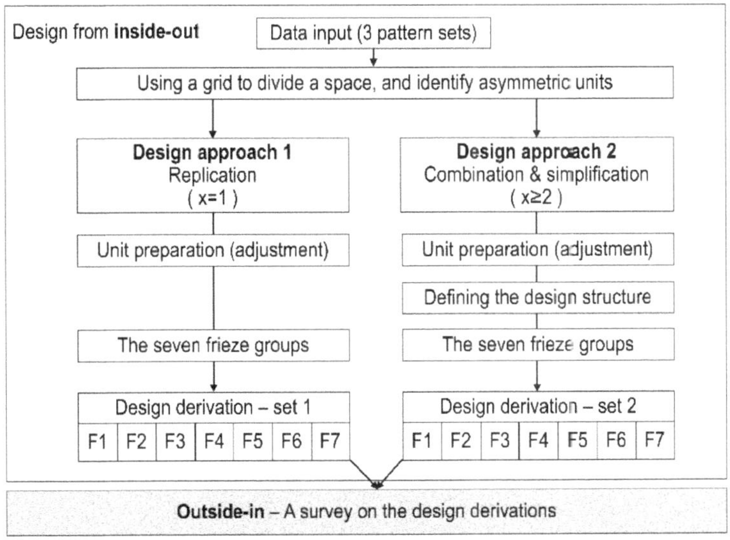

Figure 4. Research methodology. (Illustration by Kunkhet and Chudasri).

3.1. Design Process

In the first phase, our design process was an inside-out strategy that only embraced the designers' perspectives. We discussed an idea, related it to our previous research [1,2,20,21] and worked out this design experiment together from our internal resources, including a pattern booklet, capabilities and strengths. This pattern booklet (54 pages) was digitally created by Chudasri's research group in 2018 [20,21]. It displays original photographs of the 17 vintage (traditional) skirts that are exhibited in Komol Phaboraan Museum, their patterns in the hems and a replication of the 17 pattern sets shown with pixels. There was a request for decoding and visualizing these pattern sets on graph paper because these would help weavers to identify the number of threads and create the pattern elements. A graph paper is made up of fine lines that are arranged vertically and horizontally as to create many small blocks. These are comparable to (i) the interlacing of warp threads and weft threads that create pattern elements of a fabric and (ii) individual pixels in a digital picture.

The information in the original photographs of the 17 pattern sets was manually decoded and digitally recorded in computer software, i.e., Excel and Photoshop. In Excel, the column width and row height were adjusted to the smallest number (1 × 1) so that every cell represented each of the individual pixels in a picture. One could fill information in any cell (or pixel) with respect to the original patterns. This fairly small size (1 × 1) of individual pixels could conform to any shape and size of pattern elements, no matter how small or large their shapes were. Next, the pattern elements were copied to Photoshop in order to create a new file. Regarding the image quality (high or low resolution), it was necessary to "preset details" of the file at the beginning, i.e., the canvas size (e.g., 21.0 × 29.7 cm, equal to the size of A4-paper) and image resolution (e.g., 300 pixels/inch). For high-quality images, we recommend setting an image resolution for at least 300 pixels/inch at the actual size by which such information will be displayed in print, such as on A4-paper. With this setting, if such an image has to be displayed on a digital screen, it is going to be fine because a digital screen usually requires a lower resolution, e.g., 72 pixels/inch. Consequently, we obtained raster images that were save as JPG, PNG, PSD or TIFF file formats.

This method to replicate information from the original photographs is similar to a creation of pixel art. There are strong similarities between pixel art and traditional restrictive art forms, such as mosaics, cross stitch and weaving. Pixel art is commonly used in computer game development to virtually create characters or other assets in the game [22,23]. Pixel art is a form of digital art that uses computer software to create the restrictive number of pixels to form a digital picture. Such a picture may be created by different methods, such as a photograph from a digital camera and an image created by computer software. A pixel means a picture element, and it is the smallest unit of information that makes up a picture. Every pixel contains certain information that contributes to create the meaning of the whole picture [24]. In a picture, a certain shape of pixels is usually applied throughout, for example round or square. In this design experiment, we used square pixels.

Among the 17 pattern sets, we selected three pattern sets for this design experiment (Figure 5). The three images in the upper part of Figure 5 are the original photographs of the fabric patterns, whereas the other three images in the lower part were generated by computer software (explained above). Note that the three images generated by computer software have a minor difference from the original photographs. That is, the four parts of each pattern set (the supplementary part, main part, supplementary part and hem) have a common vertical reflection axis [1]. In the original photographs, the vertical reflection axes of the main part were not aligned with the two supplementary parts and the hem. Alterations to the vertical reflection axes represent a weaving method that the Long weavers employ at the present time [1].

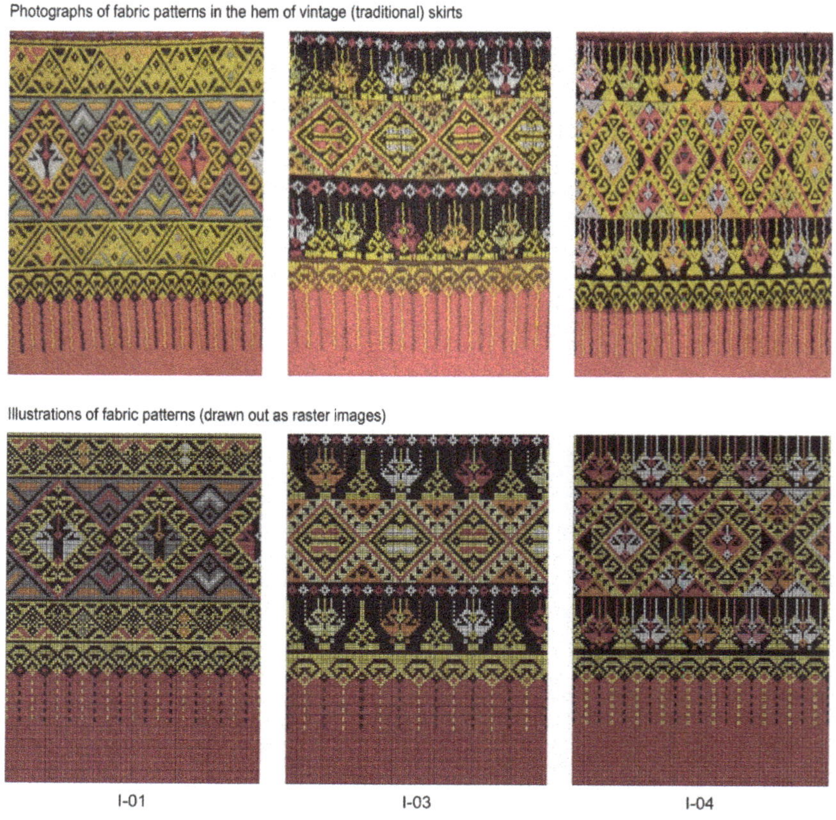

Figure 5. Three pattern sets shown with original photographs and computer-generated images (images from Chudasri).

By magnifying the three original photographs (Figure 5, the upper part), one can see the many threads that form the pattern elements, such as triangles and diamonds, which can be identified by the different colours of threads. It is obvious that the edges of the pattern elements are not smooth. Regarding the replication of information from the original photographs, one may ask why not use a vector graphic software to draw the outlines of these pattern elements as geometric shapes so that their edges are smoothened. We preferred not to do so as it is obvious that the edges of these patterns in real fabrics are not smooth. In this design experiment, we avoided any loss of information; therefore, we retained the pixelated characteristics and directly translated them into the design of tile patterns.

To construct or deconstruct two-dimensional designs on a flat surface, a designer should consider the space distribution and the relationship between grid and space [12] (p. 46). According to Josef Müller-Brockmann, as cited in [12] (pp. 35–36), a grid is a visual organization tool, which many designers use for organizing the design elements in a space. So, the design elements are presented with clarity and purpose. Relative to graph paper and the individual pixels of a picture, a grid is an arrangement of orthogonal lines that cross each other to form a network of identical squares or rectangles [12] (p. 35). Josef Müller-Brockmann, as cited in [12] (p. 46) suggests using a grid to divide a two-dimensional surface into smaller fields. Hence, we created a grid on each pattern set in order to identify asymmetric units and select the units for the design experiment. We also created two design approaches, in which these units could be translated to form tile patterns. We named these two design approaches (i) partial replication and (ii) combination and simplification. In design approach 1, only one asymmetric unit was replicated in a square, to which seven frieze groups were applied (Figure 6). This resulted in seven frieze patterns (Figure 7) (explained further in Section 4). In design approach 2, more than one asymmetric unit was selected for the design development where, the unit placement was confined to a square (Figure 8). In this square, a design structure was drawn out as guideline for the placement of the selected units, combining them and simplifying their elements (Figure 9). This resulted in a newly designed unit, which was used to generate the tile patterns (Figure 10) (explained further in Section 5). As a result of the entire design process, two sets of seven frieze patterns were illustrated (Figures 7 and 10). Although they were derived from the same origin, they looked different.

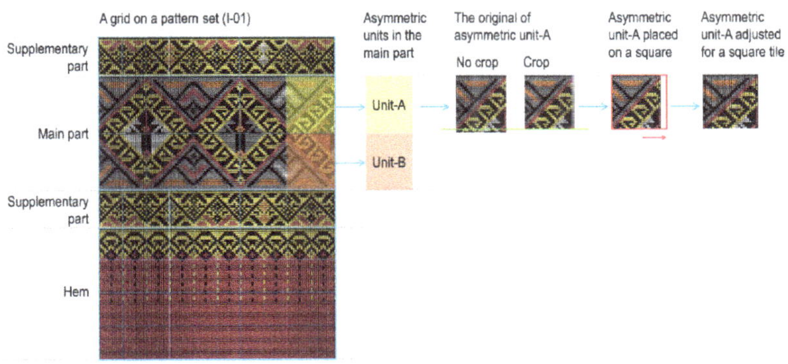

Figure 6. Asymmetric units and adjustments on a square tile. (Illustration by Kunkhet and Chudasri).

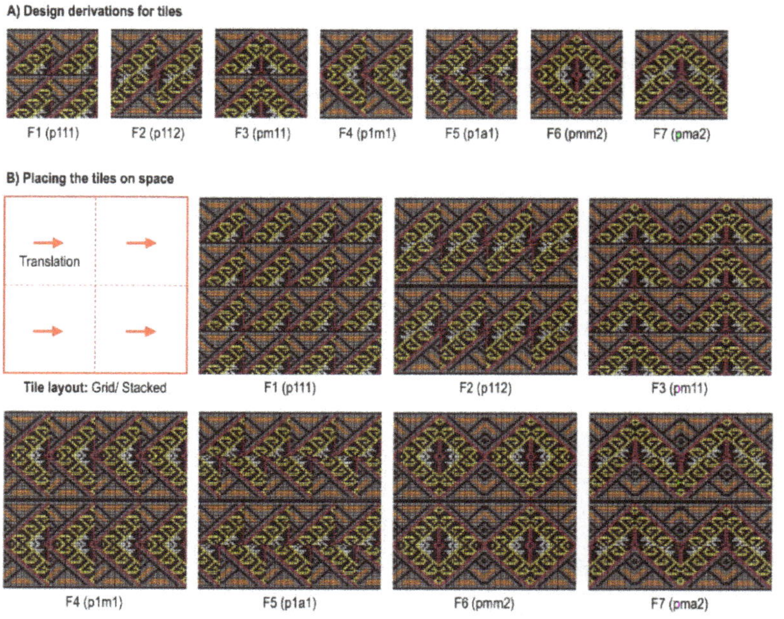

Figure 7. Design derivations from 'design approach 1'. F1–F7 represent sets of symbols for the seven frieze groups as indicated in Figure 3. Red arrows indicate tile placement in a horizontal orientation, similarly to translation. (Illustration by Kunkhet and Chudasri).

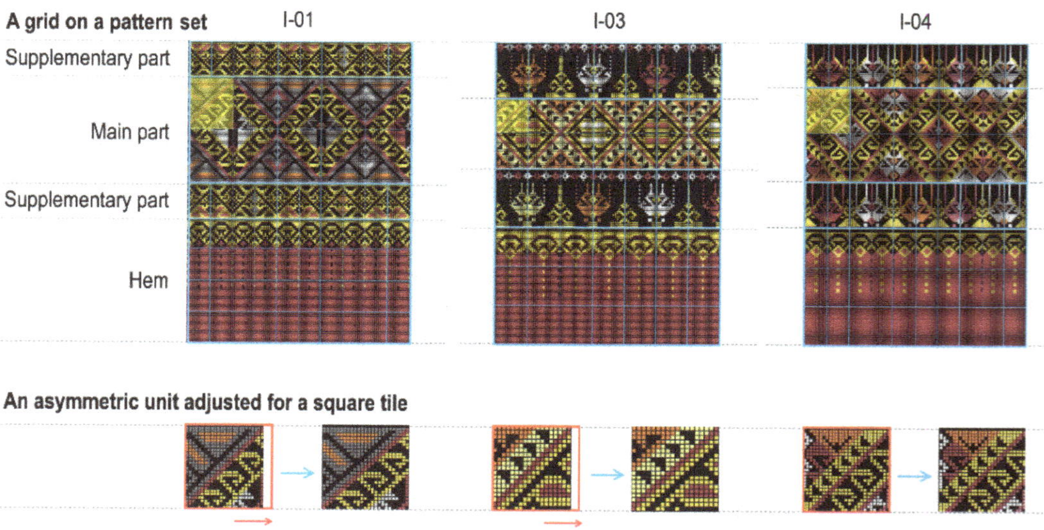

Figure 8. Asymmetric units from three main parts and adjustments on square tiles. Red arrows indicate stretching in a horizontal orientation. Blue arrows indicate the unit adjustments or compliance to the square. (Illustration by Kunkhet and Chudasri).

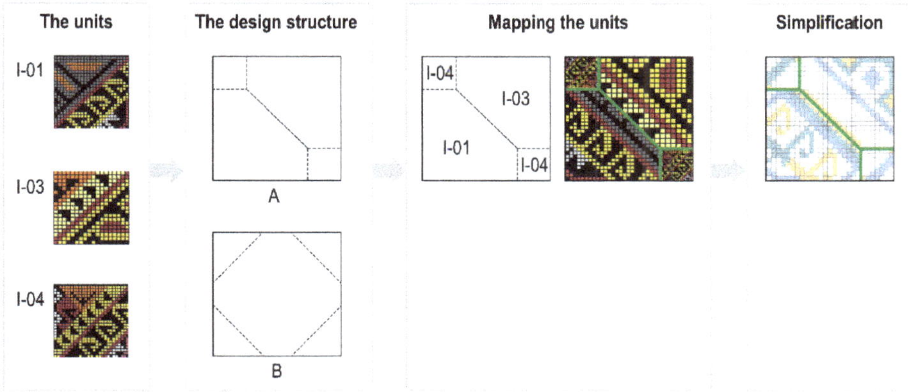

Figure 9. Mapping asymmetric units in the design structure, followed by simplification. (Illustration by Kunkhet and Chudasri).

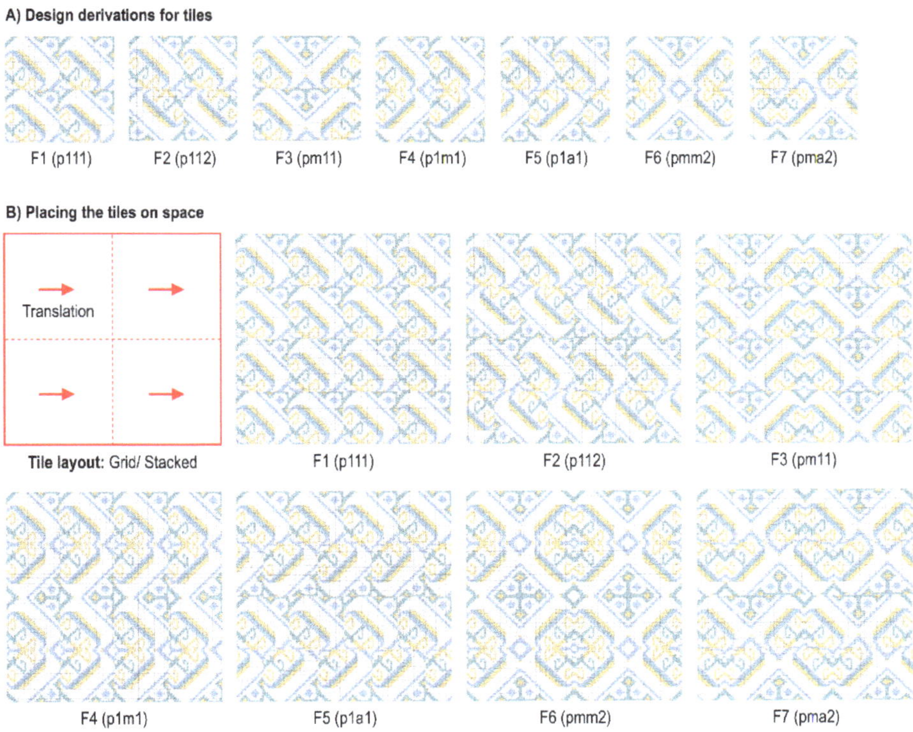

Figure 10. Design derivations from 'design approach 2'. F1–F7 represent sets of symbols for the seven frieze groups as indicated in Figure 3. Red arrows indicate tile placement in a horizontal orientation, similarly to translation. (Illustration by Kunkhet and Chudasri).

3.2. Online Survey

We intended to show to the public how the characteristics of traditional textile patterns can be translated into square tile patterns. We also hope that this design experiment will be useful for creating valuable products that are attractive to customers and raising awareness of design for cultural significance. Therefore, in the second phase, we conducted a survey with focus groups in order to gain customers' perspectives, including their preference between these two sets of design derivations. This was an outside-in strategy, which can lead to greater customer satisfaction [19]. The Thai focus groups included designers and undergraduate students (in textile design, graphic design and product design) and members of enterprises. In one week, we achieved 61 Thai respondents in total (Figure 11). This survey comprised two parts: the respondents' general information and their perspectives on pattern designs. Enquiries into their general information included gender, age, occupation, hometown and current province of residence. Their personal identities remained anonymous. There were seven questions in the second part. Survey question 1: what do you think about traditional textile patterns in design being applied to other products? Survey question 2: from the images of traditional textile patterns (Figure 5), please suggest at least three products to which these textile patterns can be applied to redesign them? Please note that in the survey questions that follow, the images of pattern designs were presented as greyscale images. This enabled the respondents to focus on the design elements in patterns rather than their different colour schemes. Survey question 3: which design approach (1 or 2) do you prefer? Respondents had to view two images (Figure 12) before selecting one. By selecting their preferred approach (Figure 13), survey question 4 presented a relevant image that showed seven frieze patterns (Figure 14 or Figure 15). The respondents were asked to choose up to three design options from amongst the seven, i.e., which option do you like? Survey question 5 asked the respondents to explain their reasons for choosing such a design approach and design options. Survey question 6 asked: do you know that two-dimensional patterns that are designed to repeat in one direction (such as tiles for interior decoration and textile patterns) can be classified based on their symmetry into "the seven frieze groups"? Survey question 7 asked if the respondents could suggest other methods for pattern design.

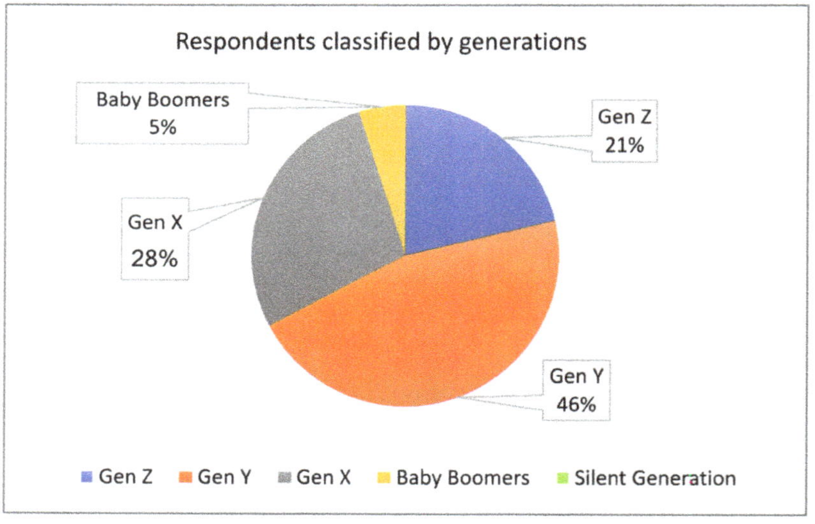

Figure 11. Numbers of respondents classified by generation, shown in percentage.

Figure 12. Two design approaches for selection. (Illustration by Kunkhet).

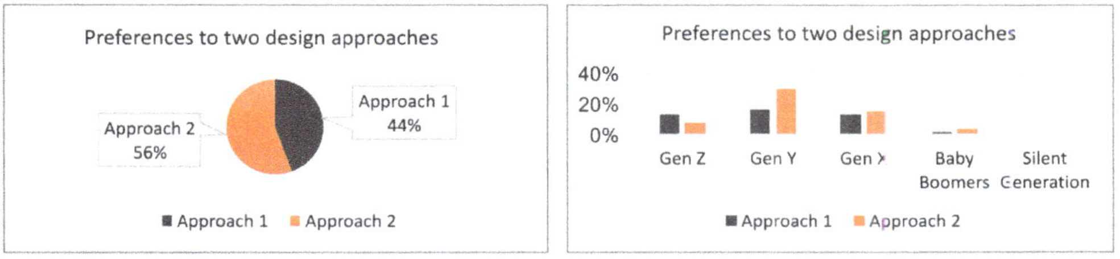

Figure 13. Percentage of respondents' preferences for two design approaches.

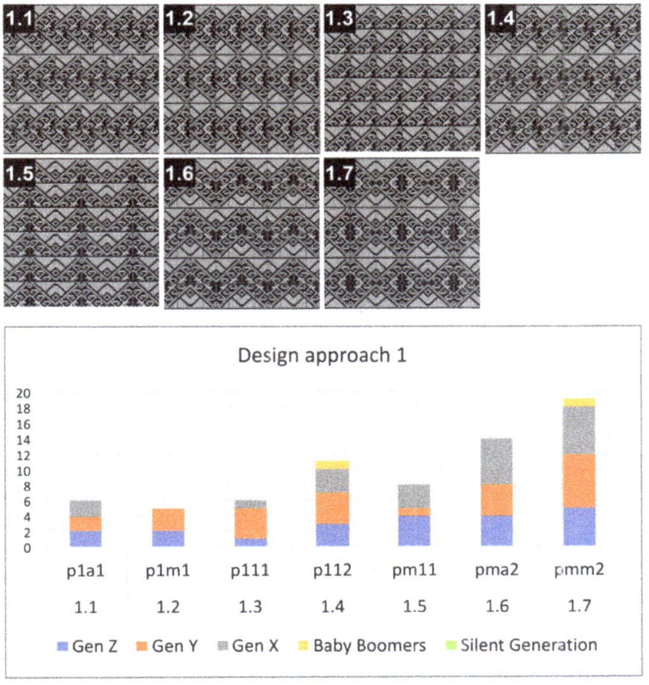

Figure 14. Design approach 1: seven frieze patterns and respondents' preferences. (Illustration by Kunkhet and Chudasri).

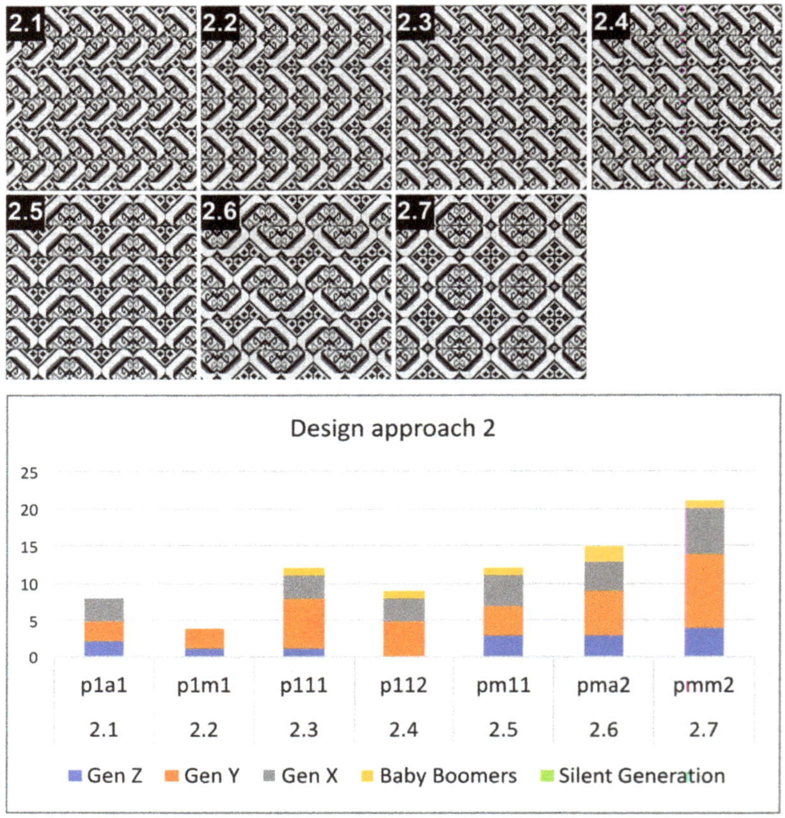

Figure 15. Design approach 2: seven frieze patterns and respondents' preferences. (Illustration by Kunkhet and Chudasri).

4. Design Approach 1: Partial Replication and Design Derivations

Design approach 1 was intended to replicate an asymmetrical unit identified from a pattern set and place it on a flat surface to create patterns on square tiles. We selected pattern set I-01 for demonstration purposes (Figure 6). In the design structure of "tin chok", the pattern in the main part is more dominant than the other three parts. So, we focused on the main part and identified two asymmetric units that were Unit-A and Unit-B, as shown in Figure 6. For the design development, a new square was needed for placing an asymmetrical unit, and we selected Unit-A for demonstration. Prior to placing an asymmetrical unit in the square, we may or may not have had to crop a boundary of an asymmetrical unit. Here, we cropped the bottom of Unit-A in order to make its height the same as the side of the square. Likewise, the width of Unit-A was not the same as the side of the square. So, we stretched the width of Unit-A to have the same distance as the side of the square. By doing so, we were aware that the "pixel aspect ratio" (PAR) of 1:1 was not preserved. The other reasons that we took into consideration were as follows. This stretch was fairly short, i.e., a distance of less than seven percent of the side of the square, which did not disturb visual characteristics of Unit-A. There was a considerable resemblance between the shape elements of the original and the adjustment of Unit-A; for example, their straight lines and directions, triangles and trapezoids. Without a mandate that every feature of shape elements must be exactly preserved, designers can freely think and come up with a creative approach that deals with a given context and thereby a design development. Consequently, the square was wholly filled up with shape elements, i.e., the adjusted Unit-A in Figure 6, which was ready for further design development at this point.

Next, the seven frieze groups were applied to the adjusted Unit-A as a set of transformation rules (Figure 3) to generate frieze patterns. There were the seven design derivations of patterns for the square tiles (Figure 7A). These patterns were ready for tiling. Amongst the several tile layouts (Figure 2), we chose a basic layout—grid/stacked—to experiment further. With this selected layout, individual tiles can be arranged in different orientations, and this will result in different tile patterns that can be repeated all over a flat surface [13] (p. 2). We were aware of this, but we preferred to placing the same individual tiles at the same distance in the same orientation all the way across a flat surface because this is convenient for builders in practice. This orientation is described as "translation" (p111) according to the four rigid motions (Figure 3). Figure 7B depicts the placements of individual tiles by translation in a space. It is noted that we also applied 90-degree rotations as different orientations of individual tiles that created different tile patterns across the space. However, in our opinion, doing so would be rather confusing for builders in practice. It would be better if such a creation was completed in the design process, for example by adjusting the tile size or dividing the space into four units using orthogonal lines.

5. Design Approach 2: Combination and Simplification—And Design Derivations

Design approach 2 was intended to combine more than two asymmetrical units in a design structure and simplify their elements to create new patterns for square tiles. We applied this concept to three pattern sets, namely I-01, I-03 and I-04, which were overlaid by grids (Figure 8). Similar to design approach 1, we focussed on patterns in the main part and identified asymmetrical units (overlaid with yellow squares or rectangles) for the design development. A new square was needed for placing the asymmetric units. It is obvious that asymmetric units identified with rectangles have only one side that fits with a square. In this experiment, the width of the rectangles was stretched to have the same distance as the side of the square. This means that the PAR of 1:1 of the original units was not preserved; however, we accepted such an adjustment with good reason (explained in Section 4). Consequently, there were three asymmetrical units resulting from the same size of the square. Yet, they comprised different sizes of square pixels: 25 × 27, 18 × 19 and 27 × 26 pixels, respectively. These were not that big issue for the design at the moment (explained below), and at this point they were ready for further design development.

The next step was to create another new square for the design structure. A design structure is the way in which the parts of something are organised or arranged into a whole or the way in which something is divided into parts. In this experiment, lines were drawn in this new square to represent the design structure. This was used to guide a combination of asymmetrical units and design elements in order to design new tile patterns. It is noted that the design structure was different from the shapes of the tiles (e.g., square, rectangle) and tile layouts (e.g., grid/stacked, checkerboard). The basic information about a design structure resides in three features: a fundamental parallelogram (FP), the design symmetry axes (DSA) and the design of rotational centres (DRC) [3] (p. 499). We conducted a visual analysis of tile patterns shown on commercial websites and observed their graphic elements in relation to a fundamental parallelogram, symmetry axes and rotational centres in the overall design. Eventually, we created two design structures (A and B), and we chose design structure A for the experimentation (Figure 9).

With the three adjusted units (I-01, I-03 and I-04, shown in Figure 9), we intended to combine and simplify their intricate elements in a way that their distinct visual characteristics were retained. As we placed the three adjusted units on the divided areas of design structure A, the different numbers of pixels in the square were not a critical issue. We also rotated, reduced or enlarged the sizing of these three units to ensure that their distinct characteristics remained in the whole design. These three units became different in size, but this was not a critical issue either. This was because in the next step there was a simplification process, in which we applied a grid to divide the whole design space to comprise

27 × 27 pixels and used it as a guideline to simplify the intricate elements. (Twenty-seven was the maximum number identified from the units after cropping or stretching). Finally, the intricate elements were simplified by retaining their distinctive lines and shapes, removing several detailed elements, adjusting parts of the lines or shapes and correlating them with the whole design. Additionally, we applied a new colour scheme to the new design unit so that it was different from the original patterns.

Next, seven frieze groups were applied as a set of transformation rules to the newly designed unit to generate frieze patterns. This process was the same as for design approach 1. Figure 10A depicts the seven design derivations. Furthermore, we arranged these tile patterns in a space in the same way as for design approach 1. Figure 10B depicts the placement of individual tiles by translation into a space.

6. Survey Findings

Currently, the people in our society can be classified into five generations based on their year of birth. (Their ages inform their generation.) Each generation has an active role in the world marketplace [25]. So, a generational breakdown was part of the data analysis in this paper. This survey comprised 61 respondents aged between 15–76 years. They were from four generations, including Gen Z (iGen or Centennials), Gen Y (or Millennials), Gen X and Baby Boomers (Figure 11). Table 1 shows the numbers of respondents classified by generation and gender identity. Thirteen respondents from Gen Z included undergraduate students (n = 8: textiles, fashion, animation, graphics, information technology), designers (n = 2: fashion, graphics), a school student (n = 1), a software developer (n = 1) and a member of an enterprise (n = 1). Twenty-eight respondents from Gen Y included designers (n = 11: fashion, textiles, graphics, product, packaging, architecture), teachers (n = 5: design, mathematics, etc.), company staff (n = 5: tile retail trade, craft enterprises), business owners (n = 4: software developers, information technology, textiles), a writer (n = 1: art and culture), a scientist (n = 1) and another person (n = 1). Seventeen respondents from Gen X included designers (n = 8: industrial design, fashion, textiles, ceramics, interior, branding, packaging, graphics, advertising), business owners (n = 4: art and crafts, textiles, real estate, etc.), teachers (n = 2: visual arts), traders (n = 2) and another person (n = 1). Lastly, there were three respondents from the Baby Boomers (one teacher, one designer and one business owner).

Table 1. Numbers of respondents classified by generation and gender identity.

Generation	Year of Birth		Age [1]	Gender			Total	
	(A.D.)	(B.E.)		Male	Female	LGBT	Respondents	Percentage
Gen Z, iGen or Centennials	1996–2015	2539–2558	7–26	4	8	1	13	21%
Gen Y or Millennials	1977–1995	2520–2538	27–45	7	17	4	28	46%
Gen X	1965–1976	2508–2519	46–57	8	9	0	17	28%
Baby Boomers	1946–1964	2489–2507	58–76	2	1	0	3	5%
Traditionalists or Silent Generation	1945 and before	2488 and before	From 77	0	0	0	0	0%
Total number of respondents							61	100%

[1] Age calculated in A.D. 2022.

6.1. Positive Responses to Applying Traditional Textile Patterns to Other Products

This section responds to survey question 1. Every respondent gave a positive response to the concept of applied design that extends traditional textile patterns to other products. In short, they said I "agree" or "support" this idea, or this is "good" or "interesting". Their overall responses suggested that this design concept could foster the preservation of traditional textile patterns while also generating new product designs that are suitable for people in the present era. They were favourable to products designed in a contemporary style, i.e., those that were trendy or modern. Around half of the respondents elaborated further. The explanations from the respondents aged between 15–26 (Gen Z) were as follows. "Traditional textile patterns are beautiful, intricate and represent the Thai identity" (ID37, 57). "Design that applies traditional textile patterns to other products can lead to added value, while traditional textile patterns are maintained" (ID43). "It would be good if someone could apply traditional textile patterns to other products that are designed to suit the lifestyles of people at the present time, and it is critical to communicate storytelling to people" (ID31). "Traditional patterns can be adjusted to look more modern [or trendy], and not be restricted to the original image only" (ID33). "This is another way in which traditional textile patterns can remain in Thai society for a long period of time" (ID59). "Traditional textile patterns can inform the identity of local peoples and their geographical areas. By applying traditional textile patterns to other products, people from other areas (outsiders) can have a better understanding of local arts" (ID27). "Local arts will be preserved and known more widely" (ID44).

The explanations from the respondents aged between 27–44 (Gen Y) were as follows. "It is about creating patterns from cultural identity and cultural roots" (ID28, 52). "Design can foster the preservation of traditional arts, which are valuable, through things that are designed for real use in the present time. This will foster adaptation that is suitable to this era. So, traditional art is retained and passed on to the society of the next generations, and there is a variety of designed products" (ID38). "In adopting these applied designs, it concerns the suitability of products and target customers" (ID40). "Design has to be adjusted to suit the present era, and target customers, especially the younger generations" (ID42). "It is good to start from something existing, and extend it further in a way that is suitable for the present time. In this sense, design is fostering cultural preservation, so traditional textile patterns will not disappear or be out of fashion" (ID26). "It is good that cultural capital is used in a design process for creating contemporary products" (ID30). "This is not only to remain local wisdom, but also to extend to new opportunities, including both cultural traditions and business opportunities" (ID16). "It is possible to apply traditional textile patterns to other products, depending on design elements, their size, arrangement and colour schemes. These should conform to the taste and wants of target customers" (ID45). "Traditional textile patterns are an intangible cultural heritage and local wisdom of Thailand. Although creators/designers can apply these patterns to other products, they should respect local wisdom with sincere understanding" (ID49). Explanations from respondents aged between 47–57 (Gen X) are as follows. "This design concept extends the cultural capital to tangible outputs" (ID23). "There will be a variety of products and alternatives for customers" (ID19, 60). "Applying traditional textile patterns to other products is an alternative way to prolong the cultural preservation of traditional textile patterns, and disseminate them to the next generations" (ID22). "This is the preservation of traditional textile patterns that also generates new products that are suitable for the present time" (ID34, 58).

6.2. Plausible Products for Pattern Designs

In survey question 2, the respondents had to view an image that comprised (i) a photograph of Long people wearing traditional skirts for a performance in a traditional ceremony, (ii) a photograph of traditional skirts displayed in Komol Phabcraan Museum and (iii) a picture of three pattern sets generated from computer software (Figure 5, lower part). The respondents were asked to suggest at least three products to which these textile patterns could be applied in order to redesign them. Besides tiles, a variety of products were suggested, including textile products. The new pattern designs may be applicable to dinnerware such as glasses, plates, bowls and placemats. A variety of products for home decoration, such as items in living rooms and bedrooms, were suggested, e.g., curtains, tablecloths, carpets, blankets, bed covers, pillowcases, bed linen, furniture, partitions, wallpaper, vases and paintings. They also suggested clothes and accessories such as skirts, T-shirts, scarves, swimming suits, handbags, shoulder bags, watch straps, shoes, hats, raincoats and long boots. Cases for digital devices can include notebooks, iPads/tablets, mouse pads and mobile phones. Other products included gift wrapping paper and packaging. Additionally, they mentioned production techniques such as printing and painting.

6.3. Preferences for Design Approaches and Frieze Patterns

In the responses to survey question 3 (Figure 12), 27 respondents (44%) preferred design approach 1, while 34 respondents (56%) preferred design approach 2 (Table 2, Figure 13). Although design approach 1 gained a lower number, in our perspective it was not very different. Both design approaches are likely to be successful in terms of design for cultural revitalisation and business opportunities.

Table 2. Preferences for two design approaches shown with a generational breakdown.

Generation	Year of Birth (A.D.)	Age [1]	The Number of Respondents			
			Raw Data		Percentage	
			Approach 1	Approach 2	Approach 1	Approach 2
Gen Z, iGen or Centennials	1996–2015	7–26	8	5	13%	8%
Gen Y or Millennials	1977–1995	27–45	10	18	16%	30%
Gen X	1965–1976	46–57	8	9	13%	15%
Baby Boomers	1946–1964	58–76	1	2	2%	3%
Traditionalists or Silent Generation	1945 and before	From 77	0	0	0%	0%
The total number of respondents			27	34	44%	56%
			61		100%	

[1] Age calculated in A.D. 2022.

Survey question 4 probed the respondents' preferences for the seven frieze patterns resulting from the selected design approach (Tables 3 and 4). Figures 14 and 15 show the seven frieze patterns (from design approaches 1 and 2, respectively) and the respondents' preferences for these patterns. Overall, they preferred the pattern designs that were connected all-around as a closed path, i.e., a visually larger object. From survey question 5, their rationale for their selection was summarised as follows. The patterns from design approach 1 (partial replication) had some original characteristics that were not changed too much, but these patterns looked trendy or contemporary and could be applicable to other products. These selected patterns showed a continuity, neatness, symmetrical balance and harmony in their designs, which were not too complex or confusing to look at. Similarly, the patterns from design approach 2 (combination and simplification) showed a connectedness and neatness in their symmetrical design. These patterns looked simple and not too complicated. They looked trendy, contemporary and classic as a result of the design layout, design elements and their sizing.

Table 3. Preferences [1] for frieze patterns from design approach 1.

Option	Symmetry Class	Gen Z	Gen Y	Gen X	Baby Boomers	Silent Generation	Frequency [2]	Popularity of Patterns in Percentage
1.1	p1a1	2	2	2	0	0	6	9%
1.2	p1m1	2	3	0	0	0	5	7%
1.3	p111	1	4	1	0	0	6	9%
1.4	p112	3	4	3	1	0	11	16%
1.5	pm11	4	1	3	0	0	8	12%
1.6	pma2	4	4	6	0	0	14	20%
1.7	pmm2	5	7	6	1	0	19	28%
		21	25	21	2	0	69	100%

[1] Each respondent was asked to select up to three options. [2] There were 27 (from 61) respondents selected design approach 1.

Table 4. Preferences [1] for frieze patterns from design approach 2.

Option	Symmetry Class	Gen Z	Gen Y	Gen X	Baby Boomers	Silent Generation	Frequency [2]	Popularity of Patterns in Percentage
2.1	p1a1	2	3	3	0	0	8	10%
2.2	p1m1	1	3	0	0	0	4	5%
2.3	p111	1	7	3	1	0	12	15%
2.4	p112	0	5	3	1	0	9	11%
2.5	pm11	3	4	4	1	0	12	15%
2.6	pma2	3	6	4	2	0	15	19%
2.7	pmm2	4	10	6	1	0	21	26%
		14	38	23	6	0	81	100%

[1] Each respondent was asked to select up to three options. [2] A total of 34 (from 61) respondents selected design approach 2.

6.4. Opportunity for Design Research and Education with Other Disciplines

There are connections between mathematics and design [14] (p. ix), [26]. We demonstrated the use of mathematics, i.e., the seven frieze groups, in the creation of the pattern designs for floor tiles. Likewise, we were curious as to whether people generally knew about this mathematical theory and how deep their knowledge was. From survey question 6, we found that the majority of the respondents did not know about the seven frieze groups, i.e., the classification of infinite discrete symmetrical groups of patterns. Fifty respondents did not know about it at all. Three respondents knew about symmetry, but they did not know how the patterns could be classified or how many symmetry groups there were. Only eight respondents knew about the seven frieze groups. They included a software developer, an entrepreneur, two teachers (maths, visual arts), two designers and two design academics. Knowing this result, we realised an opportunity to educate people, e.g., here in Thailand, about symmetry analysis and mathematical theories relate to how they can be useful in designing for cultural revitalization. Design visualization can be a vital tool to boost people's understanding of mathematics. Additionally, we must emphasise that including a design activity as part of a research process can stimulate people to acquire knowledge about different subjects of study and gain a deeper understanding through doing so. In fact, this research involved knowledge from multiple disciplines, as we included design (textiles, tiles, pattern design), mathematics (symmetry and the seven frieze groups) and a viewpoint on product design and development for business opportunities.

7. Conclusions

Traditional practices and artefacts are rich sources of knowledge (e.g., cultural and historical principles, geometric principles in practice) and a source of soft power, cultural capital and creative economy. Designers are encouraged to explore design for cultural significance and cultural revitalization [15,27]. In this article, we presented a design experiment, in which the distinct characteristics of the traditional textile patterns of "tin chok" from northern Thailand were translated through a design process into pattern designs for floor tiles. We created two design approaches ("partial replication", and "combination and simplification"), by which a mathematical theory (i.e., the seven frieze groups) was used as a set of transformation rules for generating the patterns. Consequently, we arrived at two sets of frieze patterns. Although they were derived from the same origin, they looked different. The findings from the online survey revealed that all the respondents were positive about the two design approaches and pattern designs inspired by traditional sensitivity with respect. The respondents addressed that this design activity helped to invigorate the preservation of traditional textile patterns, while it stimulated our thinking about new pattern designs and what products these patterns could be applicable to. Besides floor tiles, these patterns may be applicable to products for home decoration, packaging, digital devices and accessories, textiles and clothing accessories.

7.1. Limitations

There were four main limitations to this research. First, the design scope was restricted to the use of (i) a mathematical theory (i.e., the seven frieze groups) as a set of transformation rules for generating patterns, (ii) a square as the only tile shape for the pattern design and (iii) a grid/stacked layout for the arrangement of the square tiles on a flat surface. Second, the creation of the two design approaches ("partial replication", and "combination and simplification") was brainstormed by only two designers (the authors). Third, the online survey was conducted with 61 Thai respondents as a sample size to gain outsiders' perspectives on these design approaches and pattern designs in relation to cultural significance. This total number of respondents was sufficient for a small research project, i.e., requiring a minimum of 60 people [28]. However, the number of respondents from the Baby Boomers group did not reach a minimum of five or eight people [28] (pp. 116–117). Fourth, much of this research contained qualitative information, i.e., pattern designs and respondents' feedback. A quantitative method was not employed in the data collection; however, it was simply used to analyse parts of the information collected from the respondents' feedback.

7.2. Implications and Further Research

In the future, a scope of research can be span out, and there are several theoretical and practical implications for doing so. A design experiment may employ other mathematical theories (e.g., tessellation, the 17 wallpaper groups), other tile shapes (e.g., rectangles) and other tile layouts (e.g., checkerboard, hopscotch). Such a design experiment could invite a greater number of designers, design researchers, mathematicians and students to brainstorm, exchange their ideas and work together to develop various design approaches. The SCAMPER technique for creative thinking could be used in the design process (suggested by the respondent (ID55)), especially when brainstorming how to improve or develop products or services [29]. SCAMPER is an acronym for Substitute, Combine, Adapt, Modify/Magnify, Purpose/Put to another use, Eliminate/Minimise and Rearrange/Reverse. Regarding a survey of a small research project, a sample size of each sub-group should comprise a minimum of five or eight people to ensure that it has the potential to imply a tendency. Moreover, a survey should expand to potential groups of customers, which may go beyond just Thai people and may include foreigners. Regarding research methods, these could be adjusted to a mixed method that comprises qualitative information and quantitative information. A Theory of Change and the Futures Cone, along with relevant marketing concepts, could be incorporated to guide project planning, monitoring and evaluation. With respect to quantitative information, a mathematical model could be used as part of a transformation rule in the design process. In a survey, respondents could be asked to rank (e.g., the top three designs) or rate each design (e.g., from 0–5 stars) so their responses can be interpreted as numerical values for statistical analysis.

Author Contributions: Conceptualization, methodology, software, resources, writing—original draft preparation, visualization: A.K. and D.C.; validation, formal analysis, data curation, writing—review and editing, supervision, project administration, funding acquisition: D.C. All authors have read and agreed to the published version of the manuscript.

Funding: This research received no external funding. Publication of this article is supported by Chiang Mai University, Thailand.

Informed Consent Statement: Informed consent was obtained from every respondent involved in the online survey, and their personal identity would be remained anonymous.

Acknowledgments: This research work was partially supported by Chiang Mai University. We would like to thank Chiang Mai University, College of Arts, Media and Technology, Nattakorn Sukantamala and Supanut Chaidee (from the Department of Mathematics, Faculty of Science, Chiang Mai University). We are thankful to the respondents to the online survey and the weaving communities in Long district, Phrae province, Thailand.

Conflicts of Interest: The authors declare no conflict of interest.

References

1. Chudasri, D.; Sukantamala, N. Design research with the use of visual and symmetry analysis in indigenous woven textiles. *J. Appl. Crystallogr.* **2022**, 56. [CrossRef]
2. Kunkhet, A.; Chudasri, D.; Sukantamala, N. Developing the framework of harmonised shape grammar to regenerate traditional textile patterns from northern Thailand. *Asian J. Arts Cult.* **2022**, *22*, 1–15.
3. Valiente, J.M.; Albert, F.; Carretero, C.; Gomis, J.M. Structural Description of Textile and Tile Pattern Designs Using Image Processing. In Proceedings of the 17th International Conference on Pattern Recognition, Cambridge, UK, 26 August 2004; Volume 1, pp. 498–503.
4. Hann, M.A.; Thomson, G.M. The Geometry of Regular Repeating Pattern. *Text. Prog.* **1992**, *22*, 1–62.
5. Hann, M.A. Symmetry in Regular Repeating Patterns: Case Studies from Various Cultural Settings. *J. Text. Inst.* **1992**, *83*, 579–590. [CrossRef]
6. José María, G.; Valor, M.; Albert, F.; Contero, M. Int[egr]ated System and Methodology for Supporting Textile and Tile Pattern Design. In *International Symposium on Smart Graphics*; Springer: Berlin/Heidelberg, Germany, 2003; pp. 69–78.
7. Albert, F.; Gomis, J.M.; Valor, M.; Valiente, J.M. Methodology for Graphic Redesign Applied to Textile and Tile Pattern Design. In *International Conference on Industrial, Engineering and Other Applications of Applied Intelligent Systems*; Springer: Berlin/Heidelberg, Germany, 2004; pp. 876–885.
8. Gomis, J.M.; Albert, F.; Contero, M.; Naya, F. Calligraphic Editor for Textile and Tile Pattern Design System. In *International Symposium on Smart Graphics*; Springer: Berlin/Heidelberg, Germany, 2004; pp. 114–120.
9. Lexico. 2022. Available online: https://www.lexico.com/ (accessed on 3 September 2022).
10. Wattanasiriwech, D.; Saiton, A.; Wattanasiriwech, S. Paving Blocks from Ceramic Tile Production Waste. *J. Clean. Prod.* **2009**, *17*, 1663–1668. [CrossRef]
11. Jiang, Y. Modular Design on Tile Layout Experiment with Javascript. *Int. J. Interact. Des. Manuf.* **2022**, *16*, 1163–1173. [CrossRef]
12. Sarin, A. The Kolam Drawing: A Point Lattice System. *Design Issues.* **2022**, *38*, 34–54. [CrossRef]
13. Why Tile. Ceramic Tile Layout Pattern Guide. 2022. Available online: https://whytile.com/library/ceramic-tile-layout-pattern-guide/ (accessed on 3 September 2022).
14. Washburn, D.; Crowe, D. *Symmetries of Culture: Theory and Practice of Plane Pattern Analysis*; University of Washington Press: Seattle, DC, USA; London, UK, 1988.
15. Hann, M.A. *Symbol, Pattern & Symmetry—The Cultural Significant of Structure*; Bloomsbury: London, UK, 2013.
16. Bart, A.; Clair, B. Math and the Art of M. C. Escher. 2022. Available online: https://mathstat.slu.edu/escher/index.php/Frieze_Patterns (accessed on 23 September 2022).
17. Shubnikov, A.; Koptsik, V. *Symmetry in Science and Art*; Springer: New York, NY, USA, 1974.
18. Kopský, V.; Litvin, D.B. *International Tables for Crystallography*; Subperiodic groups; Springer: Berlin/Heidelberg, Germany, 2002; Volume E, pp. 2–28.
19. Indeed Editorial Team. Inside-Out vs. Outside-In Strategy: Benefits and Differences. 2021. Available online: https://www.indeed.com/career-advice/career-development/inside-out-vs-outside-in-strategy (accessed on 21 September 2022).
20. Chudasri, D. Co-creation through design process for reviving indigenous knowledge about traditional textile patterns. In Proceedings of the 5th International Textile and Costume Congress (ITCC), Vadodara, India, 3–5 October 2019; pp. 186–197.
21. Chudasri, D.; Walker, S.; Evans, M. Potential Areas for Design and Its Implementation to Enable the Future Viability of Weaving Practices in Northern Thailand. *Int. J. Des.* **2020**, *14*, 95–111. Available online: http://www.ijdesign.org/index.php/IJDesign/article/viewFile/3128/883 (accessed on 2 September 2022).
22. Kopf, J.; Lischinski, D. Depixelizing Pixel Art. *ACM J.* **2011**, *30*, 1–8. [CrossRef]
23. Yue, Y.; Iwasaki, K.; Chen, B.Y.; Dobashi, Y.; Nishita, T. Pixel art with refracted light by rearrangeable sticks. *Comput. Graph. Forum.* **2012**, *31*, 575–582. [CrossRef]
24. Silber, D. *Pixel Art for Game Developers*; CRC Press: Boca Raton, FL, USA; Taylor & Francis Group: Abingdon, UK, 2016; Available online: https://books.google.co.th/books?hl=th&lr=&id=OmBECgAAQBAJ&oi=fnd&pg=PP1&dq=pixel+art&ots=q0iNX4X7zX&sig=bZfcX_ZgCKsoprdxxPuTH6xzpzM&redir_esc=y#v=onepage&q=pixel%20art&f=false (accessed on 13 October 2022).
25. Center for Generational Kinetics. Generational Breakdown: Info about All of the Generations. 2022. Available online: https://genhq.com/FAQ-info-about-generations/ (accessed on 25 October 2022).
26. Journal of Mathematics and the Arts. 2022. Available online: https://www.tandfonline.com/journals/tmaa20 (accessed on 21 October 2022).
27. Walker, S.; Evans, M.; Cassidy, T.; Jung, J.; Holroyd, A.T. *Design Roots: Culturally Significant Designs, Products and Practices*; Bloomsbury: London, UK, 2018.
28. Hall, D.; Hall, I. Practical Social Research: Project Work in the Community. Macmillan Press: London, UK, 1996.
29. Elmansy, R. A Guide to the SCAMPER Technique for Creative Thinking. 2015. Available online: https://www.designorate.com/a-guide-to-the-scamper-technique-for-creative-thinking/ (accessed on 23 October 2022).

processes MDPI

Article

Enhancing Mean-Variance Mapping Optimization Using Opposite Gradient Method and Interior Point Method for Real Parameter Optimization Problems

Thirachit Saenphon, Suphakant Phimoltares * and Chidchanok Lursinsap

Advanced Virtual and Intelligent Computing Research Center (AVIC), Department of Mathematics and Computer Science, Faculty of Science, Chulalongkorn University, Bangkok 10330, Thailand
* Correspondence: suphakant.p@chula.ac.th

Abstract: The aim of optimization methods is to identify the best results in the search area. In this research, we focused on a mixture of the interior point method, opposite gradient method, and mean-variance mapping optimization, named IPOG-MVMO, where the solutions can be obtained from the gradient field of the cost function on the constraint manifold. The process was divided into three main phases. In the first phase, the interior point method was applied for local searching. Secondly, the opposite gradient method was used to generate a population of candidate solutions. The last phase involved updating the population according to the mean and variance of the solutions. In the experiments on real parameter optimization problems, three types of functions, which were unimodal, multimodal, and continuous composition functions, were considered and used to compare our proposed method with other meta-heuristics techniques. The results showed that our proposed algorithms outperformed other algorithms in terms of finding the optimal solution.

Keywords: initial population; interior point method; mean-variance mapping optimization; meta-heuristics techniques; opposite gradient method

Citation: Saenphon, T.; Phimoltares, S.; Lursinsap, C. Enhancing Mean-Variance Mapping Optimization Using Opposite Gradient Method and Interior Point Method for Real Parameter Optimization Problems. *Processes* **2023**, *11*, 465. https://doi.org/10.3390/pr11020465

Academic Editors: Jun Wei Lim and Worapon Kiatkittipong

Received: 14 December 2022
Revised: 12 January 2023
Accepted: 15 January 2023
Published: 3 February 2023

1. Introduction

Optimization methods are employed to solve real parameter optimization problems in order to obtain a vector $x = (x_1 \ldots x_D)$ that yields the optimal value of the function $f(x)$, where D is the number of dimensions of the vector. To solve optimization problems, a global solution is searched without related knowledge or the physical structure of the cost function. There are real parameter optimization problems, such as engineering optimization problems, as well as those related to scientific applications, energy savings for tissue paper mills involving energy efficiency scheduling [1], the optimal scheduling of vehicle-to-grid energy and ancillary services [2], query optimization mechanisms in cloud computing [3], energy management [4], and tools for analytics in mechanical engineering [5].

Previously, a wide variety of evolutionary algorithms have been proposed to obtain solutions from real-world continuous optimization problems. For instance, the covariance matrix adaptation evolution strategy (CMA-ES) algorithm [6] is based on correlated mutations. The covariance matrix C has been adopted to enable the proper distribution of mutations, resulting in an increased likelihood of the successful replication of the search process. However, there are several black box properties that may promptly lead to premature CMA-ES convergence, with many uncontrollable variations and uncertainties. Therefore, to increase the possibility of finding the global optimum using the restart strategy, many techniques have been proposed based on CMA-ES, such as the restart CMA evolution strategy with increasing population sizes (IPOP-CMA-ES) [7], the bi-population CMA-ES strategy (BIPOP-CMA-ES) [8], and the new bi-population CMA-ES strategy (NBIPOPaCMA-ES) [9]. Usually, a new generation is randomly sampled based on the current covariance matrix in the desired search space. Consequently, the starting point is still randomized. Therefore, the time taken to find a solution increases as the amount of data increases. Moreover, the size of the search area and the surface of the cost function have also not been considered.

Another technique that solves the optimization problem is mean-variance mapping optimization (MVMO). MVMO [10] involves an evolutionary algorithm with two interesting issues. Firstly, the MVMO process determines the range of search areas for variables within an interval [0, 1]. This process ensures that new values computed later for the optimization population are always generated within this range before the fitness value is calculated. Secondly, the statistical features used in MVMO are useful for changing the search direction. Using the mean and variance of the different solutions and a mapping function for the mutation operation yields the optimal fitness value. Additionally, for each time that the algorithm produces better fitness values, the solutions with the n-best fitness values are updated and stored in the archive for use in finding the best solution, until the specified number of iterations has been reached. MVMO is applied in chemical process applications. For example, an adaptive PID controller based on MVMO has been proposed to enhance the performance of a chemical process with variable time delay [11]. Swarm-based mean-variance mapping optimization (MVMO-S) is an extension of MVMO, which is combined with swarm intelligence. The search procedure applied to this technique begins with the particles. They are represented in the form of the consistent function of archiving and mapping to the original MVMO [12]. Another extension of MVMO is the swarm variant of hybrid mean-variance mapping optimization (MVMO-SH) [13–15], which is an evolutionary algorithm. The algorithm takes advantage of the statistical features (mean and variance) of the dynamic search function, using mapping functions for mutation and modification according to the mean and variance of the n-best solutions that are recorded in the solution archive. The process of generating new offspring from techniques based on MVMO also uses randomness, regardless of the number of populations in the actual search surface of the cost function.

In addition to the algorithms mentioned above, there are still many algorithms that use natural behavior concepts to solve optimization problems, such as the seagull and gray wolf optimization algorithms. The seagull optimization algorithm (SOA) is a new type of bioinspired optimization algorithm that is based on the characteristics of a seagull. The SOA is combined with another algorithm to solve energy problems, such as short-term wind speed forecasting problems [16]. The gray wolf optimization algorithm (GWO) [17] relies on the level of leadership and the hunting mechanism of the gray wolf population for the search process of the optimization algorithm. The algorithms, based on different animal behavior patterns, are also subject to their effectiveness, which cannot be significantly improved when modified. Valenta and Langer proposed 2D P colonies to model the gray wolf optimization algorithm [18], where the performance was compared with that of the original GWO algorithm. From the computer simulation, the 2D P colonies had good performance for optimization problems.

Seagull, gray wolf optimization, and other evolutionary algorithms have the characteristics of self-management, adaptation, and self-learning. However, the new population determination process requires random initialization, resulting in relatively low computing efficiency. Considering the advantages and disadvantages of these algorithms, facing the same problem, the efficiency can be improved by modifying the original algorithms with other initialization concepts to avoid the randomization process.

Furthermore, a philosophy-inspired algorithm, namely, yin–yang pair optimization (YYOP) [19], uses the concept of balancing between two opposite entities for exploration and exploitation. YYPO is a low-complexity method that maintains good performance.

An important step in an optimization algorithm is the population generation step. Random initialization is used to generate new candidate solutions and eliminate the solutions that yield low scores, so that the cost value can either be maximized or minimized. However, the geometric structure of the cost function is not assessed at any step of the solution when generating offspring or for the search process. We proposed the application of the opposite gradient search [20] technique to search for the solution on the surface of the cost function. The technique, namely the fast opposite gradient search (FOGS) method, refers to a surface search algorithm applied with the opposite gradient method (OGM) [21]. FOGS is different from other search methods. It is not based mainly on meta-heuristics; instead, it seeks the manifold to obtain the locations where the cost function has zero gradients and minimum values.

This article introduces methods to increase the problem-solving efficiency with a better initial population for real parameter optimization problems. The proposed method presents a combination of the interior point method, the opposite gradient for generating new offspring, and the mean-variance mapping optimization algorithm (IPOG-MVMO). The mapping optimization algorithm is applied for the mutation operation to generate a modification depending on the n-best solution's mean and variance, to obtain a candidate solution on a complex surface of the function. Therefore, this proposed research contribution involves generating new offspring and obtaining the solution from the high-dimensional space of the cost function.

The manuscript is structured as follows. Section 2 describes the concepts of the proposed technique. In Section 3, we present an experimental simulation and analyze the results. Section 4 presents the discussion. Finally, Section 5 provides the conclusions and scope for future research.

2. Proposed Concept

For the proposed method, there were three main phases. Firstly, IPM was applied for local searching. The results from the first phase were introduced as points used for the second phase to create a new population, which depended on the manifold of the cost function using OGM. Finally, the last phase aimed to obtain the best solution using MVMO.

Karmarkar [22] presented the interior point method (IPM) to solve a linear programming problem in polynomial time complexity. The number of iterations taken by IPM was not greater than 100, regardless of the size of the problem. Therefore, IPM is suitable for local searching for large problems. The repetitive path pattern of searching for IPM always walks within the domain of possible answers. Since each iteration of IPM is complicated, especially in cases where the coefficient matrix is dense, IPM spends more time on each iteration.

One important concept that has been used to find the best solution or enhance the IPM is the opposite gradient method (OGM). The OGM is used to create a new set of points from the point resulting from the IPM. MVMO then uses the new points from the OGM phase to mutate to achieve a better solution, instead of randomly generating points as the original MVMO. This proposed technique, based on a combination of IPM, OGM, and MVMO, is called IPOG-MVMO and yields a set of solutions throughout the surface of the cost function. The process of IPOG-MVMO is presented in Algorithm 1.

Algorithm 1 Proposed IPOG-MVMO algorithm

Initialize parameters.
Generate an initial population using IPM from randomized points.
Improve the population using OGM.
Update the archive using MVMO.
Return the best point in the archive.

2.1. Generating Initial Population Using Local Minima Obtained from the IPM

The proposed algorithm focused on the interior point method called the barrier method [23,24], because of the proof of convergence and its complexity. The IPM was used to find the local minima that are located near an initial population. A function $F(x)$ was defined and was differentiable in the neighborhood of a point x. The process involved formulating a rule to test a differentiable function for the local minima that satisfied the restrictions. The procedure is described in Algorithm 2, where the output of this algorithm is assigned as the initial population for the next phase.

Algorithm 2 Finding the local minima using IPM

Q: an archive storing a population
D: number of dimensions in search space
Randomize N points in the feasible domain to generate an initial population.
Set Q as an empty set.
Set parameters $t := t_0 > 0,\ \mu > 1,\ \epsilon > 0$.
FOR $\{1 \leq j \leq N\}$
Set a random point $\boldsymbol{x}_j$ as a starting point.
WHILE $(\frac{D}{t} > \epsilon)$ do
  Compute $\boldsymbol{x}_j^*$ by minimizing $tF(\boldsymbol{x}) + \varnothing(\boldsymbol{x})$, subject to $A\boldsymbol{x}_j = b$.
  Set $\boldsymbol{x}_j := \boldsymbol{x}_j^*$.
  Update new t as μt.
ENDWHILE
Insert $\boldsymbol{x}_j$ into Q.
ENDFOR
RETURN Q as an archive storing the initial population for the next phase.

According to Algorithm 2, there are three constants, $t_0 > 0,\ \mu > 1,\ \epsilon > 0$, to be defined to initiate the algorithm. In each iteration, $\boldsymbol{x}_j^*$ is computed by minimizing $tF(\boldsymbol{x}) + \varnothing(\boldsymbol{x})$ subject to constraint $A\boldsymbol{x}_j = b$ and $\varnothing(\boldsymbol{x}) = -\sum_{i=1}^{M} \log(b_i - a_i\boldsymbol{x})$. The solution is updated until $t \geq D/\epsilon$. All local minima in a set Q are used as the initial population for the next step.

2.2. Opposite Gradient Method (OGM)

This section briefly presents a concept for generating new offspring. For a vector in a D-dimensional space of the cost function, new offspring are generated at the position of the manifold by OGM, where its first derivative is zero. One of these locations is expected to be the best solution for the cost function. The proposed technique proceeded along the same line with the following observation. The first derivative of $F(\boldsymbol{x})$ is denoted by $F'(\boldsymbol{x})$ and the gradient vector field is represented as the matrix of the first derivative form. Let $\boldsymbol{x}^{(\sigma)}$ and $\boldsymbol{x}^{(\gamma)}$ be two vectors on the manifold of the cost function. The locations with zero values of the first derivatives on dimension i could be in the range between $\boldsymbol{x}^{(\sigma)}$ and $\boldsymbol{x}^{(\gamma)}$ such that $F_i\left(\boldsymbol{x}^{(\sigma)}\right) F_i\left(\boldsymbol{x}^{(\gamma)}\right) < 0$. If $\left|F_i\left(\boldsymbol{x}^{(\sigma)}\right)\right| > \left|F_i\left(\boldsymbol{x}^{(\gamma)}\right)\right|$, and the location with zero gradient is closer to $\boldsymbol{x}^{(\sigma)}$ than $\boldsymbol{x}^{(\gamma)}$. On the other hand, if $\left|F_i\left(\boldsymbol{x}^{(\sigma)}\right)\right| < \left|F_i\left(\boldsymbol{x}^{(\gamma)}\right)\right|$, the location with zero gradient is closer to $\boldsymbol{x}^{(\gamma)}$ than $\boldsymbol{x}^{(\sigma)}$. This concept can also be applied to any other dimensions.

In particular, for each pair of two vectors obtained from set Q, the output of IPM in Algorithm 2, new next-generation vectors are generated from a varying jumping distance δ related to two vectors that have different signs of gradient on each dimension, as shown in Equation (1).

$$\delta = \frac{\left|F_i\left(\boldsymbol{x}^{(\gamma)}\right)\right|}{\left|F_i\left(\boldsymbol{x}^{(\sigma)}\right)\right| + \left|F_i\left(\boldsymbol{x}^{(\gamma)}\right)\right|} \|\boldsymbol{x}^{(\gamma)} - \boldsymbol{x}^{(\sigma)}\| w \tag{1}$$

Suppose that $\boldsymbol{x}^{(\sigma)}$ and $\boldsymbol{x}^{(\gamma)}$ are selected from the output IPM with the assumption that $F_i\left(\boldsymbol{x}^{(\sigma)}\right) F_i\left(\boldsymbol{x}^{(\gamma)}\right) < 0$. Then, we have two new vectors, $\boldsymbol{x}_{new}^{(\sigma)}, \boldsymbol{x}_{new}^{(\gamma)}$, whose coefficients on dimension i are calculated from this jumping distance, as shown in Equations (2) and (3), while the coefficients on the other dimensions are still unchanged.

$$x_{new,i}^{(\sigma)} = x_i^{(\sigma)} + \delta \tag{2}$$

$$x_{new,i}^{(\gamma)} = x_i^{(\gamma)} - \delta \tag{3}$$

In Equation (1), a weight value, w, is calculated at each iteration t according to Algorithm 3 to guarantee that the new vectors are generated within the pre-specified range of the search area. In essence, the weight value is in the interval between 0 and 1 and involves the jumping distance, δ, such that the new vectors discovered outside the search space are discarded immediately.

Algorithm 3 Generating candidate vectors $x_{new}^{(\sigma)}, x_{new}^{(\gamma)}$ within the pre-specified search space

w: weight value
NP: number of iterations
Q: an archive storing the current population
Set $\mathbf{X}_{cand} = \varnothing$.
w = 0.05w(NP/t)/* Calculating an updated weight value */
Compute a jumping distance δ using Equation (1).
Create two new candidate vectors $x_{new}^{(\sigma)}, x_{new}^{(\gamma)}$ using Equations (2) and (3).
IF {$x_{new}^{(\sigma)}$ is in the range}
    Insert $x_{new}^{(\sigma)}$ into $\mathbf{X}_{cand}$.
ELSE
    Discard $x_{new}^{(\sigma)}$.
ENDIF
IF {$x_{new}^{(\gamma)}$ is in the range}
    Insert $x_{new}^{(\gamma)}$ into $\mathbf{X}_{cand}$.
ELSE
    Discard $x_{new}^{(\gamma)}$.
ENDIF
RETURN $\mathbf{X}_{cand}$

Two new vectors can be generated and located approximately between the parent vectors. Accordingly, the next generation of vectors is calculated to decrease the search area. The cost function values of these new vectors must also be within an acceptable range. If some new vectors providing better cost value are attained, these new vectors will replace some vectors in the previous generation that are stored in the archive. Subsequently, for each dimension i, all the vectors in the archive are sorted ascendingly according to their cost values and divided into two different gradient groups, which are G_i^+, a group of vectors with a positive gradient value along dimension i, and G_i^-, a group of vectors with a negative gradient value along dimension i. For each group, a vector that contains the smallest gradient is retrieved such that these vectors are used as a pair of vectors to generate the next candidate vectors in the next generation. These steps are described in Algorithm 4.

The offspring from Algorithm 4 are filled in the archive Q to store the n-best offspring. The offspring index in the archive can be arranged according to the fitness suitability sequence and can be used as a guide for conducting the search. The size of the archive Q is not changed throughout the process. Note that the archive is updated only when the new vector obtains a solution better than the existing solution in the archive.

Algorithm 4 Generating a population using the opposite gradient method (OGM)

Q: an archive storing the initial population obtained from **Algorithm 2**
NP: number of iterations
D: number of dimensions
Set count = 1
WHILE {count $\leq$ NP}
FOR $\{1 \leq i \leq D\}$
    Set w = 1
    Partition the population in Q into G_i^+ and G_i^-.
    Select $\boldsymbol{x}^{(\sigma)}$, a vector with smallest gradient from G_i^+.
    Select $\boldsymbol{x}^{(\gamma)}$, a vector with smallest gradient from G_i^-.
    Compute vectors $\boldsymbol{x}_{new}^{(\sigma)}$, $\boldsymbol{x}_{new}^{(\gamma)}$ from $\boldsymbol{x}^{(\sigma)}$ and $\boldsymbol{x}^{(\gamma)}$ using **Algorithm 3**.
    IF $\boldsymbol{x}_{new}^{(\sigma)} \in \mathbf{X}_{cand}$ and $F\left(\boldsymbol{x}_{new}^{(\sigma)}\right)$ is less than the highest cost value from vectors in Q
        Discard the vector with the highest cost value.
        IF $F_i\left(\boldsymbol{x}_{new}^{(\sigma)}\right) > 0$
            Insert $\boldsymbol{x}_{new}^{(\sigma)}$ in G_i^+.
        ELSE
            Insert $\boldsymbol{x}_{new}^{(\sigma)}$ in G_i^-.
        ENDIF
    ENDIF
    IF $\boldsymbol{x}_{new}^{(\gamma)} \in \mathbf{X}_{cand}$ and $F\left(\boldsymbol{x}_{new}^{(\gamma)}\right)$ is less than the highest cost value from vectors in Q
        Discard the vector with the highest cost value.
        IF $F_i\left(\boldsymbol{x}_{new}^{(\gamma)}\right) < 0$
            Insert $\boldsymbol{x}_{new}^{(\gamma)}$ in G_i^-.
        ELSE
            Insert $\boldsymbol{x}_{new}^{(\gamma)}$ in G_i^+.
        ENDIF
    ENDIF
ENDFOR
$Q' = \left(\cup_{i=1}^{D} G_i^+\right) \cup \left(\cup_{i=1}^{D} G_i^-\right)$
Select n best points with respect to the cost values to store in Q.
count++
ENDWHILE
RETURN Q as an archive storing the initial population for the next phase

2.3. Combining IPM, OGM, and MVMO

The combination of IPM and OGM with MVMO is presented in this section. MVMO carries out global searching and focuses on the best solution. MVMO continuously updates an archive using compact memory with a fixed storage space. The n-best offspring are stored in the archive and serve as a guide toward the search direction. A solution stored in the archive is replaced by new offspring with a lower cost function value. Once the proposed combination is conducted, the vector with the lowest cost function value represents the optimal solution of the given function.

The search procedure applied to MVMO [25] is initiated with a particular set of points with D dimensions, obtained from Algorithm 4 in this study. MVMO has a key feature in that a mapping function is used for modifying the offspring depending on the specific mean-variance of the solutions collected in an archive. The mean $\overline{x}_i$ and variance v_i of dimension i are calculated once the update of the archive for each dimension is made with Equations (4) and (5), where N is the population size stored in the archive.

$$\overline{x}_i = \frac{1}{N} \sum_{j=1}^{N} x_i(j) \quad (4)$$

$$v_i = \frac{1}{N}\sum_{j=1}^{N}(x_i(j) - \overline{x}_i)^2 \tag{5}$$

The size of the search space depends on the mapping function. Moreover, the shape of the function with respect to $\overline{x}_i$ can be controlled by two shape variables $s_{i,1}$ and $s_{i,2}$ as follows:

$$h(\overline{x}_i, s_{i,1}, s_{i,2}, x) = \overline{x}_i(1 - e^{-xs_{i,1}}) + (1 - \overline{x}_i)e^{-(1-x)s_{i,2}} \tag{6}$$

The new coefficient of each selected dimension x_i of x is calculated using h values in terms of $h_x, h_1,\ h_0$ as follows:

$$x_i = h_x + (1 - h_1 + h_0)\ x_i^r - h_0 \tag{7}$$

where x_i^r is a number obtained randomly with uniform distribution in the range of [0, 1]. It can be guaranteed that the output of Equation (7) is within the range of [–100, 100]. According to Equation (6), $h_x, h_1,\ h_0$ are calculated using $x = x_i^r$, $x = 1$, and $x = 0$, respectively.

MVMO is capable of searching the global optimum with the best mean values of the solution. Two shape variables $s_{i,1}$ and $s_{i,2}$ are originated from a variable s_i, calculated from a scaling factor f_s and variance v_i to change the shape of the function as follows:

$$s_i = -f_s \ln(v_i) \tag{8}$$

where v_i is initially set to 1 and then recalculated once the update of the archive for each dimension is made using (5). The scaling factor f_s can be used to improve the accuracy when its value is greater than 1. On the other hand, the search is coarsely conducted when the value is less than 1. The factor is initiated for the search procedure with a small value. Subsequently, the size is increased for every iteration to increase the efficiency, as denoted in Equation (9):

$$f_s = f_s^*(1 + rand) \tag{9}$$

where $rand$ is randomized within [0, 1], and f_s^* can be calculated as in Equation (10):

$$f_s^* = f_{s_i}^* + \left(\frac{j}{j_f}\right)^2 \left(f_{s_f}^* - f_{s_i}^*\right) \tag{10}$$

where j and j_f are the current iteration number and the last iteration number, respectively. In this study, the values of $f_{s_i}^*$ and $f_{s_f}^*$ were set to 1 and 25, respectively.

As aforementioned, the shape of the mapping function can be determined, which depends upon the shape variables $s_{i,1}$ and $s_{i,2}$ of x_i. To improve the efficiency of the search process, the proper shape variables can be calculated using Algorithm 5.

The parameter d_i is initially set to 1 and then continuously updated at every iteration using the increment factor Δd calculated from Equation (11).

$$\Delta d = (1 + \Delta d_0 + 2{\cdot}\Delta d_0{\cdot}(rand - 0.5)) \tag{11}$$

This factor leads to the expansion or shrinkage of the value of parameter d_i, resulting in oscillation around the current s_i. The optimal interval of Δd_0 is within [0.01, 0.4].

In essence, the original MVMO encounters two problems, causing a long-duration searching process and difficulty in finding the best solution [10,26]. The first problem is zero variance when solutions stored in the archive are located at the same position. The second problem is that the value of the variance is sometimes outside of the specified range. This research attempted to overcome the limitations of MVMO and improve the flexibility of global searching over the classical MVMO by generating the population with the opposite gradient method before proceeding to MVMO. Algorithm 6 presents the determination of the n-best solutions using MVMO within fewer generations when using the population obtained from OGM.

Algorithm 5 Calculating the shape variables $s_{i,1}$ and $s_{i,2}$ for dimension i at iteration j

```
Calculate the variance v_i using Equation (5).
Calculate the scaling factor f_s using Equations (9) and (10).
Calculate the variable s_i using Equation (8).
Set s_{i,1} = s_i and s_{i,2} = s_i.
IF s_i > 0 then
        IF s_i > d_i
                Update d_i by multiplying with Δd.
        ELSE
                Update d_i by dividing with Δd.
        ENDIF
        IF d_i > s_i
                Set α = d_i and β = s_i.
        ELSE
                Set α = s_i and β = d_i.
        ENDIF
        IF rand < 0.5
                Set s_{i,1} = α and s_{i,2} = β.
        ELSE
                Set s_{i,1} = β and s_{i,2} = α.
        ENDIF
ENDIF
RETURN s_{i,1}, s_{i,2}, d_i
```

Algorithm 6 Generating n-best offspring using MVMO

```
Q: an archive storing the initial population obtained from Algorithm 4
NP: number of iterations
D: number of dimensions
Set the parameters d_i, Δd_0, f*_{s_i}, f*_{s_f}.
WHILE {count ≤ NP}
        FOR {1 ≤ i ≤ D}
                Calculate x̄_i using Equation (4).
                Apply Algorithm 5 to obtain two shape variables s_{i,1} and s_{i,2}.
                Calculate h_x, h_1, h_0 using Equation (6).
                Generate new x_i using Equation (7).
        ENDFOR
        Calculate the cost value of x_new.
        IF F(x_new) is less than the highest cost value from vectors in Q.
                Discard the vector with the highest cost value.
                Insert x_new in Q.
        ENDIF
ENDWHILE
```

3. Experiments

The IPOG-MVMO algorithm and the other algorithms, which were NBIPOPaCMA [9], PVADE [27], and MVMO, were applied to 28 benchmark functions on real-parameter optimization [28] using a computer with a Core i7 2.10 GHz CPU and 6 GB RAM for comparative purposes. All 28 benchmark functions have been described by Liang et al. [29,30]. The functions were composed of three types: unimodal (F1–F5), multimodal (F6–F20), and composition functions (F21–F28). All algorithms began running from the same set of initial points. The global optimal point of each test function was at the origin of the vector space. The error value ε can be calculated by Equation (12):

$$\varepsilon = \left| f_j(\mathbf{0}) - f_j(\mathbf{x}^*) \right| \quad (12)$$

where $f_j(\mathbf{0})$ is the cost value at the exact optimal point of the j-th benchmark function and $f_j(\mathbf{x}^*)$ is the cost value of the optimal point obtained by a selected algorithm. The error value is rounded to zero if it is less than 0.00000001 or 1×10^{-8}.

3.1. IPOG-MVMO Parameter Set-Up

By using IPOG-MVMO, all functions have been tested with $N = 15D$ points, where D is the number of dimensions. The parameters selected for IPOG-MVMO are presented in Table 1.

Table 1. The parameters of the proposed algorithm experiment.

Parameters	Values
The number of dimensions (D) in each problem	{10, 30, 50}
Search range	$[-100, 100]^D$
$f_{s_i}^*$	1
$f_{s_f}^*$	25
d_i	1
Δd_0	0.05
N	$15D$
Size of the archive	15
NP	100,000
Number of experimental trials	50

The parameters used for IPOG-MVMO are presented in Table 1. The number of dimensions (D) in each problem, and the search range, were defined according to the evaluation criteria of CEC2013 benchmark problems. The number of points (N) varied depending on the number of dimensions, to improve the flexibility of the algorithms to deal with more difficult problems. The number of iterations (NP) and the number of experimental trials were defined to achieve reliable results. Moreover, other parameters were successfully tested in the original MVMO process.

3.2. Experimental Results

The best value, worst value, mean, and standard deviation of the error between the cost value of the solution found by an algorithm and the exact cost value were recorded over 50 trials. With these four metrics, the best value was used for the comparison and was the focus of this study. The results for 10, 30, and 50 dimensions are summarized in Tables 2–8, respectively. To clearly illustrate the comparison, only the mean of the error was selected and plotted, as is shown in Figures 1–3.

Table 2. The comparative results of the errors obtained from unimodal functions F1–F5.

Algorithms	D = 10				D = 30				D = 50			
	Best	Worst	Mean	Std.	Best	Worst	Mean	Std.	Best	Worst	Mean	Std.
F1												
NBIPOPaCMA	0	0	0	0	0	1.91×10^{-8}	0	4.30×10^{-8}	0	0	0	0
PVADE	0	0	0	0	0	0	0	0	0	0	0	0
MVMO	0	0	0	0	0	0	0	0	0	0	0	0
IPOG-MVMO	**0**	**0**	**0**	**0**	**0**	**0**	**0**	**0**	**0**	**0**	**0**	**0**
F2												
NBIPOPaCMA	0	0	0	0	0	3.45×10^{-8}	1.61×10^{-8}	3.78×10^{-9}	0	0	0	0
PVADE	0	6.57×10^{2}	1.39×10^{1}	1.12×10^{2}	1.35×10^{3}	4.23×10^{4}	9.03×10^{3}	7.40×10^{3}	3.55×10^{4}	4.15×10^{5}	2.07×10^{5}	9.44×10^{4}
MVMO	0	0	0	0	5.59×10^{-4}	9.21×10^{-4}	7.65×10^{-4}	9.49×10^{-5}	5.59×10^{-4}	9.21×10^{-4}	7.65×10^{-4}	9.49×10^{-5}
IPOG-MVMO	**0**	**0**	**0**	**0**	**0**	$\mathbf{5.27 \times 10^{-7}}$	**0**	**0**	**0**	**0**	**0**	**0**
F3												
NBIPOPaCMA	0	0	0	0	0	7.42×10^{-8}	2.21×10^{-8}	1.68×10^{-9}	1.17×10^{-5}	5.01×10^{0}	1.41×10^{0}	2.19×10^{0}
PVADE	0	4.85×10^{-3}	9.7×10^{-5}	7.15×10^{-4}	0	1.55×10^{3}	4.01×10^{1}	2.31×10^{3}	6.14×10^{5}	2.63×10^{7}	1.89×10^{7}	1.16×10^{7}
MVMO	0	0	0	0	1.00×10^{3}	1.11×10^{7}	7.95×10^{5}	1.81×10^{6}	2.88×10^{-6}	1.29×10^{6}	6.08×10^{4}	2.72×10^{5}
IPOG-MVMO	**0**	**0**	**0**	**0**	**0**	$\mathbf{1.17 \times 10^{-6}}$	$\mathbf{3.65 \times 10^{-7}}$	$\mathbf{4.62 \times 10^{-7}}$	1.75×10^{-5}	6.24×10^{-4}	2.11×10^{-4}	2.23×10^{-4}
F4												
NBIPOPaCMA	0	0	0	0	7.75×10^{-8}	1.71×10^{-8}	1.39×10^{-8}	4.81×10^{-8}	7.75×10^{-8}	1.71×10^{-8}	1.39×10^{-8}	4.81×10^{-8}
PVADE	0	2.12×10^{-1}	4.21×10^{-3}	3.13×10^{-2}	5.53×10^{-7}	1.33×10^{-3}	1.93×10^{-4}	3.31×10^{-4}	5.53×10^{-7}	1.33×10^{-3}	1.93×10^{-4}	3.31×10^{-4}
MVMO	0	1.18×10^{-8}	0	0	5.37×10^{-6}	3.91×10^{-3}	4.62×10^{-4}	7.56×10^{-4}	5.37×10^{-6}	3.91×10^{-3}	4.62×10^{-4}	7.56×10^{-4}
IPOG-MVMO	**0**	**0**	**0**	**0**	**0**	$\mathbf{1.42 \times 10^{-6}}$	$\mathbf{3.65 \times 10^{-7}}$	$\mathbf{5.64 \times 10^{-7}}$	**0**	$\mathbf{1.42 \times 10^{-6}}$	$\mathbf{3.65 \times 10^{-7}}$	$\mathbf{5.64 \times 10^{-7}}$
F5												
NBIPOPaCMA	0	1.15×10^{-8}	0	0	0	3.34×10^{-8}	1.39×10^{-8}	6.18×10^{-8}	0	0	0	0
PVADE	0	8.13×10^{-5}	1.26×10^{-5}	1.11×10^{-4}	0	3.39×10^{-8}	1.02×10^{-8}	1.14×10^{-8}	1.45×10^{-4}	1.6×10^{-4}	1.02×10^{-3}	1.11×10^{-3}
MVMO	0	4.19×10^{-8}	0	0	1.22×10^{-8}	3.01×10^{-8}	2.34×10^{-8}	2.12×10^{-8}	0	0	0	0
IPOG-MVMO	**0**	**0**	**0**	**0**	**0**	$\mathbf{2.79 \times 10^{-8}}$	$\mathbf{1.02 \times 10^{-8}}$	$\mathbf{1.46 \times 10^{-8}}$	**0**	**0**	**0**	**0**

Table 3. The comparative results of the errors obtained from multimodal functions F6–F9.

Algorithms	$D = 10$				$D = 30$				$D = 50$			
F6												
	Best	Worst	Mean	Std.	Best	Worst	Mean	Std.	Best	Worst	Mean	Std.
NBIPOPaCMA	0	0	0	0	0	2.13×10^{-8}	0	3.96×10^{-9}	0	0	0	0
PVADE	0	7.35×10^{0}	7.81×10^{0}	3.89×10^{0}	0	2.67×10^{1}	4.96×10^{-1}	3.73×10^{0}	2.57×10^{1}	1.48×10^{2}	8.52×10^{1}	3.15×10^{1}
MVMO	0	0	0	0	0	3.91×10^{1}	1.04×10^{1}	9.71×10^{0}	0	4.34×10^{1}	3.05×10^{1}	2.01×10^{1}
IPOG-MVMO	**0**	**0**	**0**	**0**	**0**	**2.81×10^{-8}**	**0**	**8.11×10^{-9}**	**0**	**5.76×10^{0}**	**1.21×10^{0}**	**2.26×10^{0}**
F7												
	Best	Worst	Mean	Std.	Best	Worst	Mean	Std.	Best	Worst	Mean	Std.
NBIPOPaCMA	0	1.62×10^{1}	1.40×10^{2}	4.80×10^{2}	5.67×10^{-3}	5.69×10^{1}	7.01×10^{0}	1.71×10^{1}	3.05×10^{-3}	1.83×10^{1}	3.94×10^{0}	8.05×10^{0}
PVADE	3.11×10^{-5}	1.10×10^{1}	3.02×10^{-1}	1.53×10^{0}	4.21×10^{-1}	2.64×10^{1}	4.60×10^{0}	5.75×10^{0}	3.92×10^{4}	4.41×10^{1}	1.89×10^{1}	9.26×10^{0}
MVMO	6.33×10^{-3}	6.34×10^{-3}	6.13×10^{-3}	0	2.48×10^{0}	1.92×10^{1}	8.17×10^{0}	3.62×10^{0}	1.52×10^{1}	4.91×10^{1}	3.04×10^{1}	7.87×10^{0}
IPOG-MVMO	**0**	**0**	**0**	**0**	**4.22×10^{-5}**	**4.62×10^{-4}**	**2.11×10^{-4}**	**2.22×10^{-4}**	**1.32×10^{-5}**	**4.81×10^{-4}**	**2.21×10^{-4}**	**2.15×10^{-4}**
F8												
	Best	Worst	Mean	Std.	Best	Worst	Mean	Std.	Best	Worst	Mean	Std.
NBIPOPaCMA	3.49×10^{-8}	1.68×10^{-8}	1.04×10^{-8}	8.57×10^{-9}	2.08×10^{1}	2.19×10^{1}	2.13×10^{1}	2.85×10^{1}	7.77×10^{-1}	2.11×10^{1}	5.87×10^{0}	9.20×10^{0}
PVADE	1.92×10^{1}	2.01×10^{1}	2.01×10^{1}	6.11×10^{-2}	2.04×10^{1}	2.14×10^{1}	2.61×10^{1}	4.08×10^{1}	2.10×10^{1}	2.12×10^{1}	2.11×10^{1}	3.49×10^{-2}
MVMO	1.94×10^{1}	2.11×10^{1}	1.98×10^{1}	6.54×10^{-2}	2.08×10^{1}	2.10×10^{1}	2.19×10^{1}	5.41×10^{2}	2.10×10^{1}	2.12×10^{1}	2.11×10^{1}	3.94×10^{-2}
IPOG-MVMO	**0**	**6.12×10^{-8}**	**1.54×10^{-7}**	**1.02×10^{-7}**	2.08×10^{1}	2.10×10^{1}	2.09×10^{1}	2.37×10^{2}	8.35×10^{2}	2.16×10^{1}	1.96×10^{1}	4.34×10^{0}
F9												
	Best	Worst	Mean	Std.	Best	Worst	Mean	Std.	Best	Worst	Mean	Std.
NBIPOPaCMA	1.92×10^{-8}	2.47×10^{1}	3.82×10^{0}	9.56×10^{-9}	2.02×10^{0}	1.78×10^{1}	4.30×10^{0}	4.81×10^{0}	0	7.41×10^{1}	1.50×10^{1}	3.30×10^{0}
PVADE	8.11×10^{-3}	3.36×10^{0}	1.34×10^{0}	9.67×10^{-1}	2.27×10^{1}	3.09×10^{1}	2.74×10^{1}	1.77×10^{0}	1.99×10^{1}	3.34×10^{1}	2.60×10^{1}	3.05×10^{0}
MVMO	4.92×10^{-1}	2.45×10^{0}	8.26×10^{-1}	6.66×10^{-1}	7.11×10^{0}	1.92×10^{1}	1.32×10^{1}	2.65×10^{0}	2.48×10^{1}	4.17×10^{1}	3.33×10^{1}	4.39×10^{0}
IPOG-MVMO	**0**	**9.45×10^{-8}**	**6.95×10^{-8}**	**4.45×10^{-8}**	**7.54×10^{-7}**	**1.19×10^{-6}**	**9.57×10^{-7}**	**2.19×10^{-7}**	4.82×10^{-2}	6.73×10^{0}	1.64×10^{0}	2.57×10^{00}

Table 4. The comparative results of the errors obtained from multimodal functions F10–F13.

Algorithms	$D = 10$				$D = 30$				$D = 50$			
	Best	Worst	Mean	Std.	Best	Worst	Mean	Std.	Best	Worst	Mean	Std.
F10												
NBIPOPaCMA	1.02×10^{-8}	6.71×10^{-8}	3.24×10^{-8}	1.84×10^{-9}	1.01×10^{-8}	3.55×10^{-8}	1.94×10^{-8}	8.32×10^{-8}	0	4.69×10^{-2}	9.37×10^{-3}	2.09×10^{-2}
PVADE	0	1.79×10^{-1}	5.01×10^{-2}	3.79×10^{-2}	1.73×10^{-2}	1.89×10^{-1}	7.64×10^{-2}	3.53×10^{-2}	4.77×10^{-2}	1.15×10^{0}	5.99×10^{-1}	3.42×10^{-1}
MVMO	9.95×10^{-3}	3.58×10^{-2}	1.58×10^{-2}	2.11×10^{-2}	7.40×10^{-3}	7.35×10^{-2}	2.78×10^{-2}	1.64×10^{-2}	0	0	0	0
IPOG-MVMO	**0**	$\mathbf{3.87 \times 10^{-8}}$	$\mathbf{2.89 \times 10^{-8}}$	$\mathbf{7.34 \times 10^{-9}}$	**0**	$\mathbf{4.01 \times 10^{-8}}$	$\mathbf{1.42 \times 10^{-8}}$	$\mathbf{1.78 \times 10^{-8}}$	**0**	$\mathbf{6.66 \times 10^{1}}$	$\mathbf{2.78 \times 10^{-3}}$	$\mathbf{3.91 \times 10^{-2}}$
F11												
NBIPOPaCMA	0	6.72×10^{-8}	3.27×10^{-8}	2.11×10^{-9}	9.95×10^{-1}	3.98×10^{0}	2.70×10^{0}	1.25×10^{0}	7.37×10^{-2}	1.89×10^{1}	3.94×10^{0}	8.36×10^{0}
PVADE	0	1.21×10^{1}	4.04×10^{0}	2.33×10^{0}	1.03×10^{0}	7.34×10^{0}	3.13×10^{0}	1.49×10^{0}	6.76×10^{1}	2.34×10^{2}	1.68×10^{2}	4.08×10^{1}
MVMO	0	6.02×10^{0}	2.32×10^{0}	1.29×10^{-2}	1.01×10^{0}	8.05×10^{0}	4.34×10^{0}	1.90×10^{0}	2.59×10^{1}	7.57×10^{1}	4.55×10^{1}	1.22×10^{1}
IPOG-MVMO	**0**	$\mathbf{1.13 \times 10^{-8}}$	$\mathbf{4.87 \times 10^{-8}}$	$\mathbf{3.13 \times 10^{-9}}$	**0**	$\mathbf{1.03 \times 10^{-7}}$	$\mathbf{2.13 \times 10^{-8}}$	$\mathbf{4.03 \times 10^{-8}}$	$\mathbf{1.10 \times 10^{-3}}$	$\mathbf{4.90 \times 10^{-2}}$	$\mathbf{1.21 \times 10^{-2}}$	$\mathbf{1.81 \times 10^{-2}}$
F12												
NBIPOPaCMA	0	1.03×10^{0}	3.43×10^{-1}	1.89×10^{-9}	6.70×10^{-8}	3.98×10^{0}	2.27×10^{0}	1.70×10^{0}	7.37×10^{-2}	1.02×10^{0}	3.30×10^{-1}	3.96×10^{-1}
PVADE	9.87×10^{-1}	1.59×10^{1}	5.86×10^{0}	3.68×10^{0}	1.59×10^{0}	3.24×10^{1}	2.13×10^{1}	3.79×10^{0}	2.20×10^{2}	3.12×10^{2}	2.56×10^{2}	2.01×10^{1}
MVMO	1.89×10^{0}	1.55×10^{1}	5.89×10^{0}	1.32×10^{0}	1.59×10^{1}	5.97×10^{1}	3.36×10^{1}	1.14×10^{1}	3.98×10^{1}	1.34×10^{2}	7.77×10^{1}	2.33×10^{1}
IPOG-MVMO	**0**	$\mathbf{3.49 \times 10^{-5}}$	$\mathbf{4.78 \times 10^{-8}}$	$\mathbf{7.56 \times 10^{-9}}$	**0**	$\mathbf{2.72 \times 10^{-8}}$	$\mathbf{8.87 \times 10^{-8}}$	$\mathbf{1.21 \times 10^{-8}}$	$\mathbf{7.74 \times 10^{-5}}$	$\mathbf{2.45 \times 10^{-3}}$	$\mathbf{9.99 \times 10^{4}}$	$\mathbf{1.05 \times 10^{-3}}$
F13												
NBIPOPaCMA	0	4.54×10^{0}	7.57×10^{-1}	1.45×10^{0}	0	3.07×10^{0}	2.77×10^{0}	1.36×10^{0}	6.46×10^{-3}	1.48×10^{1}	3.10×10^{0}	6.54×10^{0}
PVADE	0	2.08×10^{1}	8.35×10^{0}	4.34×10^{0}	7.60×10^{1}	1.56×10^{2}	1.21×10^{2}	1.86×10^{1}	8.10×10^{1}	1.62×10^{2}	1.18×10^{2}	1.69×10^{1}
MVMO	1.87×10^{0}	1.82×10^{1}	8.84×10^{0}	9.01×10^{0}	3.06×10^{1}	9.95×10^{1}	5.90×10^{1}	1.67×10^{1}	7.20×10^{1}	2.05×10^{2}	1.33×10^{2}	3.43×10^{1}
IPOG-MVMO	**0**	$\mathbf{3.58 \times 10^{-5}}$	$\mathbf{1.08 \times 10^{-5}}$	$\mathbf{1.01 \times 10^{-5}}$	**0**	$\mathbf{1.79 \times 10^{-6}}$	$\mathbf{3.23 \times 10^{-7}}$	$\mathbf{6.07 \times 10^{-7}}$	$\mathbf{6.43 \times 10^{-4}}$	$\mathbf{6.30 \times 10^{0}}$	$\mathbf{1.54 \times 10^{0}}$	$\mathbf{2.59 \times 10^{0}}$

Table 5. The comparative results of the errors obtained from multimodal functions F14–F17.

F14

Algorithms	D = 10				D = 30				D = 50			
	Best	Worst	Mean	Std.	Best	Worst	Mean	Std.	Best	Worst	Mean	Std.
NBIPOPaCMA	2.13×10^{1}	7.68×10^{2}	3.44×10^{2}	2.11×10^{2}	7.12×10^{2}	1.35×10^{3}	8.70×10^{2}	3.40×10^{2}	3.16×10^{1}	4.59×10^{1}	3.96×10^{1}	5.94×10^{0}
PVADE	4.01×10^{1}	5.13×10^{2}	1.67×10^{2}	1.07×10^{2}	2.04×10^{3}	4.01×10^{3}	3.11×10^{3}	4.70×10^{2}	2.14×10^{3}	3.90×10^{3}	3.07×10^{3}	3.80×10^{2}
MVMO	3.21×10^{0}	2.08×10^{1}	8.01×10^{0}	7.59×10^{0}	9.40×10^{1}	1.66×10^{3}	8.56×10^{2}	4.12×10^{2}	2.19×10^{3}	4.84×10^{3}	3.89×10^{3}	6.02×10^{2}
IPOG-MVMO	**2.03×10^{-5}**	**2.67×10^{-5}**	**2.39×10^{-5}**	**4.78×10^{-6}**	**2.37×10^{1}**	**6.35×10^{2}**	**3.09×10^{2}**	**3.11×10^{2}**	**9.23×10^{-5}**	**6.10×10^{1}**	**5.05×10^{1}**	**1.70×10^{0}**

F15

Algorithms	D = 10				D = 30				D = 50			
	Best	Worst	Mean	Std.	Best	Worst	Mean	Std.	Best	Worst	Mean	Std.
NBIPOPaCMA	6.87×10^{0}	2.38×10^{2}	1.17×10^{2}	8.89×10^{1}	5.53×10^{2}	1.67×10^{3}	7.65×10^{2}	2.85×10^{2}	1.27×10^{-1}	9.68×10^{1}	6.51×10^{1}	3.94×10^{1}
PVADE	3.71×10^{2}	1.02×10^{3}	7.84×10^{2}	1.68×10^{2}	2.13×10^{3}	4.94×10^{3}	3.32×10^{3}	4.08×10^{2}	4.54×10^{3}	6.17×10^{3}	5.36×10^{3}	3.54×10^{2}
MVMO	1.89×10^{2}	6.57×10^{2}	5.58×10^{2}	9.01×10^{1}	1.70×10^{3}	4.45×10^{3}	3.03×10^{3}	5.41×10^{2}	5.75×10^{3}	8.65×10^{3}	6.64×10^{3}	5.92×10^{2}
IPOG-MVMO	**2.58×10^{-5}**	**7.28×10^{-2}**	**3.21×10^{-2}**	**3.13×10^{-2}**	**2.19×10^{1}**	**1.22×10^{3}**	**2.28×10^{2}**	**2.37×10^{2}**	**3.69×10^{-3}**	**6.67×10^{1}**	**2.03×10^{1}**	**2.76×10^{1}**

F16

Algorithms	D = 10				D = 30				D = 50			
	Best	Worst	Mean	Std.	Best	Worst	Mean	Std.	Best	Worst	Mean	Std.
NBIPOPaCMA	0	1.97×10^{1}	5.68×10^{0}	7.23×10^{0}	5.68×10^{-2}	1.23×10^{0}	9.14×10^{-1}	1.86×10^{-1}	1.18×10^{-3}	3.51×10^{0}	1.20×10^{0}	1.65×10^{0}
PVADE	4.75×10^{-1}	1.26×10^{0}	8.93×10^{-1}	1.88×10^{-1}	1.42×10^{0}	3.13×10^{0}	2.32×10^{0}	3.01×10^{-1}	2.67×10^{0}	3.84×10^{0}	3.38×10^{0}	2.86×10^{-1}
MVMO	3.52×10^{-1}	6.49×10^{-1}	5.39×10^{-1}	1.62×10^{-1}	5.37×10^{-1}	1.62×10^{0}	1.08×10^{0}	2.88×10^{-1}	6.60×10^{-1}	1.70×10^{0}	1.14×10^{0}	2.37×10^{-1}
IPOG-MVMO	**0**	**7.64×10^{-1}**	**1.63×10^{-1}**	**3.18×10^{-1}**	8.08×10^{-2}	1.34×10^{0}	5.40×10^{0}	1.09×10^{1}	3.12×10^{-1}	3.25×10^{0}	2.11×10^{0}	2.24×10^{-1}

F17

Algorithms	D = 10				D = 30				D = 50			
	Best	Worst	Mean	Std.	Best	Worst	Mean	Std.	Best	Worst	Mean	Std.
NBIPOPaCMA	0	9.71×10^{1}	1.89×10^{1}	2.64×10^{1}	3.41×10^{1}	4.26×10^{1}	3.53×10^{1}	1.87×10^{0}	2.10×10^{0}	5.55×10^{1}	1.30×10^{1}	2.37×10^{1}
PVADE	1.15×10^{1}	1.75×10^{1}	1.39×10^{1}	1.39×10^{1}	7.44×10^{1}	1.17×10^{2}	9.52×10^{1}	1.10×10^{1}	1.44×10^{2}	3.17×10^{2}	2.37×10^{2}	1.10×10^{1}
MVMO	1.03×10^{1}	1.18×10^{1}	1.11×10^{1}	7.65×10^{-1}	4.38×10^{1}	8.14×10^{1}	6.72×10^{1}	8.50×10^{0}	8.89×10^{1}	1.43×10^{2}	1.11×10^{2}	1.34×10^{1}
IPOG-MVMO	**0**	**3.07×10^{1}**	**9.85×10^{0}**	**8.85×10^{0}**	**3.25×10^{1}**	**6.60×10^{1}**	**4.31×10^{1}**	**4.89×10^{0}**	4.50×10^{0}	6.66×10^{1}	3.48×10^{1}	3.30×10^{1}

Table 6. The comparative results of the errors obtained from composition functions F18–F20.

Algorithms	D = 10				D = 30				D = 50			
	Best	Worst	Mean	Std.	Best	Worst	Mean	Std.	Best	Worst	Mean	Std.
F18												
NBIPOPaCMA	0	5.74×10^{1}	1.58×10^{1}	1.51×10^{1}	3.86×10^{1}	1.72×10^{2}	6.38×10^{1}	4.45×10^{1}	8.60×10^{-2}	9.86×10^{1}	2.54×10^{1}	4.13×10^{1}
PVADE	1.13×10^{1}	3.26×10^{1}	2.40×10^{1}	3.90×10^{0}	1.40×10^{2}	1.89×10^{2}	1.66×10^{2}	1.12×10^{1}	3.43×10^{2}	4.22×10^{2}	3.65×10^{2}	1.12×10^{1}
MVMO	1.51×10^{1}	2.01×10^{1}	1.72×10^{1}	2.58×10^{0}	4.58×10^{1}	7.88×10^{1}	5.94×10^{1}	7.59×10^{0}	8.06×10^{1}	1.63×10^{2}	1.07×10^{2}	1.73×10^{1}
IPOG-MVMO	**0**	$\mathbf{1.01 \times 10^{-1}}$	$\mathbf{3.27 \times 10^{-2}}$	$\mathbf{3.71 \times 10^{-2}}$	$\mathbf{1.88 \times 10^{1}}$	$\mathbf{3.38 \times 10^{1}}$	$\mathbf{2.94 \times 10^{1}}$	$\mathbf{9.87 \times 10^{0}}$	$\mathbf{1.20 \times 10^{-2}}$	$\mathbf{4.18 \times 10^{0}}$	$\mathbf{1.34 \times 10^{0}}$	$\mathbf{1.92 \times 10^{0}}$
F19												
NBIPOPaCMA	0	7.38×10^{-1}	4.34×10^{-1}	2.01×10^{-1}	1.14×10^{0}	2.54×10^{0}	2.40×10^{0}	4.41×10^{-1}	1.60×10^{-1}	2.85×10^{0}	8.65×10^{-1}	1.16×10^{0}
PVADE	3.27×10^{-1}	9.69×10^{-1}	5.98×10^{-1}	1.25×10^{-1}	3.10×10^{0}	1.06×10^{1}	6.49×10^{0}	1.74×10^{0}	1.06×10^{1}	3.10×10^{1}	2.12×10^{1}	4.74×10^{0}
MVMO	3.52×10^{-1}	6.07×10^{-1}	5.19×10^{-1}	1.44×10^{-1}	1.10×10^{0}	3.13×10^{0}	1.93×10^{0}	4.09×10^{-1}	2.79×10^{0}	7.55×10^{0}	4.82×10^{0}	1.24×10^{0}
IPOG-MVMO	**0**	$\mathbf{1.90 \times 10^{-1}}$	$\mathbf{2.59 \times 10^{-2}}$	$\mathbf{6.02 \times 10^{-2}}$	$\mathbf{3.90 \times 10^{-1}}$	$\mathbf{2.88 \times 10^{0}}$	$\mathbf{1.06 \times 10^{0}}$	$\mathbf{4.73 \times 10^{-1}}$	$\mathbf{1.52 \times 10^{-1}}$	$\mathbf{3.28 \times 10^{0}}$	$\mathbf{1.53 \times 10^{0}}$	$\mathbf{1.52 \times 10^{0}}$
F20												
NBIPOPaCMA	2.11×10^{0}	4.85×10^{0}	3.26×10^{0}	9.87×10^{-1}	1.01×10^{1}	1.56×10^{1}	1.29×10^{1}	5.98×10^{-1}	7.85×10^{-1}	2.53×10^{1}	7.01×10^{0}	1.05×10^{0}
PVADE	9.96×10^{-1}	3.52×10^{0}	2.12×10^{0}	5.61×10^{-1}	1.04×10^{1}	1.50×10^{1}	1.34×10^{1}	1.89×10^{-1}	1.95×10^{1}	2.55×10^{1}	2.36×10^{1}	1.90×10^{0}
MVMO	1.83×10^{0}	2.78×10^{0}	2.34×10^{0}	4.76×10^{-1}	8.81×10^{0}	1.19×10^{1}	1.04×10^{1}	5.85×10^{-1}	1.87×10^{1}	2.23×10^{1}	2.01×10^{1}	6.84×10^{-1}
IPOG-MVMO	1.12×10^{0}	2.74×10^{0}	1.95×10^{0}	4.01×10^{-1}	$\mathbf{6.84 \times 10^{0}}$	$\mathbf{1.28 \times 10^{1}}$	$\mathbf{1.01 \times 10^{1}}$	$\mathbf{6.23 \times 10^{-1}}$	2.10×10^{0}	1.78×10^{1}	2.30×10^{0}	4.57×10^{-1}

Table 7. The comparative results of the errors obtained from composition functions F21–F24.

Algorithms	D = 10				D = 30				D = 50			
	Best	Worst	Mean	Std.	Best	Worst	Mean	Std.	Best	Worst	Mean	Std.
F21												
NBIPOPaCMA	3.64×10^{0}	1.42×10^{1}	1.05×10^{1}	4.47×10^{0}	2.00×10^{2}	2.00×10^{2}	2.00×10^{2}	0	1.00×10^{2}	2.00×10^{2}	1.91×10^{2}	1.43×10^{1}
PVADE	2.53×10^{2}	3.81×10^{2}	2.90×10^{2}	4.06×10^{1}	2.00×10^{2}	4.43×10^{2}	3.17×10^{2}	6.22×10^{1}	8.36×10^{2}	1.22×10^{3}	9.56×10^{2}	1.44×10^{2}
MVMO	3.35×10^{-5}	3.90×10^{2}	2.26×10^{2}	9.73×10^{1}	2.00×10^{2}	4.43×10^{2}	2.95×10^{2}	1.07×10^{2}	1.00×10^{2}	1.13×10^{3}	2.45×10^{2}	1.91×10^{2}
IPOG-MVMO	**2.68×10^{0}**	**1.50×10^{1}**	**1.22×10^{1}**	**4.11×10^{0}**	**2.00×10^{2}**	**4.45×10^{2}**	**2.57×10^{2}**	**1.15×10^{1}**	**1.00×10^{2}**	**5.24×10^{2}**	**2.29×10^{2}**	**3.48×10^{1}**
F22												
NBIPOPaCMA	3.31×10^{0}	1.00×10^{1}	8.81×10^{0}	3.05×10^{0}	3.54×10^{2}	1.08×10^{3}	8.03×10^{2}	3.12×10^{2}	1.89×10^{2}	3.86×10^{3}	1.45×10^{3}	6.01×10^{2}
PVADE	2.34×10^{1}	5.07×10^{2}	2.57×10^{2}	1.11×10^{2}	1.72×10^{3}	3.38×10^{3}	2.49×10^{3}	3.86×10^{2}	6.11×10^{3}	1.01×10^{4}	7.72×10^{3}	8.44×10^{2}
MVMO	5.02×10^{0}	1.00×10^{2}	3.27×10^{1}	1.91×10^{1}	2.35×10^{2}	2.00×10^{3}	8.19×10^{2}	4.28×10^{2}	1.17×10^{3}	4.94×10^{3}	2.76×10^{3}	8.38×10^{2}
IPOG-MVMO	**3.00×10^{0}**	**1.76×10^{1}**	**6.20×10^{0}**	**2.48×10^{2}**	**2.00×10^{2}**	**2.83×10^{3}**	**6.26×10^{2}**	**2.55×10^{2}**	3.92×10^{2}	4.68×10^{3}	2.35×10^{3}	6.94×10^{2}
F23												
NBIPOPaCMA	3.59×10^{0}	4.14×10^{0}	3.96×10^{0}	2.01×10^{1}	3.52×10^{2}	3.59×10^{3}	1.29×10^{3}	1.35×10^{3}	1.37×10^{3}	1.31×10^{4}	4.29×10^{3}	5.00×10^{3}
PVADE	1.18×10^{1}	1.12×10^{2}	9.14×10^{1}	1.78×10^{1}	4.74×10^{3}	7.25×10^{3}	5.81×10^{3}	5.04×10^{2}	7.91×10^{3}	1.54×10^{4}	1.17×10^{4}	1.48×10^{3}
MVMO	5.63×10^{1}	8.71×10^{2}	4.95×10^{2}	2.03×10^{2}	1.69×10^{3}	4.45×10^{3}	3.09×10^{3}	5.28×10^{2}	6.17×10^{3}	1.12×10^{4}	8.64×10^{3}	1.32×10^{3}
IPOG-MVMO	**3.54×10^{0}**	**3.14×10^{0}**	**3.89×10^{0}**	**3.01×10^{-1}**	4.89×10^{2}	4.28×10^{3}	2.79×10^{3}	9.41×10^{2}	**4.97×10^{2}**	**7.29×10^{3}**	**5.79×10^{3}**	**9.41×10^{2}**
F24												
NBIPOPaCMA	3.59×10^{0}	3.70×10^{0}	3.74×10^{0}	5.01×10^{-2}	1.39×10^{2}	3.01×10^{2}	2.42×10^{2}	8.00×10^{1}	2.35×10^{2}	3.86×10^{2}	3.27×10^{2}	7.60×10^{1}
PVADE	1.01×10^{2}	2.11×10^{2}	1.94×10^{2}	1.30×10^{1}	2.01×10^{2}	2.61×10^{2}	2.03×10^{2}	1.39×10^{0}	2.40×10^{2}	3.28×10^{2}	2.78×10^{2}	1.83×10^{1}
MVMO	1.01×10^{2}	2.07×10^{2}	1.83×10^{2}	3.74×10^{1}	2.04×10^{2}	2.20×10^{2}	2.11×10^{2}	3.77×10^{0}	2.28×10^{2}	2.61×10^{2}	2.45×10^{2}	8.05×10^{0}
IPOG-MVMO	**9.85×10^{-1}**	**4.42×10^{0}**	**2.07×10^{0}**	**1.44×10^{-2}**	**1.05×10^{2}**	**4.62×10^{2}**	**2.17×10^{2}**	**1.46×10^{0}**	**1.05×10^{2}**	**4.62×10^{2}**	**2.17×10^{2}**	**1.46×10^{1}**

Table 8. The comparative results of the errors obtained from composition functions F25–F28.

F25												
Algorithms	$D = 10$				$D = 30$				$D = 50$			
	Best	Worst	Mean	Std.	Best	Worst	Mean	Std.	Best	Worst	Mean	Std.
NBIPOPaCMA	7.31×10^{0}	7.49×10^{0}	7.44×10^{0}	5.01×10^{-2}	2.05×10^{2}	3.01×10^{2}	2.66×10^{2}	4.26×10^{1}	2.34×10^{2}	3.86×10^{2}	3.23×10^{2}	7.86×10^{1}
PVADE	1.90×10^{2}	2.13×10^{2}	2.04×10^{2}	3.77×10^{0}	2.00×10^{2}	2.56×10^{2}	2.30×10^{2}	2.04×10^{1}	3.17×10^{2}	3.92×10^{2}	3.54×10^{2}	1.72×10^{1}
MVMO	1.01×10^{2}	2.01×10^{2}	1.94×10^{2}	2.25×10^{1}	2.00×10^{2}	2.71×10^{2}	2.51×10^{2}	9.39×10^{0}	3.00×10^{2}	3.50×10^{2}	3.26×10^{2}	1.13×10^{1}
IPOG-MVMO	**7.09×10^{0}**	**7.57×10^{0}**	**7.37×10^{0}**	**4.88×10^{-2}**	2.09×10^{2}	2.75×10^{2}	2.55×10^{2}	2.84×10^{1}	2.36×10^{2}	2.77×10^{2}	2.57×10^{2}	8.88×10^{1}
F26												
Algorithms	$D = 10$				$D = 30$				$D = 50$			
	Best	Worst	Mean	Std.	Best	Worst	Mean	Std.	Best	Worst	Mean	Std.
NBIPOPaCMA	4.04×10^{0}	4.29×10^{0}	4.14×10^{0}	1.41×10^{-1}	1.55×10^{2}	3.26×10^{2}	2.42×10^{2}	7.92×10^{1}	1.95×10^{2}	4.84×10^{2}	3.43×10^{2}	1.42×10^{2}
PVADE	1.04×10^{2}	2.00×10^{2}	1.84×10^{2}	3.33×10^{1}	2.00×10^{2}	3.11×10^{2}	2.18×10^{2}	4.01×10^{1}	2.00×10^{2}	3.96×10^{2}	3.47×10^{2}	6.01×10^{1}
MVMO	1.02×10^{2}	1.15×10^{2}	1.08×10^{2}	3.52×10^{0}	2.00×10^{2}	2.00×10^{2}	2.00×10^{2}	4.47×10^{-3}	2.00×10^{2}	2.00×10^{2}	2.00×10^{2}	3.21×10^{-3}
IPOG-MVMO	**3.94×10^{0}**	**4.46×10^{0}**	**4.12×10^{0}**	**1.29×10^{-1}**	1.89×10^{2}	3.11×10^{2}	2.12×10^{2}	2.29×10^{1}	**1.94×10^{2}**	**2.96×10^{2}**	**2.12×10^{2}**	**2.29×10^{1}**
F27												
Algorithms	$D = 10$				$D = 30$				$D = 50$			
	Best	Worst	Mean	Std.	Best	Worst	Mean	Std.	Best	Worst	Mean	Std.
NBIPOPaCMA	1.13×10^{1}	1.26×10^{1}	1.22×10^{1}	5.70×10^{-1}	4.00×10^{2}	5.85×10^{2}	4.75×10^{2}	6.75×10^{1}	4.00×10^{2}	2.15×10^{3}	1.48×10^{3}	9.01×10^{2}
PVADE	2.95×10^{2}	5.35×10^{2}	3.64×10^{2}	8.98×10^{1}	3.06×10^{2}	3.61×10^{2}	3.26×10^{2}	1.14×10^{1}	8.27×10^{2}	1.36×10^{3}	1.11×10^{3}	1.85×10^{2}
MVMO	2.48×10^{2}	2.88×10^{2}	2.60×10^{2}	4.03×10^{0}	3.43×10^{2}	6.89×10^{2}	4.74×10^{2}	9.19×10^{1}	9.09×10^{2}	1.31×10^{3}	1.08×10^{3}	1.10×10^{2}
IPOG-MVMO	**5.75×10^{0}**	**1.11×10^{1}**	**7.34×10^{0}**	**2.45×10^{-1}**	3.95×10^{2}	5.17×10^{2}	4.34×10^{2}	2.45×10^{1}	5.95×10^{2}	1.17×10^{3}	7.34×10^{2}	2.45×10^{2}
F28												
Algorithms	$D = 10$				$D = 30$				$D = 50$			
	Best	Worst	Mean	Std.	Best	Worst	Mean	Std.	Best	Worst	Mean	Std.
NBIPOPaCMA	3.60×10^{0}	1.11×10^{1}	9.63×10^{0}	3.31×10^{0}	3.00×10^{2}	3.00×10^{2}	3.00×10^{2}	0	4.00×10^{2}	4.00×10^{2}	4.00×10^{2}	0
PVADE	8.45×10^{1}	3.01×10^{2}	2.38×10^{2}	4.52×10^{1}	3.00×10^{2}	3.00×10^{2}	3.00×10^{2}	2.23×10^{-5}	4.00×10^{2}	3.54×10^{5}	4.64×10^{2}	4.39×10^{2}
MVMO	9.87×10^{1}	2.90×10^{2}	1.88×10^{2}	1.00×10^{2}	3.00×10^{2}	3.00×10^{2}	3.00×10^{2}	5.36×10^{-4}	4.00×10^{2}	4.00×10^{2}	4.00×10^{2}	1.29×10^{-2}
IPOG-MVMO	**3.00×10^{0}**	**8.72×10^{0}**	**4.49×10^{0}**	**2.37×10^{-1}**	**3.00×10^{2}**	**3.00×10^{2}**	**3.00×10^{2}**	**2.37×10^{-5}**	**4.00×10^{2}**	**4.00×10^{2}**	**4.00×10^{2}**	**1.13×10^{-2}**

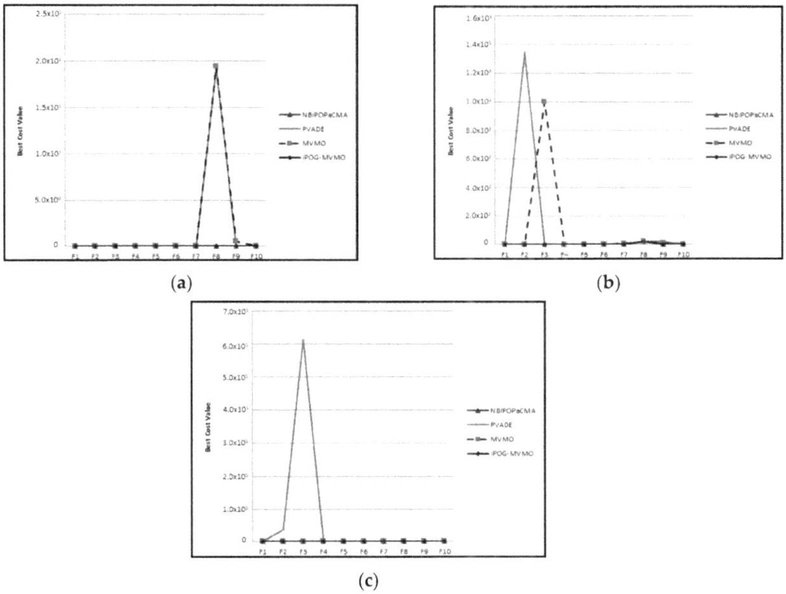

Figure 1. Comparison of the best cost value for F1–F10 at (**a**) 10 dimensions, (**b**) 30 dimensions, and (**c**) 50 dimensions.

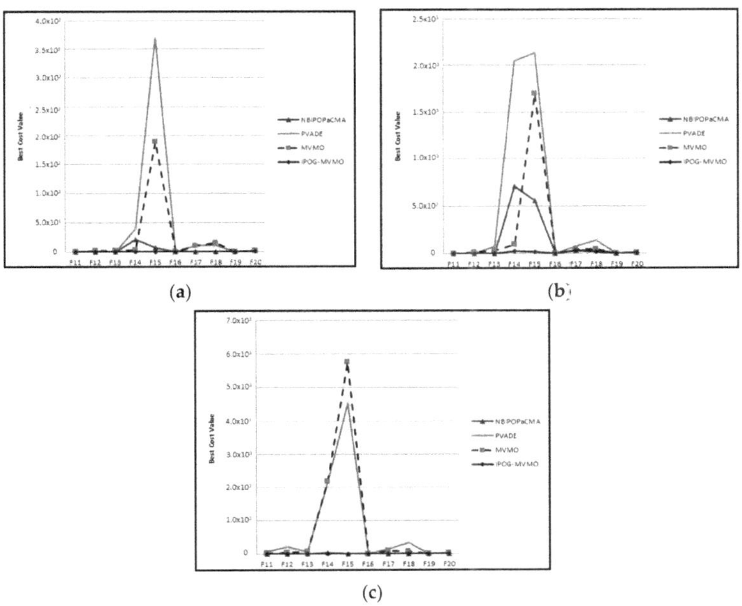

Figure 2. Comparison of the best cost value for F11–F20 at (**a**) 10 dimensions, (**b**) 30 dimensions, and (**c**) 50 dimensions.

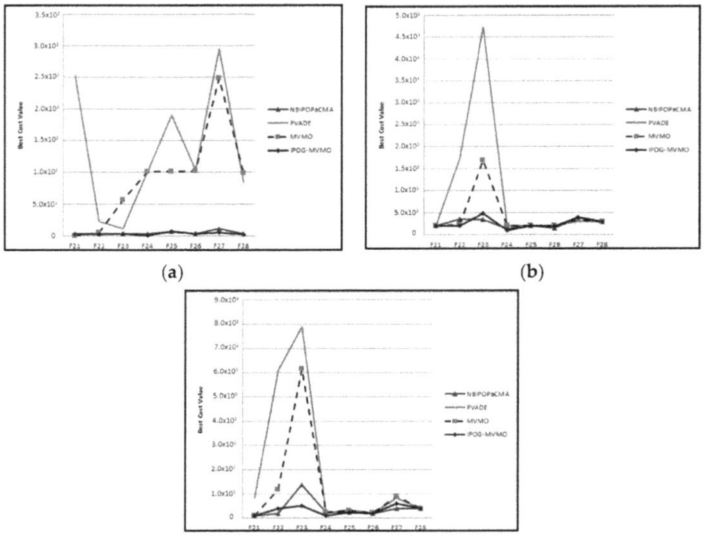

Figure 3. Comparison of the best cost value for F21–F28 at (**a**) 10 dimensions, (**b**) 30 dimensions, and (**c**) 50 dimensions.

To solve unimodal functions F1–F5, the proposed algorithm was able to search the solution without error in every case except F3 (D = 50), where NBIPOPaCMA yielded the better solution, as is shown in Table 2. Moreover, when considering the mean of the error, IPOG-MVMO yielded a zero mean in 10 out of 15 cases. In other words, in these 10 cases, IPOG-MVMO guaranteed the best solution for every trial in every case. It seems that IPOG-MVMO is a suitable technique for solving unimodal problems.

The multimodal functions F6–F20 were analyzed in three groups based on their complexity. To solve multimodal functions in the D = 10 group with the lowest complexity, IPOG-MVMO attained results with no error, except for F14, F15, and F20. Nevertheless, these results were still satisfactory when compared with other algorithms. For the D = 30 group, IPOG-MVMO performed worse than the other algorithms only in two cases, which were F8 and F16. Moreover, IPOG-MVMO also attained error-free results in F6 and F10–F13. For the D = 50 group with the highest complexity, IPOG-MVMO yielded lower performance than the other algorithms in only five cases, which were F8, F9, F16, F17, and F20.

To solve the composition functions F21–F28, the proposed algorithm also outperformed the other algorithms in all cases of D = 10. The algorithm was able to attain better results than the other algorithms in F22 and F24 for D = 30, and in F23, F24, and F26 for D = 50, respectively. Note that this type of function could not provide zero errors in all cases due to its complexity.

The superior results for each problem are highlighted in bold where the best value of IPOG-MVMO was obtained and was better than the other methods; although sometimes, more than one method provided the best value. Clearly, among the 28 × 3 cases, there were 68 cases in which IPOG-MVMO yielded the best result. Likewise, when considering the mean value, IPOG-MVMO yielded the best result in 59 out of the 28 × 3 cases. The results indicated that applying IPM and OGM to MVMO provided a solution that was closer to the optimal solution.

Some examples for the performance comparison were selected, as presented in Figure 4. We also present the average error with respect to the number of iterations (NP) in Figure 4. The average error was computed over 50 trials. The results showed that applying IPM and

OGM to MVMO can accelerate the search process to achieve a solution that was closer to the optimal solution within a smaller number of iterations.

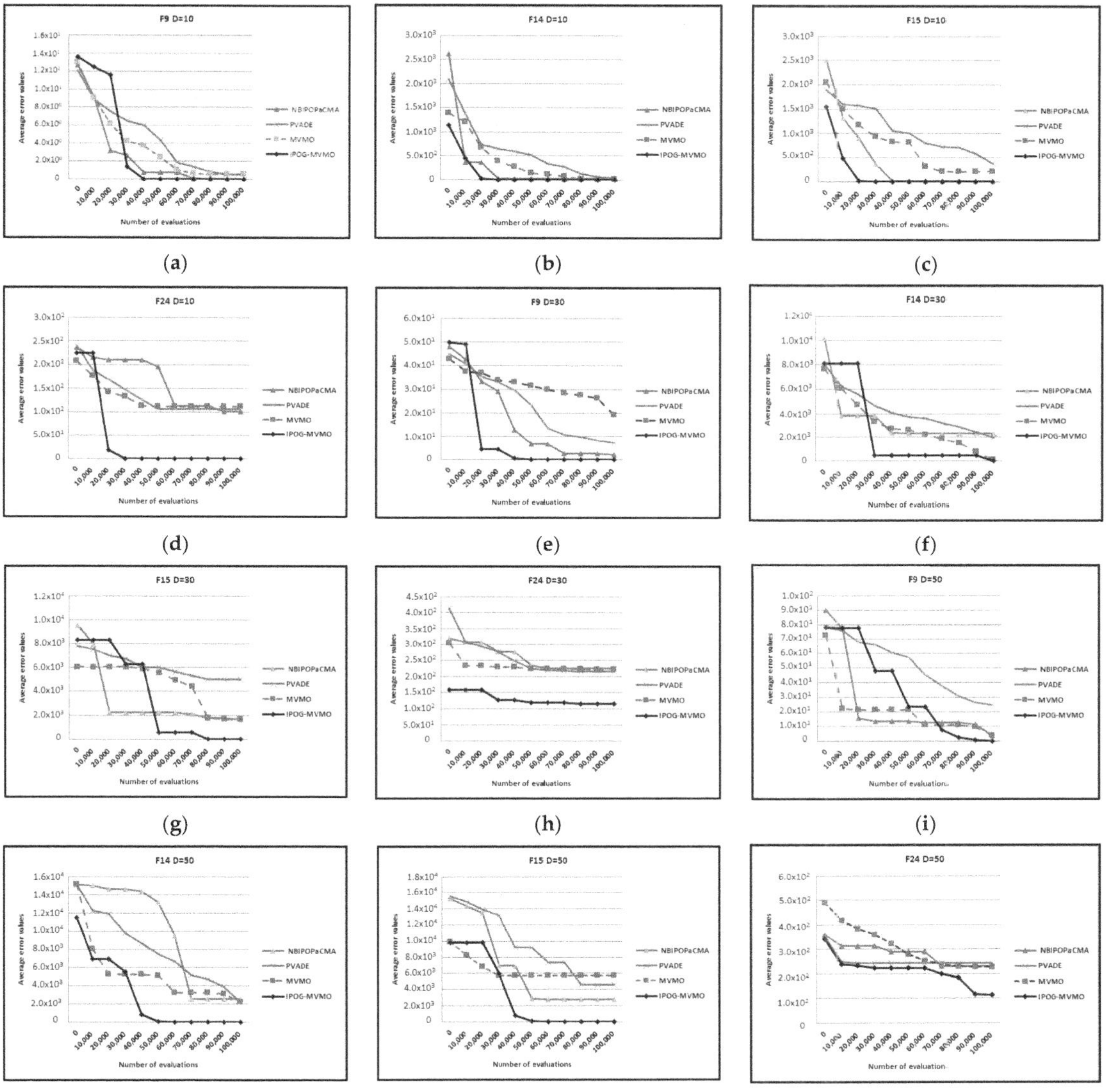

Figure 4. Examples of the performance comparison among IPOG-MVMO, NBIPOPaCMA, PVADE, and MVMO. F9, F14, F15, and F24 at 10, 30, and 50 dimensions were selected. The vertical axis represents the averaged error, while the horizontal axis represents the number of iterations. All were averaged over 50 trials. (**a**–**d**) F9, F14, F15, and F24 with $D = 10$; (**e**–**h**) F9, F14, F15, and F24 with $D = 30$; and (**i**–**l**) F9, F14, F15, and F24 with $D = 50$, respectively.

4. Discussion

4.1. Optimal Solutions

Twenty-eight different functions of three types were taken to verify our proposed method. The experimental results in terms of error are shown in Tables 2–8. When considering the best solution out of all the methods, it was found that IPOG-MVMO achieved the best result in 68 cases out of 84 cases, or 80.95%. Next, three types of functions were further analyzed separately. For the unimodal functions F1–F5, the proposed method offered the best result in 14 out of 15 cases, or 93.33%. This is quite acceptable when applying our method to this type of function. In the case of the multimodal functions F6–F20, there were 45 cases in total. Among these cases, IPOG-MVMO yielded the best result in 37 cases, or 82.22%. This was slightly lower than for the unimodal function due to the complexity of the problem. The remaining functions were the composition functions, F21–F28. The proposed method yielded the best result in 17 out of 24 cases, or 70.83%, which was lower than for unimodal and multimodal functions. Subsequently, the results can be discussed from a different perspective, where we mainly focused on the number of dimensions. There were 28 cases for each dimension and our method yielded the best result in 27, 22, and 19 cases, or 96.43%, 78.57%, and 67.86%, respectively.

In essence, zero error indicates whether the algorithm can reach the exact solution within the pre-defined number of iterations. IPOG-MVMO provided zero error in 33 out of 84 cases. Table 9 presents the likelihood that IPOG-MVMO and four comparative methods could reach the exact solution in terms of function type and the number of dimensions. YYPO, a method inspired by the philosophy of creating a balance between two concepts, was also conducted and is included in this table. For unimodal functions, there was only one case at $D = 50$ in which the error was non-zero, whereas the percentage dropped when the number of dimensions was higher in multimodal functions. Moreover, NBIPOPaCMA presented a higher likelihood of obtaining zero error at $D = 50$; however, IPOG-MVMO still provided the best results at $D = 10$ and $D = 30$. For the composition function, the best solution over 50 trials did not reach the exact solution for any algorithm. Since the function type and the number of dimensions corresponded to the problem difficulty and complexity, it can be inferred that although our IPOG-MVMO method outperformed the other comparative methods, the performance gradually decreased as the problem difficulty and complexity rose. However, using different parameter settings might result in greater performance for solving black box problems.

Table 9. Likelihood of obtaining zero error in 50 trials.

Algorithms	Unimodal Functions			Multimodal Functions			Composition Functions		
	D	D	D	D	D	D	D	D	D
NBIPOPaCMA	100	80	60	33.33	13.33	20	0	0	0
PVADE	100	60	20	26.67	6.67	0	0	0	0
MVMO	100	20	40	13.33	6.67	13.33	0	0	0
YYPO	40	40	20	6.67	6.67	6.67	0	0	0
IPOG-MVMO	100	100	80	80	33.33	13.33	0	0	0

When considering the mean value of error in Tables 2–8, it was found that IPOG-MVMO yielded the best result in 59 out of 84 cases, or 70.24%. Typically, the zero mean rarely occurs because the method must reach the exact solution in every trial. Our proposed method achieved zero mean in 13 cases, although most were unimodal functions. According to the analysis of the function type, IPOG-MVMO yielded the best result in 12 cases, or 80%, for the unimodal functions, 34 cases, or 75.56%, for the multimodal functions, and 13 cases, or 54.17%, for the composition functions. It can be seen that the type of function also affected the performance in terms of the mean value. In the case of the number of dimensions, IPOG-MVMO yielded the best result in 25 cases, or 89.29%, 18 cases, or 64.29%, and 16 cases, or 57.14%, in 10, 30, and 50 dimensions, respectively. The number of dimensions was another factor that impacted our algorithm.

Finally, the worst value of error presented a solution to be expected over a specific number of trials. IPOG-MVMO achieved the best result in 48 cases, or 57.14%, which was higher than the other methods. Again, the method still worked well in 11 cases, or 73.33%, of unimodal functions, while it yielded the best result in 29 cases, or 64.44%, and eight cases, or 33.33%, in multimodal and composition functions, respectively. In terms of the number of dimensions, our method achieved the best result in 20 out of 28 cases in 10 dimensions. For the cases with 30 dimensions, our method yielded the best result in 12 cases, or 42.86%, and the percentage became higher when the number of dimensions was 50, with a 57.14% rate of obtaining the best result. More parameter settings and experiments are required to identify the relationship between the number of dimensions and the optimal solutions in the worst-case scenario.

4.2. Stability

The standard deviation of error can be used to measure the stability of the proposed method compared with the other methods. Zero standard deviation means that the method can provide the same solution over a limited number of trials. Under the experiments, IPOG-MVMO obtained the best result in 46 out of 84 cases, or 54.76%, and there were 12 cases of zero standard deviation, 10 of which were uniform functions. Based on the function type, the unimodal function was the type with the highest stability, and the best result was obtained in 11 out of 15 cases, whereas the multimodal functions yielded the best result in 26 out of 45 cases. Unfortunately, the composition functions method yielded the best result only in nine out of 24 cases, which was greater than the other methods, but most of the cases were in 10 dimensions. For the number of dimensions, the method yielded the best result in 20 cases (71.43%), 11 cases (39.29%), and 15 cases (53.57%), in 10, 30, and 50 dimensions, respectively. Without taking the composition functions into consideration, it seems that the number of dimensions had a minor impact on the stability of the method, because, as previously mentioned, the composition method worked well only in 10 dimensions.

5. Conclusions

In this study, IPOG-MVMO, a combination of the interior point method (IPM), the opposite gradient method (OGM), and mean-variance mapping optimization (MVMO), was proposed to identify a solution for the continuous real-world optimization problem. IPM and OGM were combined with the original MVMO in the area of an IPM local search and then a new population was created using the opposite gradient concept. This procedure ensured that the newly created population was located somewhere close to the minimum value so that a better solution could be obtained using MVMO. For experiments, creating a new population close to the minimum value to obtain an accurate and fast solution by using the local search strength of IPM and OGM's solution convergence was a key step in this study. When these two approaches were combined with traditional MVMO, the best cost was achieved, which was better than using a single technique, i.e., MVMO. It was important to generate a new population near the minimum in the optimization problem, which can be applied to optimization problems in many areas, such as energy saving for tissue paper mills through energy efficiency scheduling, and the optimal scheduling of energy and ancillary services. Future research will focus on the hybridization of IPOG-MVMO and other techniques, and improving the algorithms to ensure compatibility with combinatorial real-world problems.

Author Contributions: Conceptualization, T.S., S.P. and C.L.; methodology, T.S., S.P. and C.L.; software, T.S.; validation, T.S. and S.P.; formal analysis, T.S. and S.P.; investigation, S.P.; resources, T.S. and C.L.; data curation, T.S.; writing—original draft preparation, T.S.; writing—review and editing, S.P.; visualization, T.S.; supervision, S.P. and C.L.; project administration, T.S. and S.P.; funding acquisition, T.S. All authors have read and agreed to the published version of the manuscript.

Funding: This research was funded by the Officer of the Higher Education Commission, Ministry of Education, Thailand, through Silpakorn University under Grant Number 5/2555.

Institutional Review Board Statement: Not applicable.

Informed Consent Statement: Not applicable.

Data Availability Statement: The data presented in this study are available in [31].

Acknowledgments: The authors would like to gratefully thank the Faculty of Information and Communication Technology, Silpakorn University, and the Officer of the Higher Education Commission, Ministry of Education, Thailand, for supporting this work through a grant funded under the Higher Education Research Promotion and National Research University Project of Thailand.

Conflicts of Interest: The authors declare no conflict of interest. The funders had no role in the design of the study; the collection, analyses, or interpretation of data; the writing of the manuscript; or the decision to publish the results.

References

1. Zeng, Z.; Chen, X.; Wang, K. Energy saving for tissue paper mills by energy-efficiency scheduling under time-of-use electricity tariffs. *Processes* **2021**, *9*, 274. [CrossRef]
2. Sortomme, E.; El-Sharkawi, M.A. Optimal scheduling of vehicle-to-grid energy and ancillary services. *IEEE Trans. Smart Grid* **2012**, *3*, 351–359. [CrossRef]
3. Azhir, E.; Jafari Navimipour, N.; Hosseinzadeh, M.; Sharifi, A.; Darwesh, A. Deterministic and non-deterministic query optimization techniques in the cloud computing. *Concurr. Comput. Pract. Exp.* **2019**, *31*, e5240. [CrossRef]
4. Foroozandeh, Z.; Ramos, S.; Soares, J.; Vale, Z. Energy management in smart building by a multi-objective optimization model and pascoletti-serafini scalarization approach. *Processes* **2021**, *9*, 257. [CrossRef]
5. Belfiore, N.P.; Rudas, I.J. Applications of computational intelligence to mechanical engineering. In Proceedings of the IEEE 15th International Symposium on Computational Intelligence and Informatics (CINTI), Budapest, Hungary, 9–21 November 2014.
6. Hansen, N.; Sibylle, D.M.; Koumoutsakos, P. Reducing the time complexity of the derandomized evolution strategy with covariance matrix adaptation (CMA-ES). *Evol. Comput.* **2003**, *11*, 1–18. [CrossRef] [PubMed]
7. Auger, A.; Hansen, N. A restart CMA evolution strategy with increasing population size. In Proceedings of the IEEE Congress on Evolutionary Computation (CEC), Edinburgh, Scotland, UK, 2–5 September 2005.
8. Hansen, N. Benchmarking a BI-population CMA-ES on the BBOB-2009 function tested workshop. In Proceedings of the GECCO Genetic and Evolutionary Computation Conference, Montreal, QC, Canada, 8–12 July 2009.
9. Loshchilov, I. CMA-ES with restarts for solving CEC 2013 benchmark problems. In Proceedings of the IEEE Congress on Evolutionary Computation (CEC), Cancun, Mexico, 20–23 June 2013.
10. Erlich, I.; Venayagamoorthy, G.K.; Nakawiro, W. A mean-variance optimization algorithm. In Proceedings of the 2010 IEEE World Congress on Computational Intelligence, Barcelona, Spain, 18–23 July 2010.
11. Salazar, E.; Herrera, M.; Camacho, O. An Application of MVMO Based Adaptive PID Controller for Process with Variable Delay. In *Systems and Information Sciences*; Botto-Tobar, M., Zamora, W., Larrea Plúa, J., Bazurto Roldan, J., Santamaría Philco, A., Eds.; Springer: Cham, Switzerland, 2021; Volume 1273, pp. 1–6. [CrossRef]
12. Rueda, J.L.; Erlich, I. Short-term transmission expansion planning by using swarm mean-variance mapping optimization. In Proceedings of the 17th International Conference on Intelligent System Applications to Power Systems, Tokyo, Japan, 1–4 July 2013.
13. Rueda, J.L.; Erlich, I. Hybrid mean-variance mapping optimization for solving the IEEE-CEC 2013 competition problems. In Proceedings of the IEEE Congress on Evolutionary Computation (CEC), Cancun, Mexico, 20–23 June 2013.
14. Khoa, T.H.; Vasant, P.M.; Singh, M.S.B.; Dieu, V.N. Swarm based mean-variance mapping optimization for solving economic dispatch with cubic fuel cost function. *Lect. Notes Comput. Sci* **2015**, *9012*, 3–12. [CrossRef]
15. Mori, H.; Ikegami, H. An efficient MVMO-SH method for optimal capacitor allocation in electric power distribution systems. In *Advances in Swarm Intelligence*; Tan, Y., Takagi, H., Shi, Y., Niu, B., Eds.; Lecture Notes in Computer Science; Springer: Cham, Switzerland, 2017; Volume 10386, pp. 1–10. [CrossRef]
16. Chen, X.; Li, Y.; Zhang, Y.; Ye, X.; Xiong, X.; Zhang, F. A novel hybrid model based on an improved seagull optimization algorithm for short-term wind speed forecasting. *Processes* **2021**, *9*, 387. [CrossRef]
17. Sun, X.; Fan, Z.; Ji, Y.; Wang, S.; Yan, S.; Wu, S.; Fu, Q.; Ghazali, K.H. Robust multi-user detection based on hybrid gray wolf optimization. *Concurr. Comput. Pract. Exp.* **2021**, *33*, e5273. [CrossRef]
18. Valenta, D.; Langer, M. On numerical 2D P colonies modelling the gray wolf optimization algorithm. *Processes* **2021**, *9*, 330. [CrossRef]
19. Punnathanam, V.; Kotecha, P. Yin-yang-pair optimization: A novel lightweight optimization algorithm. *Eng. Appl. Artif. Intell.* **2016**, *54*, 62–79. [CrossRef]
20. Saenphon, T.; Lursinsap, C. Fast evolutionary solution finding for optimization using opposite gradient movement. In Proceedings of the International Conference on Natural Computation (ICNC), Shanghai, China, 26–28 July 2011.

21. Saenphon, T.; Phimoltares, S.; Lursinsap, C. Combining new fast opposite gradient search with ant colony optimization for solving travelling salesman problem. *Eng. Appl. Artif. Intell* **2014**, *35*, 324–334. [CrossRef]
22. Narendra, K.K. A new polynomial-time algorithm for linear programming. *Combinatorica* **1984**, *4*, 373–395. [CrossRef]
23. Ohmori, S.; Yoshimoto, K. A Primal-Dual Interior-Point Method for Facility Layout Problem with Relative-Positioning Constraints. *Algorithms* **2021**, *14*, 60. [CrossRef]
24. Boyd, S.; Vandenberghe, L. *Convex Optimization*; Cambridge University Press: Cambridge, UK, 2004; pp. 561–620. [CrossRef]
25. Rueda, J.L.; Erlich, I. Evaluation of the mean-variance mapping optimization for solving multimodal problems. In Proceedings of the IEEE Symposium Series on Computational Intelligence, Singapore, 16–19 April 2013.
26. Cepeda, J.C.; Rueda, J.L.; Erlich, I.; Korai, A.W.; Gonzalez-Longatt, F.M. Mean-variance mapping optimization algorithm for power system applications in DIgSILENT power factory. In *PowerFactory Applications for Power System Analysis*; Springer: Cham, Switzerland, 2014; pp. 267–296. [CrossRef]
27. Coelho, L.D.; Ayala, H.V.H.; Freire, R.Z. Population's variance-based adaptive differential evolution for real parameter optimization. In Proceedings of the IEEE Congress on Evolutionary Computation (CEC), Cancun, Mexico, 20–23 June 2013.
28. Tianjun, L.; Stutzle, T. Benchmark results for a simple hybrid algorithm on the CEC 2013 benchmark set for real-parameter optimization. In Proceedings of the IEEE Congress on Evolutionary Computation (CEC), Cancun, Mexico, 20–23 June 2013.
29. Liang, J.; Qin, A.K.; Suganthan, P.N.; Baskar, S. Comprehensive learning particle swarm optimizer for global optimization of multimodal functions. *IEEE Trans. Evol. Comput.* **2006**, *10*, 281–295. [CrossRef]
30. Liang, J.; Qu, B.Y.; Suganthan, P.N.; Hernández-Díaz, A.G. *Problem Definitions and Evaluation Criteria for the CEC 2013 Special Session and Competition on Real-Parameter Optimization*; Computational Intelligence Laboratory, Zhengzhou University: Zhengzhou, China; Nanyang Technological University: Singapore, 2013.
31. Liang, J.J.; Qu, B.Y.; Suganthan, P.N. *Zhengzhou China and Technical Report*; Nanyang Technological University: Singapore, 2013.

processes

Article

Experimental Evaluation of Chemical Reactions Involved in Ultrasonic-Assisted Absorption of Bulk CO_2

Fatemeh Shokrollahi [1], Kok Keong Lau [1,*] and Behzad Partoon [2]

[1] CO2 Research Center (CO2RES), Department of Chemical Engineering, Universiti Teknologi PETRONAS, Seri Iskandar 32610, Perak, Malaysia; fatemeh_g03717@utp.edu.my

[2] Biological and Chemical Engineering Department, Faculty of Technical Science, Aarhus University, Nørreborgade 44, 8000 Aarhus, Denmark; behzad.partoon@bce.au.dk

* Correspondence: laukokkeong@utp.edu.my

Abstract: As the most mature natural gas sweetening process, absorption has always been improved to meet the separation requirement. Recently, ultrasonic irradiation has been proposed as a technique that can intensify CO_2 absorption. However, further studies are still required, particularly focusing on the sonochemical effect. Since the influence of the sonochemical effect on the reaction pathway is still debatable, attention must be given to verifying the influence of ultrasonic irradiation on the chemical reactions of CO_2 absorption. Hence, this work aims to evaluate the influence of $\dot{O}H$ radicals generated by the sonochemical effect on the chemical reactions involved during CO_2 absorption using promoter-free methyldiethanolamine (MDEA). For the evaluation, various samples under irradiated and non-irradiated conditions are analyzed using the HPLC characterization technique. The results show that the hypothesis of changing the reaction pathway due to the presence of the sonochemical effect is invalid. However, it can accelerate the generation of hydroxyl radicals ($\dot{O}H$) via water sonolysis. Thus, the origin of sonochemistry in aqueous solutions is defined as water sonolysis. The analysis of the CO_2 absorption rate also demonstrates the presence of accelerated chemical reactions (contributed by the $\dot{O}H$ radicals), which could potentially make the slow kinetic MDEA more practical for industrial application.

Keywords: absorption; ultrasonic irradiation; sonochemical effect; HPLC characterization; reaction pathway

Citation: Shokrollahi, F.; Lau, K.K.; Partoon, B. Experimental Evaluation of Chemical Reactions Involved in Ultrasonic-Assisted Absorption of Bulk CO_2. *Processes* **2023**, *11*, 3266. https://doi.org/10.3390/pr11123266

Academic Editor: Vladimir S. Arutyunov

Received: 28 October 2023
Revised: 12 November 2023
Accepted: 16 November 2023
Published: 22 November 2023

1. Introduction

The demand for natural gas is predicted to rise by more than 60% between now and 2040. Depending on the geological condition of the reservoir, the raw natural gas has a wide range of gases, including methane, heavier hydrocarbons such as ethane, propane, isobutene, normal butane, and a substantial amount of impurities such as CO_2, H_2S, and CO [1]. Removing acidic impurities from natural gas, especially CO_2, is a vital step toward further utilization of this energy source. Such CO_2 elimination in both on/offshore natural gas wells can be achieved by developing carbon capture and storage processes that permit the mitigation and utilization of CO_2.

Currently, absorption, especially chemical absorption, is the most developed separation process widely used to capture CO_2 from raw natural gas [2]. Nevertheless, this process still suffers from various drawbacks, including high energy consumption for solvent regeneration, solvent loss, equipment corrosion, and large column sizes [3,4]. These disadvantages pose various technical and financial challenges to the CO_2 absorption process, necessitating the search for intensification techniques. Such process intensification techniques can reduce the size of conventional absorption columns, decrease the equipment capital cost, and make CO_2 absorption more feasible. Available techniques to intensify the mass transfer of the absorption process include packed bed columns (PB), rotating packed bed columns (RPB), hollow fiber membrane contactors (HFMC), bubble columns (BC), microchannel reactors (MC), and ultrasonic irradiation [5,6]. Among them, ultrasonic irradiation is a novel technique requiring extensive investigations before industrial scaleup.

In general, ultrasonic irradiation is expressed as sound waves with frequencies higher than the upper limit of human hearing ability, ranging from 20 kHz to 500 MHz. Depending on the applied frequency and the presence of the sonophysical and sonochemical effects, this technique has the potential to be used in different research areas, including food technology [7], cleaning [8], medical treatment [9], crude oil upgrading [10], synthesis of nanomaterials [11,12], and enhancement of multiphase reactions [13,14]. However, using high-frequency ultrasonic irradiation in the natural gas purification field is limited, particularly as a potential intensification technique for CO_2 absorption. The concept of using ultrasonic irradiation in the CO_2 absorption process was first proved by Tay et al. (2017) [15–17]. They investigated the effect of high-frequency ultrasonic irradiation on the CO_2 absorption performance of a batch lab-scale reactor. Three different solvents were tested, including water, monoethanolamine (MEA), and potassium carbonate (K_2CO_3). The results showed a remarkable rise in the CO_2 absorption rate by utilizing ultrasonic irradiation with a frequency of 1.7 MHz. They stated that ultrasonic irradiation can improve CO_2 absorption performance because of the sonochemical and sonophysical effects. However, particular attention was given to the sonophysical effects. After proving the concept, the focus switched to the continuous mode. Hence, Yusof et al. (2019) developed a novel continuous high-frequency ultrasonic contactor system for the CO_2 absorption process [18]. They investigated the effect of different operating parameters on the ultrasonic-assisted CO_2 absorption from natural gas using water. They mentioned that the continuous ultrasonic contactor system for CO_2 absorption performed better than other contactors. According to their results, the continuous ultrasonic contactor system showed the highest mass transfer coefficient (0.037 s^{-1}) between the other contactors, including the bubble column, stirred vessel, and counter-current packed column. They did not, however, address the significance of sonophysical and sonochemical effects on such improvements [18]. To suit the industrial mode, they also evaluated the performance of the proposed continuous system with MEA. The influence of various operating parameters on the overall mass transfer coefficient was evaluated. They reported that at the optimized conditions for the CO_2-MEA system, the high-frequency ultrasonic absorption system could intensify the mass transfer up to 12 times compared to the conventional packed bed column. They concluded that the sonophysical effects are mainly responsible for the enhanced CO_2 absorption process in the continuous high-frequency ultrasonic reactor [19]. The effects of different absorbents, including diethanolamine (DEA), sodium hydroxide (NaOH), and sodium carbonate (Na_2CO_3), on ultrasound-assisted CO_2 absorption were investigated in the study conducted by Marjanian et al. (2021) [20]. They stated that the extent of improving CO_2 absorption performance in the ultrasonic-assisted absorption system depends on the absorbent type.

As a way forward to maturing this technique, more fundamental research on CO_2 absorption performance utilizing other commercial chemical absorbents and focusing on sonochemical effects was required. Therefore, in our previous research [21], attention was given to MDEA as a chemical absorbent with a high inherent CO_2 absorption capacity but slow absorption kinetics with CO_2. The influence of different operating parameters on CO_2 absorption performance in an ultrasonic-assisted batch reactor was evaluated using MDEA in single and blended forms. The potential of the ultrasonic-assisted reactor was then further elucidated through a comparative analysis with magnetic stirring and non-irradiated conditions. The findings demonstrated that the CO_2 absorption rate in the ultrasonic-assisted reactor was approximately 28 and 52 times higher than in the stirred reactor and the non-irradiated condition, respectively, when 50% promoted-free MDEA was utilized. Even with the addition of 5% piperazine (PZ), the ultrasonic-assisted CO_2 absorption system outperformed other systems. Compared to the magnetic stirred reactor, the CO_2 absorption rate in this case was 17 times higher. Furthermore, compared to the non-irradiated condition, the CO_2 absorption rate was 54 times higher. Both sonochemical and sonophysical effects were responsible for the increase in the CO_2 absorption rate [21]. However, as the role of the sonochemical effect on the improvement of the CO_2 absorption process in the ultrasonic-assisted reactor was not yet elucidated, our recent study aimed to parametrically quantify the sonochemical effect during CO_2 absorption in the ultrasonic-assisted reactor using promoter-free MDEA. A novel indirect method was implemented to identify $\dot{O}H$ radicals as the representative of the sonochemical effect. Terephthalic acid (TA) was selected as a chemical probe for detecting and scavenging $\dot{O}H$ radicals during the CO_2 absorption using promoter-free MDEA. Subsequently, the quantification was performed via scavenging of $\dot{O}H$ radicals using TA. Despite the parametric quantification, the role of the sonochemical effect on the chemical reactions involved in CO_2 absorption has not yet been investigated.

2. Theory

In general, the CO_2 absorption process contains three key steps, including vapor-liquid mass transfer, liquid-liquid mass transfer, and liquid chemical reaction [22]. By applying high-frequency ultrasonic irradiation, all these steps can be influenced due to the presence of sonophysical and sonochemical effects. Figure 1 illustrates the sonophysical and sonochemical effects induced by high-frequency ultrasonic irradiation during the CO_2 absorption process.

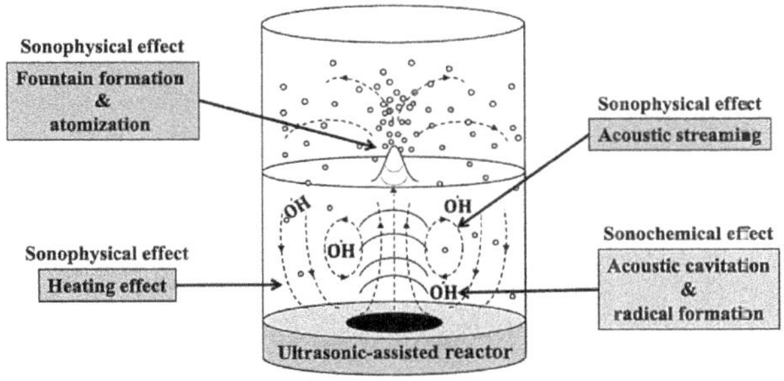

Figure 1. Different effects induced by high-frequency ultrasonic irradiation.

According to Figure 1, atomization, acoustic streaming, liquid fountain formation, and heating are the main sonophysical effects induced by high-frequency ultrasonic irradiation during the CO_2 absorption process. The sonophysical effects can improve mass transfer by adequately mixing, lowering mass transfer resistance in the liquid phase, and increasing the effective interfacial areas between the gas and liquid phases [15,16]. In contrast, the sonochemical effect is caused by a phenomenon known as acoustic cavitation [19]. Cavitation involves various steps. The three main steps of cavitation are formation (nucleation), bubble growth (expansion) during the cycles until reaching a critical size, and violent collapse in the liquid bulk. These steps are repeated consistently. In the formation step, if the ultrasonic intensity is high enough to surpass the liquid's tensile strength, there comes a point at which intermolecular forces can no longer hold the molecular structure together. Cavitation nuclei form in the liquid phase as a result of this point [23,24]. The formed microbubbles may grow rapidly if the ultrasonic intensity is sufficiently strong. This step is recognized as an expansion [25]. The implosion step happens when the cavitation bubble becomes so dense that it cannot continue absorbing energy to sustain itself. As a result, the surrounding liquid rushes into it, causing an implosion [23,26]. Each bubble functions as a core when implosion occurs, producing energy that increases the temperatures and pressure on a microscale. Due to inducing extremely high local temperature and pressure, the collapsing bubbles can create an unusual environment for chemical reactions by forming highly reactive radical species [23,27].

According to the hotspot theory, a chemical reaction happening in an ultrasonic-assisted system is a heterogeneous reaction in which reactive species are generated from the cavitation bubble. In this theory, three regions are anticipated for the incidence of chemical reactions, including (1) a hot gaseous nucleus or thermolytic zone, (2) an interfacial area having a radial temperature gradient and a local radical density, and (3) the solution bulk at ambient temperature. These three regions are shown in Figure 2.

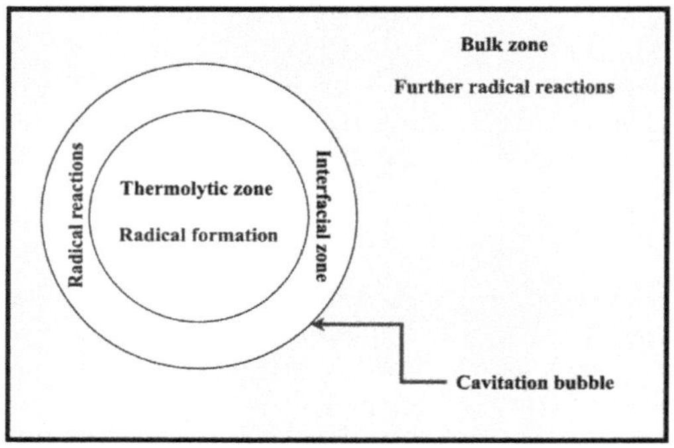

Figure 2. Reaction zones according to hotspot theory.

In the first region, which is a hot gaseous nucleus, the extreme conditions created by collapse trigger bond breakage of the presented vapors and gases within the bubble nucleus, leading to the generation of free radicals, particularly $O\dot{H}$ radicals. The formed $O\dot{H}$ radicals can either react with each other to form new molecules and radicals or diffuse into the liquid bulk [24]. The second region is the liquid shell around the imploding cavity, which is estimated to heat up to approximately 2000 K during implosion. Combustion and radical reactions (including $O\dot{H}$ generated from H2O decomposition) occur in this solvent layer around the hot bubble. In the liquid bulk, as the last region, there is no direct sonochemical activity, but subsequent reactions with ultrasonically produced intermediates may occur. Furthermore, a limited number of radicals generated in the cavities or at the interface may move into the bulk liquid phase and react with the substrate to form new products [24,28]. Hence, there might be some differences in the reaction pathways in the ultrasonic-assisted system compared to the conventional system. Due to the generation of $O\dot{H}$ radicals, there is a hypothesis that the sonochemical effect is capable of speeding up or changing the reaction pathway of the CO_2 absorption process in the high-frequency ultrasonic-assisted system. The generated $O\dot{H}$ radicals could be the cause of the ultrasonic-assisted system's improved performance, as summarized in Table 1. However, the role of the sonochemical effect on the chemical reaction of CO_2 absorption has not been discussed. Thus, further studies are substantial to investigate the role of the sonochemical effect on the reaction pathway during ultrasonic-assisted CO_2 absorption.

Table 1. Absorption rate comparison of the various MDEA-based CO_2 absorption techniques.

	Absorbent	MDEA	MDEA	MDEA	MDEA	MDEA	MDEA
	Promoter	MEA	DEA	DEA	PZ	PZ	-
	Target Gas	CO_2	CO_2	CO_2	CO_2	CO_2	CO_2
	Technique	WWC	WWC	WWC	DC	HFUA	HFUA
Operating conditions	C_{MDEA}	1.5 kmol·m^{-3}	1.5 kmol·m^{-3}	1.5 kmol·m^{-3}	2.636 kmol·m^{-3}	50 wt%	50 wt%
	$C_{Promoter}$	0.1 kmol·m^{-3}	0.1 kmol·m^{-3}	0.4 kmol·m^{-3}	0.364 kmol·m^{-3}	5 wt%	0 wt%
	T	40 °C	40 °C	40 °C	60 °C	70 °C	70 °C
	P	0.239 bar	0.335 bar	0.4 bar	0.8 bar	11 bar	11 bar
Absorption rate		0.12 mol·h^{-1}	0.15 mol·h^{-1}	0.25 mol·h^{-1}	0.12 mol·h^{-1}	15.23 mol·h^{-1}	4.20 mol·h^{-1}
Reference		[29]	[30]		[31]	[21]	

3. Methodology

3.1. Materials

The materials used in this work and the specifications provided by their suppliers are presented in Table 2.

Table 2. The used materials and their respective information.

Material	Formula	Supplier	Purity	Application
Carbon dioxide	CO_2	Air Products, Kuala Lumpur, Malaysia	99.99%	Target absorbing gas
Distilled water	H_2O	-	100%	Preparation of aqueous solutions
Methyldiethanolamine	$C_5H_{13}NO_2$	Revlogi Materials, Puchong, Malaysia	≥99%	Main chemical absorbent
Terephthalic acid	$C_8H_6O_4$	Sigma Aldrich, Burlington, MA, USA	≥99%	$\dot{O}H$ radical scavenger
Sodium hydroxide	NaOH	Sigma Aldrich, Burlington, MA, USA	≥99%	HPLC mobile phase
Disodium hydrogen phosphate anhydrous	Na_2HPO_4	Merck, Darmstadt, Germany	≥99%	HPLC mobile phase

3.2. Experimental Setup

The schematic of the ultrasonic-assisted CO_2 absorption system is shown in Figure 3. The reactor was fabricated using stainless steel, operating in the batch absorption configuration. The reactor consisted of a top cover and a cylindrical body. The volume of the reactor, inner diameter, and height were 250 mL, 5 cm, and 17.8, respectively. The temperature sensor, pressure sensor, gas inlet and outlet, and pH probe were on the top cover in five different ports. For direct contact with the solution, the transducer, which had a 1.5 cm diameter and a frequency of 1.7 MHz, was positioned at the bottom of the ultrasonic reactor. It should be noted that the transducer converts the electrical power into ultrasonic irradiation using the calorimetric method [32]. The procedure for conducting the CO_2 absorption experiments is comprehensively described in our previous research paper [21].

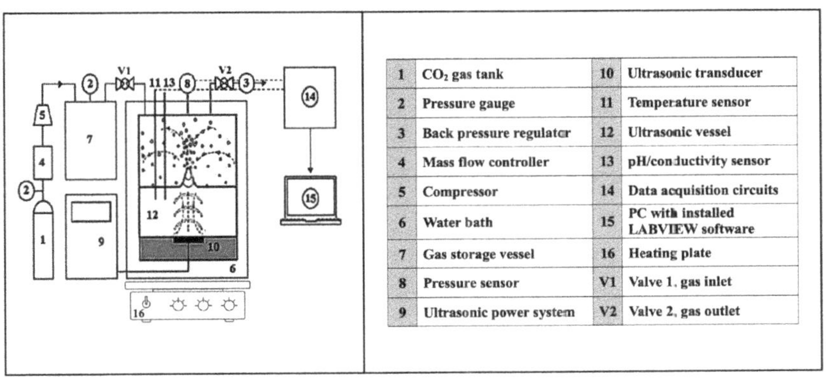

Figure 3. Schematic of the ultrasonic-assisted CO_2 absorption system.

3.3. Procedure of Reaction Pathway Evaluation

Various samples needed to be analyzed to provide sufficient analytical data for identifying the role of the sonochemical effect, in particular $\dot{O}H$ radicals, on the reaction pathway. The HPLC characterization technique was implemented for analysis. The analysis was performed by evaluating the significance of $\dot{O}H$ radicals during the CO_2 absorption process using promoter-free MDEA. In line with this aim, in-vitro detection of $O\dot{H}$ radicals was performed. TA was used as the $O\dot{H}$ radical scavenger. The optimum mass concentration of TA to be used in the subsequent experiments was 10 wt%. It should be noted that the optimum mass concentration of TA is defined as the concentration necessary to trap all the generated $O\dot{H}$ radicals from the reaction mixture [33,34], which can subsequently inhibit the CO_2 absorption process without affecting the reaction pattern. The samples were divided into four groups, with each group being tested three times under two different conditions, including irradiated and non-irradiated. The samples are listed in Table 3. For each sample, the HPLC chromatograms were analyzed to identify the stable chemical compounds formed under different conditions. The identification of chemical compounds was performed based on the standard verification method. Subsequently, the concentration of the identified chemical compounds was calculated based on the HPLC peak area and HPLC calibration curves. Finally, to verify the mechanism, the reaction pathway was evaluated by comparing concentration variations between irradiated and non-irradiated samples.

Table 3. List of samples subjected to HPLC analysis.

Group No.	Sample Description
A	50 wt% aqueous MDEA
B	50 wt% aqueous MDEA + 10 wt% TA
C	50 wt% aqueous MDEA + CO_2
D	50 wt% aqueous MDEA + 10 wt% TA + CO_2

3.4. *HPLC Equipment and Chromatographic Conditions*

As mentioned in Section 3.3, the analysis of the scavenged $O\dot{H}$ radicals were performed by the HPLC instrument (Model Agilent series 1100). A YMC-Pack Polymer C18 reverse phase column (250 mm × 6.0 mm ID) was selected. The column was carefully washed with a mixture of water and acetonitrile (60:40) 1 h before and 2 h after use. The eluent was a mixture of 100 mM Na_2HPO_4 and 100 mM NaOH (ratio 60:40 at a pH of 12). The eluent flow rate was 0.6 mL·min^{-1}, and the analysis was conducted at 20 °C. The injection volume of the sample was 20 μL, and the UV light at a wavelength of 215 nm was used as the detector. For calibration purposes, two sets of standards with different concentrations were freshly prepared. The first set of calibration standards consisted of an aqueous solution of MDEA with concentrations of 10, 20, 30, 40, and 50 wt%. Since TA is poorly soluble in water, the second set of standards consisted of various mass concentrations of TA dissolved in the 50% aqueous MDEA solution. TA concentrations were 1, 3, 5, 7, 9, and 11 wt%. The different sets of the prepared standards were then injected into the HPLC system. For each set of calibration standards, the relationship between the peak area of the compound and its concentration was evaluated. The calibration curve for each set of standards was then plotted. Subsequently, the coefficient of determination (R2) was obtained from the regression line. The calibration curves for MDEA and TA with the respective R2 values of 0.99 and 0.98 are presented in Figure 4.

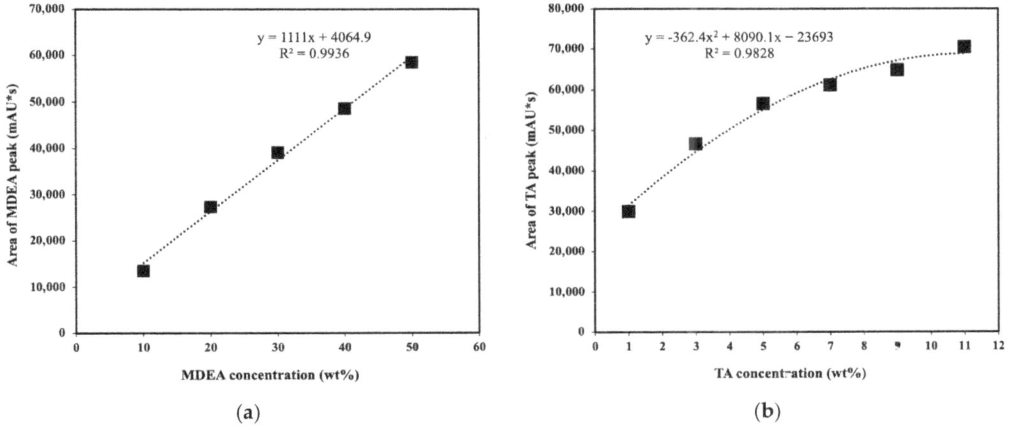

Figure 4. HPLC calibration curve and calibration equation for (**a**) MDEA, (**b**) TA.

4. Results and Discussion

As mentioned, four different groups of liquid-phase samples were chosen for analysis. Each group was examined under irradiated and non-irradiated conditions. Then the samples were characterized by the HPLC analytical technique with UV detection. According to the obtained HPLC data, stable chemical compounds formed under irradiated and non-irradiated conditions were identified. The final concentrations of the identified chemical compounds were calculated using the HPLC peak area and the calibration curves. The reaction pathway was finally evaluated by comparing the detected peaks, which represent the chemical compounds and their concentration variations between the two testing conditions. The results of the HPLC-UV characterization of the selected samples are as follows.

4.1. HPLC Analysis of Aqueous MDEA Solution

Two samples containing 50 wt% MDEA in the aqueous form were injected into the HPLC system. The ultrasonic irradiation was applied to the first sample (A1) but not to the second (A2). The ultrasonic power and the frequency were 12.36 W and 1.7 MHz, respectively. The chromatogram of these samples is shown in Figure 5. Firstly, the straight baseline was observed in both chromatograms, indicating the accuracy of the results. Moreover, one peak was observed in each chromatogram, representing the MDEA. The identity of the detected peak was validated by comparing it to previously tested MDEA standards. The obtained HPLC data is summarized in Table 4. The final concentrations of MDEA in both samples were determined based on the peak area and the HPLC calibration curve for MDEA, as shown in Figure 6. The same aqueous MDEA solution with the approximate initial concentration of 50 wt% was used for both samples. The results show insignificant changes in the final concentration and indicate the stability of MDEA molecules once irradiated by high-frequency ultrasonic irradiation. Such an insignificant rise in the concentration of the irradiated sample might be due to the possibility of minor evaporation of the water content of the aqueous solution during the high-frequency ultrasonic irradiation [35,36]. In particular, the evaporation is induced by the heating effect of ultrasonic irradiation. When the ultrasonic wave transmits to the liquid phase, some of the ultrasound energy is converted to heat energy, causing the liquid phase to heat up and evaporate. However, external cooling systems like an air fan, cooling water jacket, cooling coil, or water bath can maintain the liquid temperature and minimize evaporation [37].

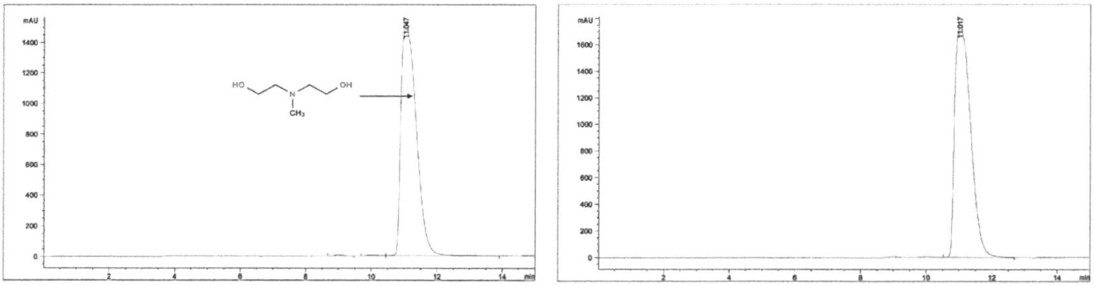

(a) (b)

Figure 5. HPLC chromatograms of 50 wt% aqueous MDEA solution under different conditions (**a**) Irradiated, (**b**) Non-irradiated.

Table 4. Chromatographic data of aqueous MDEA solution.

Ultrasonic Power: 12.36 and 0 W, Initial [MDEA]: 50 wt%, Temperature: 70 °C Identification Method: HPLC-UV				
Sample No.	**Condition**	**Observed RT of Peak [min]**	**Peak Area × 10^4 [mAU × s]**	**Identity**
A1	Irradiated	11.047	6.14	MDEA
A2	Non-irradiated	11.017	5.98	MDEA

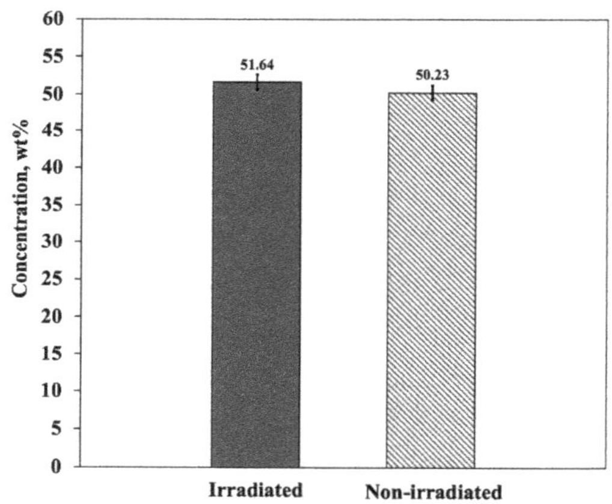

Figure 6. Concentration variation of aqueous MDEA solution under different testing conditions.

4.2. HPLC Analysis of Aqueous MDEA Solution Blended with TA

Similar to the previous step, two samples in the aqueous form containing 50 wt% MDEA blended with 10 wt% of TA were prepared for HPLC analysis. The ultrasonic irradiation was applied to the first sample (B1) but not to the second (B2). The HPLC chromatogram of these samples is shown in Figure 7.

According to Figure 7, two peaks were observed in each chromatogram, identified as TA and MDEA for the irradiated sample. However, it was expected to observe one more peak representing 2-hydroxyterephthalic acid (HTA), but due to the low concentration, the peak shape is a bit flat and cannot be observed. The concentrations of the identified peaks were determined based on the peak area, as mentioned in Table 5, and the HPLC calibration curve for MDEA and TA. The results are presented in Figure 8. Since the same peaks were detected in both samples and no additional peaks were observed, it can be stated that ultrasonic irradiation did not change the structure of the mixture. Moreover, according to the results, no significant changes in the concentrations of the MDEA compound in both irradiated and non-irradiated samples were observed. However, the concentration of TA compound in the irradiated sample was lower than in the non-irradiated sample, indicating the presence of more $\dot{O}H$ radicals under the irradiated condition. These findings confirmed the stability of MDEA molecules; thus, it can be stated that the possibility of breaking the chemical bond of the MDEA molecule with high-frequency ultrasonic irradiation is negligible. In other words, the generation of $\dot{O}H$ radicals scavenged by TA might be due to the dissociation of the water content of the aqueous solution [38,39].

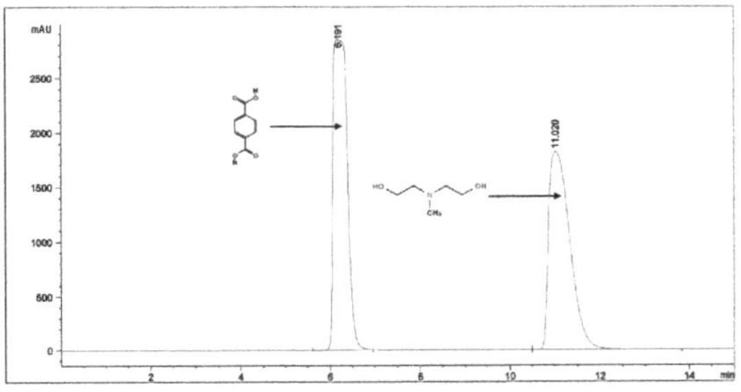

(a)

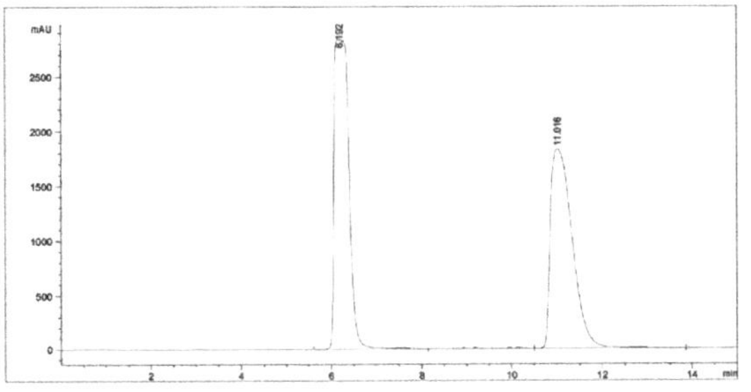

(b)

Figure 7. HPLC chromatograms of 50 wt% aqueous MDEA solution blended with 10 wt% TA under different conditions (**a**) Irradiated, (**b**) Non-irradiated.

Table 5. Chromatographic data of aqueous MDEA solution blended with TA.

Ultrasonic Power: 12.36 & 0 W, Initial [MDEA]: 50 wt%, Temperature: 70 °C Identification Method: HPLC-UV				
Sample No.	**Condition**	**Observed RT of Peak [min]**	**Peak Area × 10^4 [mAU × s]**	**Identity**
B1	Irradiated	6.191	6.32	TA
		11.020	6.12	MDEA
B2	Non-irradiated	6.192	6.81	TA
		11.016	5.98	MDEA

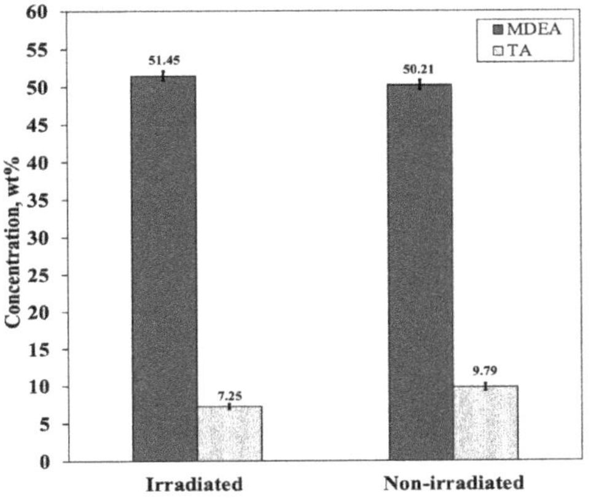

Figure 8. Concentration variation of aqueous MDEA solution blended with TA under different testing conditions.

The above-mentioned assumption can also be discussed from a molecular point of view. According to the molecular structure of MDEA, there are two O-H functional groups in the 1 and 1′ positions, as shown in Figure 9. In each position, the O-H functional group is attached to a carbon atom, which is sterically hindered. Additionally, when atoms are close together, the energy of the covalent bonds that keep the atoms together in the molecule increases [40,41]. Due to the increased energy, breaking the chemical bond of the O-H functional groups of the MDEA molecule for generating $\dot{O}H$ radicals is more complex than in water, which is not considered a steric hindrance. Thus, the formation of $\dot{O}H$ radicals can be mainly attributed to the dissociation of water rather than MDEA.

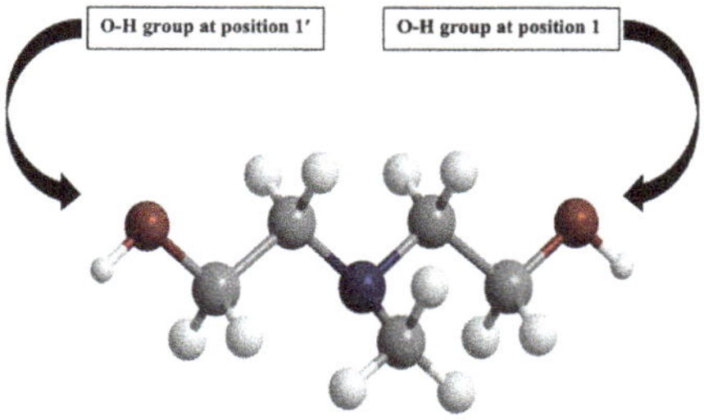

Figure 9. MDEA molecular structure (Gray, white, red, and blue balls represent C, H, O, and N atoms, respectively).

4.3. HPLC Analysis of the CO_2 Absorbed into the Aqueous MDEA Solution

The CO_2 absorption into the aqueous MDEA solution was also analyzed by HPLC under two different conditions to provide sufficient information for elucidating the role of the sonochemical effect on the chemical reactions. In the first approach, the CO_2 absorption process using promoter-free MDEA was performed in an ultrasonic-assisted reactor (C1). Whereas in the second approach, the CO_2 absorption process was carried out under the non-irradiated condition (C2). For both approaches, the experimental conditions were the optimized conditions obtained in our previous research work [21]. After the absorption process, the samples were injected into the HPLC equipment for analysis. The HPLC chromatogram and information of detected peaks are presented in Figure 10, and Table 6, respectively. Based on the retention time of the MDEA standards, one peak was identified as MDEA, but the other remained unidentified. The unknown peak might be any stable intermediate formed during the CO_2 absorption process utilizing MDEA aqueous solution, which can be identified further using the GC-MS characterization technique. However, the unknown peak was detected at a similar retention time in both testing samples, indicating that no change had happened to the sample after inducing ultrasonic irradiation. Therefore, due to the similarity of the presented peaks, it can be stated that no unexpected changes in terms of forming new chemical compounds occurred during CO_2 absorption into an aqueous MDEA solution in an ultrasonic-assisted reactor. According to the obtained data, the concentration of MDEA, as the identified peak, was calculated. Figure 11 shows the MDEA concentration changes for irradiated and non-irradiated samples after absorbing 11 bars of CO_2. The results showed no significant difference in the MDEA concentration of both samples, indicating that a similar amount of absorbent was consumed during the CO_2 absorption process.

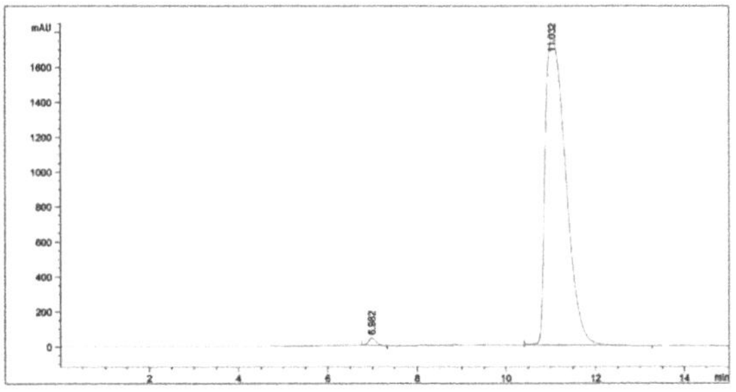

(**a**)

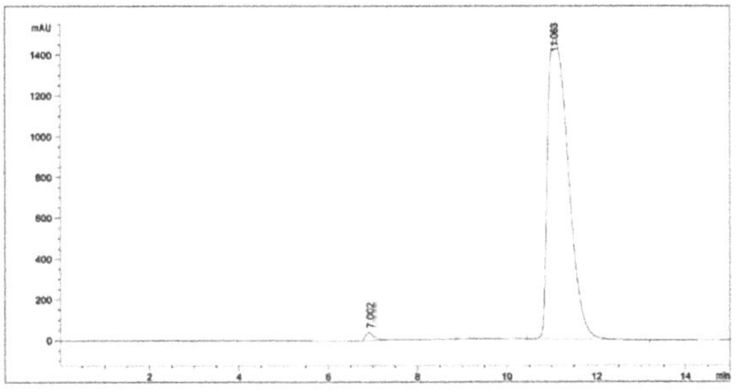

(**b**)

Figure 10. HPLC chromatograms of CO_2 absorption into aqueous MDEA solution under different conditions (**a**) Irradiated, (**b**) Non-irradiated.

Table 6. Chromatographic data of the CO_2 absorbed into aqueous MDEA solution.

Ultrasonic Power: 12.36 and 0 W, Initial [MDEA]: 50 wt%, Temperature: 70 °C, CO_2 Pressure: 10 bar Identification Method: HPLC-UV				
Sample No.	**Condition**	**Observed RT of Peak [min]**	**Peak Area × 10^4 [mAU × s]**	**Identity**
C1	Irradiated	11.032	5.64	MDEA
		6.982	0.052	Unknown
C2	Non-irradiated	11.063	5.43	MDEA
		7.002	0.055	Unknown

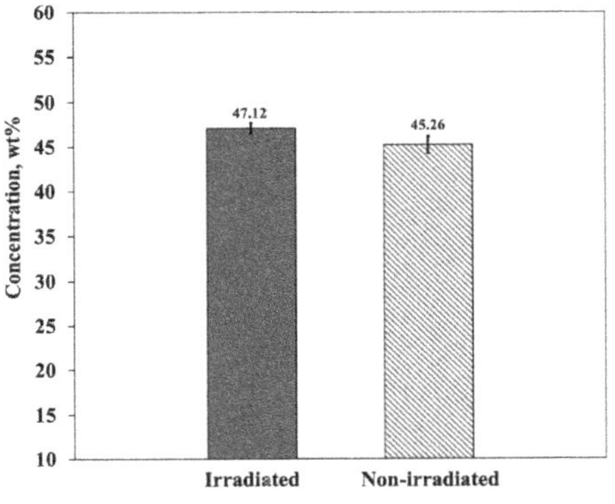

Figure 11. Concentration variation of aqueous MDEA solution after CO_2 absorption under different testing conditions.

4.4. HPLC Analysis of the CO_2 Absorbed into the Aqueous MDEA Solution Blended with TA

The CO_2 absorption process utilizing an aqueous MDEA solution was further investigated by HPLC in the presence of TA. The tests were performed under irradiated (D1) and non-irradiated (D2) conditions using a 50 wt% aqueous solution of MDEA mixed with 10 wt% TA. After the CO_2 absorption process, the samples were injected into the HPLC system. As previously stated, the operating conditions were set at the optimized experimental conditions for promoter-free MDEA. The HPLC chromatogram of the samples and the information about the detected peaks are presented in Figure 12 and Table 7, respectively.

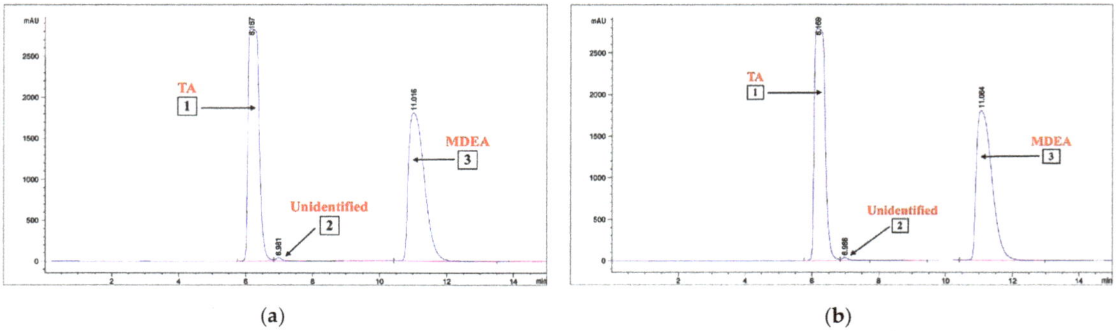

Figure 12. HPLC chromatograms of CO_2 absorption into aqueous MDEA solution mixed with TA under different conditions (**a**) Irradiated, (**b**) Non-irradiated.

Table 7. Chromatographic data of the CO_2 absorbed into aqueous MDEA solution mixed with TA.

Ultrasonic Power: 12.36W, [MDEA]: 50 wt%, [TA]: 10 wt%, Temperature: 70 °C, Pressure: 10 bar Identification Method: HPLC-UV					
Sample No.	**Condition**	**Peak Number**	**Observed RT of Peak [min]**	**Peak Area × 10^4 [mAU × s]**	**Identity**
D1	Irradiated	1	6.167	6.43	TA
		2	6.981	0.06	Unknown
		3	11.018	5.84	MDEA
D2	Non-irradiated	1	6.169	6.81	TA
		2	6.986	0.05	Unknown
		3	11.064	5.55	MDEA

As can be seen, in each chromatogram, three peaks were observed. According to the retention time and referring to the calibration standards, the MDEA and TA peaks were identified. But the second peak in both samples could not be identified. Nevertheless, the retention time of the unknown peak was almost similar in both samples, indicating the formation of an identical compound. Due to the similarity of the observed peaks in both irradiated and non-irradiated conditions, it can be stated that the CO_2 absorption performed in almost the same way. In both conditions, the concentrations of the identified peaks, MDEA and TA, were calculated based on the respective calibration curves. Figure 13 shows the concentration variation of the identified peaks for irradiated and non-irradiated samples. For the sample tested under the non-irradiated condition, the concentration of TA is approximately equal to the initial concentration of 10 wt%, indicating that the formation of $\dot{OH}$ radicals is not feasible under the non-irradiated condition. Conversely, the variation in TA concentration in the irradiated sample confirmed the presence of $\dot{OH}$ radicals and their involvement during the CO_2 absorption process. Furthermore, there were no significant differences in MDEA concentrations for the irradiated sample compared to the scavenger-free CO_2 absorption process (C1). The insignificant concentration variations of MDEA support the idea of MDEA molecular stability versus water molecules under high-frequency ultrasonic irradiation. Thus, the scavenged $\dot{OH}$ radicals during the CO_2 absorption originated insignificantly from the breakage of the MDEA chemical bonds. In other words, the formation of $\dot{OH}$ radicals seems to be predominantly caused by the dissociation of water molecules, which may eventually assist in the CO_2 absorption process.

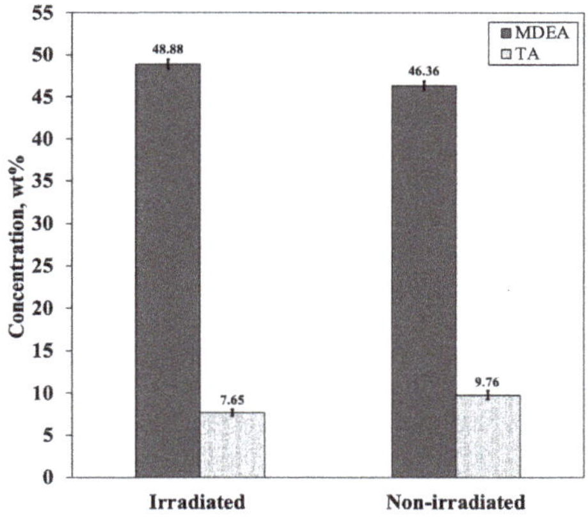

Figure 13. Concentration variation of MDEA and TA under differert conditions after CO_2 absorption.

4.5. Evaluation of the Reaction Pathway

According to the HPLC characterization results of the samples, no new chemical compounds were detected during the CO_2 absorption process utilizing MDEA in the high-frequency ultrasonic-assisted reactor. Thus, the hypothesis of changing the reaction pathway due to the presence of the sonochemical effect is invalid. The stability of MDEA molecules compared to water molecules explains the non-alteration in the reaction pathway despite inducing high frequency and high power ultrasonic irradiation. It means that the applied ultrasonic energy can readily break the water molecules but not the MDEA molecules. Even if the MDEA molecules partially break down and release $\dot{O}H$ radicals, they would immediately react with the available $\dot{O}H$ radicals and reestablish themselves since they are in an irradiated aqueous solution with abundant $\dot{O}H$ radicals. Thus, it can be stated that despite the presence of the sonochemical effect, the reaction of CO_2 absorption into aqueous MDEA solution can still be explained by the conventional reaction pathway shown in Figure 14.

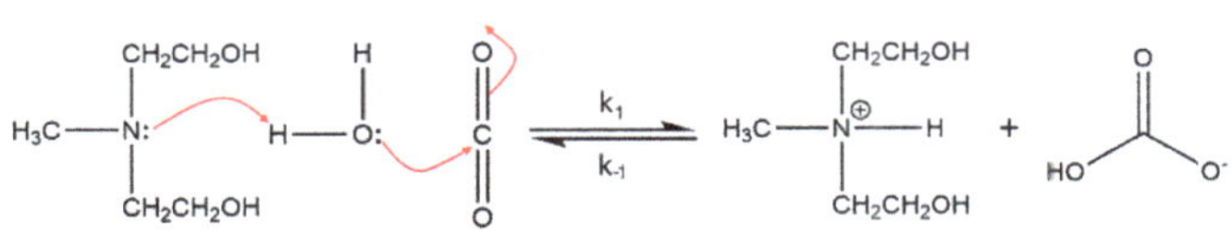

Figure 14. The reaction mechanism of CO_2 absorption into aqueous MDEA solution.

According to the conventional reaction pathway, MDEA, as a tertiary alkanolamine, cannot directly react with CO_2 as it has no hydrogen atom attached to the nitrogen atom. It is worth noting that the presence of hydrogen atoms is essential for a direct reaction with CO_2. Due to the lack of hydrogen and, consequently, proton, MDEA can provide a sink for the hydrogen ions produced when CO_2 hydrolysis in water to form bicarbonate. The complete mechanism includes three reactions [42]:

$$CO_2 + H_2O \leftrightarrow HCO_3^- + H^+ \tag{1}$$

$$CO_2 + OH^- \leftrightarrow HCO_3^- \tag{2}$$

$$MDEA + CO_2 + H_2O \leftrightarrow MDEAH^+ + HCO_3^- \tag{3}$$

Therefore, while the sonochemical effect cannot change the reaction pathway, in the aqueous solutions, it can accelerate the generation of $\dot{O}H$ radicals via water sonolysis. In particular, for the above reactions, such acceleration could alter the slow formation of bicarbonate, resulting in a faster rate of CO_2 absorption.

4.6. Significance of Ultrasonic Irradiation on CO_2 Absorption Rate

The sonochemical effect can significantly influence the rate of CO_2 absorption. To clarify its influence, the CO_2 absorption rate was assessed in the absence and presence of TA (as the scavenger for $\dot{O}H$ radicals) operating under irradiated and non-irradiated settings. The basis of rate determination is explained in our previous research work [21]. The findings are shown in Figure 15. The absorption rate under the non-irradiated condition remained approximately unchanged in both the presence and absence of TA. Whereas for the irradiated condition, it changed considerably. In an irradiated setting and without TA, the CO_2 absorption rate was about 4.20 mol/h. However, the CO_2 absorption rate was reduced to 0.75 mol/h by adding TA. The 5.6-fold reduction can explain the significance of the sonochemical effect on absorption rate enhancement. Thus, despite the non-alteration of the reaction pathway, the generated $\dot{O}H$ radicals induced by the sonochemical effect can significantly influence the CO_2 absorption rate.

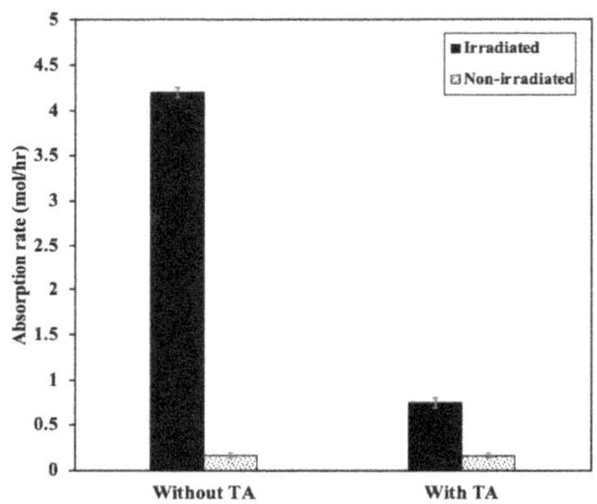

Figure 15. Experimental absorption rate of different systems using 0 and 12.36 W ultrasonic power, CO_2 initial pressure of 11 bar, temperature of 70 °C, MDEA concentration of 50 wt%, and TA concentration of 10 wt%.

5. Conclusions

Ultrasonic irradiation is a novel technique that can intensify the CO_2 absorption process because of the sonochemical and sonophysical effects. The fundamental principles of sonophysical and sonochemical effects in assisting CO_2 absorption have been proven recently. Nevertheless, more detailed research is still needed to light up the path of commercial use of ultrasonic-assisted CO_2 absorption systems. Particularly, it is crucial to investigate the significance of the sonochemical effects on the chemical reactions of the CO_2 absorption process. It is hypothesized that the sonochemical effect might affect the chemical reactions either by accelerating or altering them. Therefore, to assess the above-mentioned assumptions, the role of the sonochemical effect on the reaction pathway was evaluated by analyzing various samples. The sample analysis was performed by evaluating the importance of $\dot{O}H$ radicals during the CO_2 absorption process using promoter-free MDEA. Moreover, TA was used as the potential $\dot{O}H$ radical scavenger. For each sample, chemical components from the liquid phase were analyzed by HPLC and compared under irradiated and non-irradiated conditions. The comparison aimed to find possible evidence of the sonochemical effect's capability to change the reaction pathway. The similarity of the comparative results claimed that the sonochemical effect could not change the reaction pathway. Thus, the admitted origin of sonochemistry in aqueous solutions was water sonolysis, which could generate the $\dot{O}H$ radicals and subsequently accelerate the CO_2 absorption process.

Author Contributions: Conceptualization, K.K.L. and F.S.; writing—original draft preparation, K.K.L. and F.S.; writing—review and editing, K.K.L. and B.P.; supervision, K.K.L.; funding acquisition, K.K.L. All authors have read and agreed to the published version of the manuscript.

Funding: The research was funded by the YUTP-PRG project (015PBC-013).

Data Availability Statement: Data are contained within the article.

Acknowledgments: The authors wish to acknowledge the technical support provided by CO2 Research Centre (CO2RES), Universiti Teknologi PETRONAS, Malaysia.

Conflicts of Interest: The authors declare no conflict of interest.

Nomenclature

bar	Pressure unit
BC	Bubble column
CO	Carbon monoxide
CO_2	Carbon dioxide
DC	Disk column
DEA	Diethanolamine
HFMC	Hollow fiber membrane contactor
HFUA	High-frequency ultrasonic assisted
HPLC	High-performance liquid chromatography
H_2S	Hydrogen sulfide
ID	Inner diameter
K	Temperature unit, Kelvin
K_2CO_3	Potassium carbonate
mAU	milli-Absorbance Unit
M	Molar
MC	Microchannel
MEA	Monoethanolamine
MHz	Megahertz, frequency unit
MDEA	Methyl diethanolamine
NaOH	Sodium hydroxide
Na_2CO_3	Sodium carbonate
nm	Nanometer, wavelength unit
$\dot{O}H$	Hydroxyl radical
PB	Packed bed
RPB	Rotating packed bed
RT	Retention time
TA	Terephthalic acid
T	Temperature
UV	Ultraviolet spectroscopy
W	Watt, ultrasonic power unit
WWC	Wetted wall column

References

1. Shokrollahi, F.; Lau, K.K.; Partoon, B.; Smith, A.M. A review on the selection criteria for slow and medium kinetic solvents used in CO_2 absorption for natural gas purification. *J. Nat. Gas Sci. Eng.* **2022**, *98*, 104390. [CrossRef]
2. Rufford, T.E.; Smart, S.; Watson, G.C.; Graham, B.F.; Boxall, J.; Da Costa, J.D.; May, E.F. The removal of CO 2 and N 2 from natural gas: A review of conventional and emerging process technologies. *J. Pet. Sci. Eng.* **2012**, *94–95*, 123–154. [CrossRef]
3. Kenarsari, S.D.; Yang, D.; Jiang, G.; Zhang, S.; Wang, J.; Russell, A.G.; Wei, Q.; Fan, M. Review of recent advances in carbon dioxide separation and capture. *RSC Adv.* **2013**, *3*, 22739–22773. [CrossRef]
4. MacDowell, N.; Florin, N.; Buchard, A.; Hallett, J.; Galindo, A.; Jackson, G.; Adjiman, C.S.; Williams, C.K.; Shah, N.; Fennell, P. An overview of CO_2 capture technologies. *Energy Environ. Sci.* **2010**, *3*, 1645–1669. [CrossRef]
5. Tay, W.H.; Lau, K.K.; Lai, L.S.; Shariff, A.M.; Wang, T. Current development and challenges in the intensified absorption technology for natural gas purification at offshore condition. *J. Nat. Gas Sci. Eng.* **2019**, *71*, 102977. [CrossRef]
6. Tan, L.S.; Shariff, A.M.; Lau, K.K.; Bustam, M.A. Factors affecting CO 2 absorption efficiency in packed column: A review. *J. Ind. Eng. Chem.* **2012**, *18*, 1874–1883. [CrossRef]

7. Bhargava, N.; Mor, R.S.; Kumar, K.; Sharanagat, V.S. Advances in application of ultrasound in food processing: A review. *Ultrason. Sonochem.* **2021**, *70*, 105293. [CrossRef] [PubMed]
8. Aktij, S.A.; Taghipour, A.; Rahimpour, A.; Mollahosseini, A.; Tiraferri, A. A critical review on ultrasonic-assisted fouling control and cleaning of fouled membranes. *Ultrasonics* **2020**, *108*, 106228. [CrossRef]
9. Miller, D.L.; Smith, N.B.; Bailey, M.R.; Czarnota, G.J.; Hynynen, K.; Makin, I.R.S. Overview of therapeutic ultrasound applications and safety considerations. *J. Ultrasound Med.* **2012**, *31*, 623–634. [CrossRef]
10. Ghahremani, H.; Nasri, Z.; Eikani, M.H. Ultrasound-assisted oxidative desulfurization (UAOD) of Iranian heavy crude oil: Investigation of process variables. *J. Pet. Sci. Eng.* **2021**, *204*, 108709. [CrossRef]
11. Karthik, K.; Nikolova, M.P.; Phuruangrat, A.; Pushpa, S.; Revathi, V.; Subbulakshmi, M. Ultrasound-assisted synthesis of V_2O_5 nanoparticles for photocatalytic and antibacterial studies. *Mater. Res. Innov.* **2020**, *24*, 229–234. [CrossRef]
12. Zhang, Z.; Wang, K.; Xu, C.; Zhang, Y.; Wu, W.; Lu, C.; Liu, W.; Rao, Y.; Jiang, C.; Xu, C.; et al. Ultrasound enhancing the mass transfer of droplet microreactor for the synthesis of $AgInS_2$ nanocrystals. *Chem. Eng. J.* **2022**, *435*, 134948. [CrossRef]
13. Gogate, P.R.; Sutkar, V.S.; Pandit, A.B. Sonochemical reactors: Important design and scale up considerations with a special emphasis on heterogeneous systems. *Chem. Eng. J.* **2011**, *166*, 1066–1082. [CrossRef]
14. Wilhelm, A.-M.; Laugier, F.; Kidak, R.; Ratsimba, B.; Delmas, H. Ultrasound to Enhance a Liquid–Liquid Reaction. *J. Chem. Eng. Jpn.* **2010**, *43*, 751–756. [CrossRef]
15. Tay, W.H.; Lau, K.K.; Shariff, A.M. High frequency ultrasonic-assisted chemical absorption of CO_2 using monoethanolamine (MEA). *Sep. Purif. Technol.* **2017**, *183*, 136–144. [CrossRef]
16. Tay, W.H.; Lau, K.K.; Shariff, A.M. High frequency ultrasonic-assisted CO_2 absorption in a high pressure water batch system. *Ultrason. Sonochem.* **2016**, *33*, 190–196. [CrossRef]
17. Tay, W.H.; Lau, K.K.; Shariff, A.M. High performance promoter-free CO_2 absorption using potassium carbonate solution in an ultrasonic irradiation system. *J. CO_2 Util.* **2017**, *21*, 383–394. [CrossRef]
18. Yusof, S.M.M.; Lau, K.K.; Shariff, A.M.; Tay, W.H.; Mustafa, N.F.A.; Lock, S.S.M. Novel continuous ultrasonic contactor system for CO_2 absorption: Parametric and optimization study. *J. Ind. Eng. Chem.* **2019**, *79*, 279–287. [CrossRef]
19. Yusof, S.M.M.; Shariff, A.M.; Tay, W.H.; Lau, K.K.; Mustafa, N.F.A. Mass transfer intensification of CO_2 absorption in monoethanolamine using high frequency ultrasonic technology in continuous system. *Int. J. Greenh. Gas Control* **2020**, *102*, 103157. [CrossRef]
20. Marjanian, M.M.; Shahhosseini, S.; Ansari, A. Investigation of the ultrasound assisted CO_2 absorption using different absorbents. *Process Saf. Environ. Prot.* **2021**, *149*, 277–288. [CrossRef]
21. Shokrollahi, F.; Lau, K.K.; Partoon, B.; Lai, L.S. Elucidation of Operating Parameters Influencing the Ultrasonic-Assisted Absorption of Bulk CO 2 Using Unpromoted and Promoted Methyldiethanolamine. *Ind. Eng. Chem. Res.* **2023**, *62*, 2843–2865. [CrossRef]
22. Karlsson, H.; Svensson, H. Rate of Absorption for CO_2 Absorption Systems Using a Wetted Wall Column. *Energy Procedia* **2017**, *114*, 2009–2023. [CrossRef]
23. Pokhrel, N.; Vabbina, P.K.; Pala, N. Sonochemistry: Science and Engineering. *Ultrason. Sonochem.* **2016**, *29*, 104–128. [CrossRef] [PubMed]
24. Adewuyi, Y.G. Sonochemistry: Environmental science and engineering applications. *Ind. Eng. Chem. Res.* **2001**, *40*, 4681–4715. [CrossRef]
25. Bhangu, S.K.; Ashokkumar, M. Theory of Sonochemistry. In *Sonochemistry: From Basic Principles to Innovative Applications*; Springer: Cham, Switzerland, 2017; pp. 1–28. [CrossRef]
26. Savun-Hekimoğlu, B. A Review on Sonochemistry and Its Environmental Applications. *Acoustics* **2020**, *2*, 766–775. [CrossRef]
27. Wu, T.Y.; Guo, N.; Teh, C.Y.; Hay, J.X.W. Theory and Fundamentals of Ultrasound. In *Advances in Ultrasound Technology for Environmental Remediation*; Springer: Dordrecht, The Netherlands, 2013; pp. 5–12. [CrossRef]
28. Pham, T.D.; Shrestha, R.A.; Virkutyte, J.; Sillanpää, M. Recent studies in environmental applications of ultrasound. *J. Environ. Eng. Sci.* **2013**, *8*, 403–412. [CrossRef]
29. Liao, C.H.; Li, M.H. Kinetics of absorption of carbon dioxide into aqueous solutions of monoethanolamine+N-methyldiethanolamine. *Chem. Eng. Sci.* **2002**, *57*, 4569–4582. [CrossRef]
30. Lin, C.Y.; Soriano, A.N.; Li, M.H. Kinetics study of carbon dioxide absorption into aqueous solutions containing N-methyldiethanolamine + diethanolamine. *J. Taiwan Inst. Chem. Eng.* **2009**, *40*, 403–412. [CrossRef]
31. Zhang, X.; Zhang, C.F.; Qin, S.J.; Zheng, Z.S. A kinetics study on the absorption of carbon dioxide into a mixed aqueous solution of methyldiethanolamine and piperazine. *Ind. Eng. Chem. Res.* **2001**, *40*, 3785–3791. [CrossRef]
32. Shokrollahi, F.; Lau, K.K.; Tay, W.H.; Lai, L.S. Power measurement by calorimetric method using water infrequency range between 1.7 Mhz to 3 Mhz. *Int. J. Eng. Technol.* **2018**, *7*, 106–109. [CrossRef]
33. Charbouillot, T.; Brigante, M.; Mailhot, G.; Maddigapu, P.R.; Minero, C.; Vione, D. Performance and selectivity of the terephthalic acid probe for •OH as a function of temperature, pH and composition of atmospherically relevant aqueous media. *J. Photochem. Photobiol. A Chem.* **2011**, *222*, 70–76. [CrossRef]
34. Barreto, J.C.; Smith, G.S.; Strobel, N.H.P.; McQuillin, P.A.; Miller, T.A. Terephthalic acid: A dosimeter for the detection of hydroxyl radicals in vitro. *Life Sci.* **1994**, *56*, PL89–PL96. [CrossRef] [PubMed]

35. Bjorndalen, N.; Islam, M.R. The effect of microwave and ultrasonic irradiation on crude oil during production with a horizontal well. *J. Pet. Sci. Eng.* **2004**, *43*, 139–150. [CrossRef]
36. Kudo, T.; Sekiguchi, K.; Sankoda, K.; Namiki, N.; Nii, S. Effect of ultrasonic frequency on size distributions of nanosized mist generated by ultrasonic atomization. *Ultrason. Sonochem.* **2017**, *37*, 16–22. [CrossRef]
37. Tamidi, A.M.; Lau, K.K.; Khalit, S.H. A review of recent development in numerical simulation of ultrasonic-assisted gas-liquid mass transfer process. *Comput. Chem. Eng.* **2021**, *155*, 107498. [CrossRef]
38. Miyaji, A.; Kohno, M.; Inoue, Y.; Baba, T. Hydroxyl radical generation by dissociation of water molecules during 1.65 MHz frequency ultrasound irradiation under aerobic conditions. *Biochem. Biophys. Res. Commun.* **2017**, *483*, 178–182. [CrossRef]
39. Merouani, S.; Hamdaoui, O.; Rezgui, Y.; Guemini, M. Computational engineering study of hydrogen production via ultrasonic cavitation in water. *Int. J. Hydrogen Energy* **2016**, *41*, 832–844. [CrossRef]
40. Luo, Q.; Zhou, Q.; Feng, B.; Li, N.; Liu, S. A Combined Experimental and Computational Study on the Shuttle Mechanism of Piperazine for the Enhanced CO_2Absorption in Aqueous Piperazine Blends. *Ind. Eng. Chem. Res.* **2022**, *61*, 1301–1312. [CrossRef]
41. Laurence, C.; Berthelot, M. Observations on the strength of hydrogen bonding. *Perspect. Drug Discov. Des.* **2000**, *18*, 39–60. [CrossRef]
42. Maleki, A.; Irani, V.; Tavasoli, A.; Vahidi, M. Enhancement of CO_2 solubility in a mixture of 40 wt% aqueous N-Methyldiethanolamine solution and diethylenetriamine functionalized graphene oxide. *J. Nat. Gas Sci. Eng.* **2018**, *55*, 219–234. [CrossRef]

MDPI
St. Alban-Anlage 66
4052 Basel
Switzerland
www.mdpi.com

Processes Editorial Office
E-mail: processes@mdpi.com
www.mdpi.com/journal/processes

www.ingramcontent.com/pod-product-compliance
Lightning Source LLC
Chambersburg PA
CBHW040819250926
51850CB00007BA/1212
* 9 7 8 3 0 3 6 5 9 9 9 3 9 *